FENUGREEK

Improvement in Yield Potential Through Plant Growth Regulators (PGRs) and Depolymerized Polysaccharides

FENUGREEK

Improvement in Yield Potential Through Plant Growth Regulators (PGRs) and Depolymerized Polysaccharides

Tariq Ahmad Dar, PhD
Bilal Ahmad, PhD
Moinuddin, PhD
Khalid Rehman Hakeem, PhD
Editors

AAP | APPLE ACADEMIC PRESS

First edition published 2026

Apple Academic Press Inc.
1265 Goldenrod Circle, NE,
Palm Bay, FL 32905 USA

760 Laurentian Drive, Unit 19,
Burlington, ON L7N 0A4, Canada

CRC Press
2385 NW Executive Center Drive,
Suite 320, Boca Raton FL 33431

4 Park Square, Milton Park,
Abingdon, Oxon, OX14 4RN, UK

Library and Archives Canada Cataloguing in Publication

Title: Fenugreek : improvement in yield potential through plant growth regulators (PGRs) and depolymerized polysaccharides / Tariq Ahmad Dar, PhD, Bilal Ahmad, PhD, Moinuddin, PhD, Khalid Rehman Hakeem, PhD, editors.
Names: Dar, Tariq Ahmad, editor.
Description: First edition. | Includes bibliographical references and index.
Identifiers: Canadiana (print) 20250183617 | Canadiana (ebook) 20250183633 | ISBN 9781774919163 (hardcover) | ISBN 9781003593676 (ebook)
Subjects: LCSH: Fenugreek. | LCSH: Fenugreek—Growth.
Classification: LCC SB317.F44 F46 2025 | DDC 633.8/8363—dc23

Library of Congress Cataloging-in-Publication Data

CIP data on file with US Library of Congress

ISBN: 978-1-77491-916-3 (hbk)
ISBN: 978-1-00359-367-6 (ebk)

Dedication

To my mother, Hajrah Begum, in Heaven: Thank you for always loving me and guiding me. Even if you are not physically present, I can still feel your love guiding me. You will never leave my thoughts. I miss you terribly, and I love you. I pray to Allah (SWT) to bless your soul and grant you Jannat-ul-Firdous (the highest level of heaven).

> *"O Allah, forgive and have mercy upon her, excuse her and pardon her, and make honorable her reception. Expand her entry, cleanse her with water, snow, and ice, and purify her of sin as a white robe is purified of filth. Exchange her home for a better home. Admit her into the Garden; protect her from the punishment of the grave and the torment of the Fire."*

—Prof. (Dr.) Khalid Rehman Hakeem, PhD

> *To my new born daughter MAHNOOR, a precious baby born after almost 7 years of our marriage. We are highly thankful to ALLAH SWT for blessing us with this precious gift. May ALLAH protect her from evil, bless her with his choicest blessings and increase her iman. I wish you a life filled with good health, happiness, and peace of mind. May you treat others with kindness, compassion, and understanding. My dearest daughter may you always find the strength and wisdom to navigate life's challenges. Ameen*

—Dr Tariq Ahmad Dar, PhD

About the Editors

Tariq Ahmad Dar, PhD
Assistant Professor, Department of Botany, Sri Pratap College, Cluster University Srinagar (Jammu & Kashmir), India

Tariq Ahmad Dar, PhD, earned his MSc in Botany from the P.G. Department of Botany, University of Kashmir, Hazratbal, Srinagar, in 2010, and his PhD in Botany from Aligarh Muslim University, Aligarh, India, in 2015. Dr. Tariq Ahmad Dar has 20 publications to his credit. He has published his research findings in journals of international repute, such as *Environmental and Experimental Botany, Industrial Crops and Products, Carbohydrate Polymers, Journal of Applied Research on Medicinal and Aromatic Plants, Journal of Radiation Research and Applied Sciences, Plant Physiology Reports, Journal of Essential Oil Research, Frontiers in Plant Science,* and *Journal of Plant Growth Regulation,* among others. Dr. Dar has presented his papers at national and international conferences and has attended about 11 national and international conferences. Moreover, he is a life member of the Indian Botanical Society and the Indian Science Congress Association. He has guided the dissertations of 10 MSc students. He is actively involved in enhancing the growth, yield, and active constituents of medicinal and aromatic plants using plant growth regulators like jasmonates, gibberellins, and depolymerized polysaccharides. Currently, Dr. Dar is focusing on the use of marine polysaccharides as plant growth regulators and increasing the yield of active constituents of medicinal plants. Additionally, the use of radiation-processed depolymerized polysaccharides like sodium alginate, chitosan, and carrageenan to alleviate heavy metal stress and different abiotic stress conditions like salinity, drought, and chilling are the focus areas. Dr. Dar has eight years of teaching experience at the undergraduate level and over two years at the postgraduate level.

Bilal Ahmad, PhD

Assistant Professor, Department of Botany, Govt Degree College, KhanSahib, Budgam, affliated with University of Kashmir Hazratbal Srinagar, India

Bilal Ahmad, PhD, completed his doctorate at Aligarh Muslim University, Uttar Pradesh, India, in 2018. During his research, Dr. Ahmad studied the influence of various plant growth regulators, nanoparticles, and radiation-processed polysaccharides on medicinally important plants under normal and stressful conditions, including heavy metal stress. He is currently an assistant professor and has been teaching since 2018. Dr. Ahmad has authored 35 articles in reputed international journals, including the *International Journal of Biological Macromolecules, Plant Growth Regulation, Current Plant Biology, Journal of Plant Growth Regulation, Plant Physiology and Biochemistry, and Scientia Horticulturae,* among others. He has also contributed to edited volumes of books published by international publishers. Dr. Ahmad has presented his research at national and international conferences and has received several awards. Additionally, he serves as a reviewer for various reputed journals.

Moinuddin, PhD

Professor in Botany, Women's College, Aligarh Muslim University, Aligarh, India

Moinuddin, PhD, is a Professor of Botany at Women's College at Aligarh Muslim University (AMU), Aligarh, India, from which he earned his PhD in Botany. He is conducting research on the mineral nutrition of triticale. He conducted his postdoctoral work on the drought-stress physiology of wheat (at the International Maize and Wheat Improvement Center [CIMMYT], Mexico) and various other crops (Indian Agricultural Research Institute, New Delhi, and Institute of Pulses Research (PRII), Gurgaon, India. He has been working on the effects of plant growth regulator (PGRs), nanoparticles, and marine polysaccharides for more than 13 years at Aligarh Muslim University (AMU), Aligarh, under normal and water-stress conditions. In his account, he has 108 research papers, review articles, and book chapters (total impact factor value: 124) in addition to 42 conference presentations of national and international repute. He has also guided several MPhil and PhD students at AMU.

Khalid Rehman Hakeem, PhD
Professor, King Abdulaziz University, Jeddah, Saudi Arabia

Khalid Rehman Hakeem, PhD, is working as a Professor at King Abdulaziz University, Jeddah, Saudi Arabia. He is currently the Head of the Plant Biotechnology Research Unit at the Princess Dr. Najlaa Bint Saud Al Saud Center for Distinguished Research in Biotech, King Abdulaziz University, Jeddah. He is also working as a Visiting Professor at Daffodil International University, Dhaka, Bangladesh. After completing his PhD in Botany, specializing in Plant Ecophysiology and Molecular Biology, from Jamia Hamdard, New Delhi, India, in 2011, he initially worked as an Assistant Professor at the University of Kashmir, Srinagar. Later, he joined Universiti Putra Malaysia, Selangor, Malaysia, and worked there as a Postdoctoral Fellow in 2012 and as a Fellow Researcher (Associate Professor) from 2013 to 2016, respectively. Dr. Hakeem joined King Abdulaziz University in August 2016 as an Associate Professor and was promoted to Professor in 2019.

Prof. Hakeem has more than 14 years of teaching and research experience in plant eco-physiology, biotechnology and molecular biology, medicinal plant research, plant-microbe-soil interactions, as well as environmental studies. He is the recipient of several fellowships at both national and international levels. He was recently elected as a Fellow of the Royal Society of Biology, London. He has served as a visiting scientist at Fatih University, Istanbul, Turkey, as well as at Jinan University, Guangzhou, China. He is also involved with several international research projects with different government organizations. To date, he has authored and edited more than 95 books with international publishers, including Springer Nature, Academic Press (Elsevier), CRC Press, etc. He has also to his credit more than 185 research publications in peer-reviewed international journals and 80 book chapters in edited volumes with international publishers.

At present, he is serving as an editorial board member and reviewer for several high-impact international scientific journals from Elsevier, Springer Nature, Taylor & Francis, Cambridge Press, and John Wiley Publishers. He is included in the advisory board of Cambridge Scholars Publishing, UK. He is also a Fellow of the Plantae Group of the American Society of Plant Biologists, a member of the World Academy of Science, a member of the International Society for Development and Sustainability, Japan, and a member of the Asian Federation of Biotechnology, Korea. He has been listed in *Marquis Who's Who in the World* from 2014 to 2021. Currently, he is engaged in studying plant processes at both eco-physiological and molecular levels.

Contents

Contributors

Bilal Ahmad
Assistant Professor, Department of Botany, Govt Degree College, KhanSahib, Budgam, affliated with University of Kashmir Hazratbal Srinagar, India

Barket Ali
Department of Botany, Sri Pratap College Cluster University Srinagar, Jammu & Kashmir, India

Khan Bilal Mukhtar Ahmed
Centre of Agricultural Education, faculty of Agricultural Sciences, AMU-Aligarh, India

M. Akanksha
Osmania University, Department of Botany, Hyderabad, Telangana, India

Shakir Ali
Division of Fruit Science, Sher-e Kashmir University of Agricultural Science and Technology, Kashmir, India

Anmol
Department of Horticulture, School of Agriculture, Lovely Professional University, Phagwara, Punjab, India

Abdurrahman Ay
Department of Soil Science and Plant Nutrition, Faculty of Agriculture, Ondokuz Mayis University, Samsun, Türkiye

Themmeichon Chamroy
Department of Horticulture, School of Agriculture, Lovely Professional University, Phagwara, Punjab, India

Tariq Ahmad Dar
Department of Botany, Sri Pratap College, Cluster University Srinagar, Jammu & Kashmir, India

Piverjeet Kaur Dhillon
Assistant Professor (Home Science), Krishi Vigyan Kendra, Tarn Taran, Guru Angad Dev Veterinary and Animal Sciences University, Ludhiana, Punjab, India

Muhammad Mughees Ud Din
Institute of Soil and Environmental Sciences, University of Agriculture, Faisalabad, Pakistan

Vinod Kumar Dubey
Department of Genetics & Plant Breeding, College of Agriculture, Acharya Narendra Deva University of Agriculture and Technology, Ayodhya, Uttar Pradesh, India

Mohd. Gulfishan
School of Agricultural Sciences, Glocal University, Saharanpur, Uttar Pradesh, India

Tushar Gupta
Amity Institute of Biotechnology, Amity University, Noida, Uttar Pradesh, India

Khalid Rehman Hakeem
Department of Biological Sciences, Faculty of Science, King Abdulaziz University, Jeddah, Saudi Arabia

Ajmat Jahan
Advanced Plant Physiology Section, Department of Botany, Aligarh Muslim University, Aligarh, Uttar Pradesh, India

Ulfat Jan
Division of Agriculture Entomology SKUAST Kashmir, Jammu and Kashmir, India
Department of Electrical Engineering, Chandigarh University, Jammu and Kashmir, India

Kounser Javid
Sher-e Kashmir University of Agricultural Science and Technology, Jammu & Kashmir, India

Deepika Jeengar
Department of Entomology, College of Agriculture, Govind Ballabh Pant University of Agriculture and Technology, Pantnagar, Uttarakhand, India

A. C. Jnanesha
CSIR–Central Institute of Medicinal and Aromatic Plants, Research Center, Boduppal, Hyderabad, Telangana, India

Harjinder Kaur
Department of Horticulture, School of Agriculture, Lovely Professional University, Phagwara, Punjab, India

Rupesh Kaushik
Department of Horticulture, School of Agriculture, Lovely Professional University, Phagwara, Punjab, India

M. Masroor A. Khan
Advanced Plant Physiology Section, Department of Botany, Aligarh Muslim University, Aligarh, Uttar Pradesh, India

Afiya Khursheed
Division of Vegetable Science, Sher-e Kashmir University of Agricultural Science and Technology, Jammu & Kashmir, India

Lyubka Koleva
Agricultural University, Department of Plant Physiology, Biochemistry, and Genetics, Plovdiv, Bulgaria

M. Nithya Kruthi
Amity Institute of Biotechnology, Amity University, Noida, Uttar Pradesh, India

A. Laxman Kumar
Sri Konda Laxman Telangana State Horticultural University, Mulugu, Siddipet, Telangana, India

Ranjith Kumar
CSIR–Central Institute of Medicinal and Aromatic Plants, Research Center, Boduppal, Hyderabad, Telangana, India

S. Bharath Kumar
Sri Konda Laxman Telangana State Horticultural University, Mulugu, Siddipet, Telangana, India

Sanjeev Kumar
Department of Fruit Science, Dr. Yashwant Singh Parmar University of Horticulture and Forestry, Nauni, Solan, Himachal Pradesh, India

Amit Layek
Department of Agricultural Entomology, Bidhan Chandra Krishi Viswavidyalaya, Mohanpur, Nadia, West Bengal, India

Madhu
Assistant Professor (Home Science), Kashi Naresh Government PG College, Gyanpur, Bhadohi, Uttar Pradesh, India

B. Maruthi
Bidhan Chandra Krishi Viswavidyalaya, Mohanpur, Nadia, West Bengal, India

Anis A. Mirza
Department of Horticulture, School of Agriculture, Lovely Professional University, Phagwara, Punjab, India

Filza Mustafa
Atta-Ur-Rahman School of Applied Biosciences, National University of Sciences and Technology, Islamabad, Pakistan

Priyanka Narad
Amity Institute of Biotechnology, Amity University, Noida, Uttar Pradesh, India

Arindam Pal
Department of Entomology, Post-Graduate College of Agriculture, Dr. Rajendra Prasad Central Agricultural University, Pusa, Samastipur, Bihar, India

Shruti Pandey
Amity Institute of Biotechnology, Amity University, Noida, Uttar Pradesh, India

Kaushik Pramanik
Department of Agricultural Entomology, Bidhan Chandra Krishi Viswavidyalaya, Mohanpur, Nadia, West Bengal, India

Priya
Department of Entomology, Post-Graduate College of Agriculture, Dr. Rajendra Prasad Central Agricultural University, Pusa, Samastipur, Bihar, India

Batulla Rashid
Sher-e Kashmir University of Agricultural Science and Technology, Jammu & Kashmir, India

Unjman Rashid
Department of Botany, Jamia Hamdard University, New Delhi, India

Kondle Ravi
Department of Horticulture, School of Agriculture, Lovely Professional University, Phagwara, Punjab, India

Rehan
Department of Horticulture, School of Agriculture, Lovely Professional University, Phagwara, Punjab, India

Yawar Sadiq
Advanced Plant Physiology Section, Department of Botany, Aligarh Muslim University, Aligarh, Uttar Pradesh, India

Ajay Saroha
Department of Horticulture, School of Agriculture, Lovely Professional University, Phagwara, Punjab, India

Debarati Seal
Department of Agronomy, Bidhan Chandra Krishi Viswavidyalaya, Mohanpur, Nadia, West Bengal, India

Abhishek Sengupta
Amity Institute of Biotechnology, Amity University, Noida, Uttar Pradesh, India

Sher Muhammad Shahzad
Department of Soil and Environmental Sciences, College of Agriculture, University of Sargodha, Pakistan

Naina Sharma
Amity Institute of Biotechnology, Amity University, Noida, Uttar Pradesh, India

Priyanka Sharma
Amity Institute of Biotechnology, Amity University, Noida, Uttar Pradesh, India

Amir Hussain Sheir
Department of Electrical Engineering, Chandigarh University, Jammu & Kashmir, India

Priya Singla
Academic Counsellor, Indira Gandhi National Open University, Nirman Campus of Education, Research, and Training, Sunam (Sangrur), Punjab, India

Vyshnavi Sunil
Department of Entomology, College of Agriculture, Kerala Agricultural University, Vellayani, Trivandrum, Kerala, India

Muhammad Danish Toor
Department of Soil Science and Plant Nutrition, Faculty of Agriculture, Ondokuz Mayis University, Samsun, Türkiye; Department of Agrochemistry and Soil Science, Agricultural University, Plovdiv, Bulgaria

Parikshit Uchai
Department of Entomology, Post-Graduate College of Agriculture, Dr. Rajendra Prasad Central Agricultural University, Pusa, Samastipur, Bihar, India

S. Venugopal
CSIR–Central Institute of Medicinal and Aromatic Plants, Research Center, Boduppal, Hyderabad, Telangana, India

Ab Waheed Wani
Department of Horticulture, School of Agriculture, Lovely Professional University, Phagwara, Punjab, India

Zarina
Division of Vegetable Science, Sher-e Kashmir University of Agricultural Science and Technology, Jammu, India

Preface

The four most common noncommunicable ailments prevailing in humans are diabetes, cardiovascular diseases, chronic respiratory diseases, and cancers. Currently, diabetes is the most prevalent disease in the human population worldwide. The WHO estimated about 830 million diabetic patients in the world in 2022, with the prevalence rising more rapidly in low- and middle-income countries than in high-income countries. According to the WHO, there was a 3% increase in diabetes-mediated mortality rates between 2000 and 2019. In 2021 alone, diabetes and diabetes-driven kidney diseases caused two million deaths. Therefore, the scientific community is striving hard to combat this deadly disease by developing effective treatments and exploring new measures. Some allopathic drugs are available in the market for treating the disease; however, due to some side effects, these drugs cannot be continued for longer periods of time. Hence, the research focus is shifting to the use of herbal drugs in this regard. Of the several herbal drugs available in the market, the most common and easily available herb for the disease is fenugreek (*Trigonella foenum-graecum*). It is highly important in the management of diabetes and contributes greatly to alternative systems of medicine.

Fenugreek, belonging to the Fabaceae family, is a miraculous medicinal herb. The plant is rich in different nutritional and pharmaceutical compounds like polyphenols, galactomannans, polysaccharides, steroidal sapogenins, triterpenoids, and several alkaloids. The major alkaloids present in the fenugreek seed are trigonelline and diosgenin. Trigonelline is believed to counter diabetes mellitus (DM) by stimulating the secretion of insulin, modulating/regenerating beta cells, and controlling the activity of glucose metabolism-related enzymes. Fenugreek seed acts as a dietary supplement. In addition to reducing blood cholesterol concentration, the use of fenugreek seeds significantly decreases the symptoms of diabetes, such as polydipsia, polyuria, urine sugar, renal hypertrophy, and glomerular filtration rate. The plant is also rich in fiber content.

Keeping in view the tremendous medicinal applications of fenugreek, it is highly desirable to increase the seed yield as well as the active constituents of fenugreek. The present book highlights the nutritional and medicinal applications of fenugreek, apart from different methods of enhancing its yield and quality, by improving the growth, physiological, and biochemical parameters of the plant. The methods include the use of conventional mineral nutrients, plant growth regulators, nanoparticles, polyacrylamide, and depolymerized polysaccharides like sodium alginate, chitosan, and carrageenan. Besides, the book covers the aspect of stress (salinity, heavy metal, and biotic stress) and its alleviation strategies with regard to the fenugreek crop.

The book consists of 14 chapters, including both review and research articles, written by experts on the subject from around the globe. We are confident that the present book will be useful to researchers who are working in the concerned field. The book will also be useful to undergraduate and postgraduate students, research scholars, teachers, ethnobotanists, pharmacologists, and herbal growers.

We are greatly thankful to Apple Academic Press for the prompt acceptance and sufficient assistance in the compilation of this scientific task. Sincere thanks are also expressed to the team members of Apple Academic Press for their dedication and cooperation in producing this valuable book.

We extend our sincere thanks to all the contributors for their research contributions and consistent patience.

Last but not least, thanks are also due to the family members of the editors for their moral support, blessings, and inspiration in the compilation of this book.

CHAPTER 1

Origin and Geographical Distribution of Fenugreek

DEBARATI SEAL,[1] AMIT LAYEK,[2] and KAUSHIK PRAMANIK[2]

[1]*Department of Agronomy, Bidhan Chandra Krishi Viswavidyalaya, Mohanpur, Nadia, West Bengal, India*

[2]*Department of Agricultural Entomology, Bidhan Chandra Krishi Viswavidyalaya, Mohanpur, Nadia, West Bengal, India*

ABSTRACT

Fenugreek (*Trigonella foenum-graecum* L.), a plant from the Fabaceae family with hard, light brown pointed seeds, is known as Trigonella, meaning "little triangle" (Latin word), for its typical flower's (triangular) shape. It is a self-pollinated legume grow annually and also a common aromatic and spice crop, found to be grown as wild or cultivated across the world. Its natural geographical range extends from Iran to Northern parts of India, although it is now grown in various parts of the earth as well. Generally referred to as "Methi" in Hindi and "Vendhayam" in Tamil, fenugreek is an ancient medicinal plant which includes active ingredients like flavonoids, alkaloids, saponins, steroids, and other substances like trigonelline, protein, fixed oils, mucilage, and so forth. Fenugreek is habitually used as the most significant usual, versatile herbal remedy in Chinese, Indian, Tibetan, and Iranian medical practices. This crop has long been regarded as both traditional food and medicine. With its antioxidant and antimicrobial properties. This plant is used to lower blood lipids and sugar levels in diabetics and non-diabetics. It has been discovered to be a useful medicinal plant with the potential for several applications and a fount for creating diosgenin, the raw constituent employed in pharmaceutical products, particularly steroidal

Fenugreek: Improvement in Yield Potential Through Plant Growth Regulators (PGRs) and Depolymerized Polysaccharides. Tariq Ahmad Dar, Bilal Ahmad, Moinuddin, and Khalid Rehman Hakeem (Eds.)

hormones. This plant can thrive in a broad spectrum of environments and may even be cultivated profitably on marginal grounds since it is relatively resistant of salt and drought. These qualities and the potentiality for heavy metal remediation make fenugreek a good choice for a variety of cropping systems. Fenugreek has recently farmed in several nations from Asia, Africa, and Europe. India is the world's top producer. The major states cultivating fenugreek in India include Haryana, Gujarat, Rajasthan, Madhya Pradesh, West Bengal, and Uttaranchal. Rajasthan has the largest area and production in India, accounting for approximately 37% and 33% of the total area and production, respectively. Thus, this chapter focuses on discussing the origin and distribution of fenugreek.

1.1 INTRODUCTION

Fenugreek is one of the earliest known aromatic and therapeutic herbs contains plenty of several important phytochemicals. In addition to "Methi" in Hindi, Urdu, Punjabi as well as in Marathi, it is also known as "hayseed" in English, "Uluva" in Malayalam, "Vendhayam" in Tamil, "Shoot" in Hebrew, "Hulba" in Arabic, "Dari" in Persian and "Moshoseitaro" in Greek. It is a fragrant annual leguminous species from the Fabaceae family that is diploid in status and free of aneuploidy. It has a close genetic resemblance with the widely known forage plant lucerne (*Medicago sativa* L.), sometimes also referred as the "Queen of Forages."

For many centuries, this plant has been utilized as a significant ancient, multifunctional remedial plant widely used in the Indian Ayurveda, alongside its use in the treatment of mankind and animal patients in ancient Tibetan and Chinese therapeutic practices (Dangi et al., 2016). It may also be used as a spice (seeds) and an herbal medicine (either fresh or dried out leaves; Zandi et al., 2015). Robust pods and delicate leaflets of fenugreek are utilized for variety of medical treatments and as a flavoring ingredient into the Indian cuisine (Baliga et al., 2017). This legume plant is cultivated extensively as a flavoring spice in many regions of Southern Asia, the Middle East, Mediterranean Europe and Northern Africa. In the popular East Indian curries, it is act as a major ingredient and as a curry powder.

The scientific word "*foenum-graecum*" implies "Greek hay" suggests that fenugreek was widely recognized as the source of fodder since ancient times (Basu et al., 2008, 2009; Zandi et al., 2011). It was historically cultivated as a source of fodder in areas such as Mediterranean Eurasia, the Middle East,

Northern Africa and Russia before spreading to various regions of the planet, such as the extensive Indian subcontinent as well as to significant portions of the Southern Asia (Dangi et al., 2016).

As a self-pollinating crop exhibiting an unpredictable growing pattern, fenugreek appears to grow continuously over the course of the growing season (Solorio-Sánchez et al., 2014). As a result, fenugreek keeps blooming in the field as long as occurrence of frost or a special desiccant is used. Delicate branches and sub-branches that bears light-green colored compound trifoliate leaflets with long stalks are generated by fenugreek and pinnate structure have triangular stipules (Figure 1.1). It also produces white to yellowish white tint small papilionaceous flowers with 5 petals appears either in terminal or axillary position. Its axillary flowers are hermaphrodite, insect-pollinated, and produces 2 to 8 pods of 15 cm length (Ahmad et al., 2016).

The pods of fenugreek are oblique and 10–18 cm long and 3.5–5 mm wide, having short hair. The greenish to reddish pods become brownish when they are mature (Petropoulos, 2002). The embryo of the fenugreek seed is yellow and is encircled by a corneous layer of endosperm that is white and partially translucent (Wani and Kumar, 2018). Fenugreek is generally 30 to 60 cm long, having a long cylindrical stem of pinkish color, and a root with a finger-like shape. Plants which are grown under dry land (rainfed) condition are shorter in height as compared to the irrigated ones. Fenugreek takes 5–10 days for germination, and then takes 5–8 days to develop first trifoliate leaves. Its seeds are firm, smooth, tiny (5 mm length), and yellowish to golden-yellow in color. It grows quickly and takes 4 to 7 months to achieve maturity. Midsummer, between June and August, is the time of blossoming, and late summer is when the seeds ripen. It may also flourish in a tropical environment with moderate winters and cool summers (Ahmad et al., 2016).

Fenugreek is used in various medical purposes, and since it includes necessary levels of proteins, carbs, lipids, and amino acids for good health, it may also be used as a dietary supplement for people. The seeds and leaves of this crop are consumed across various parts around the globe for a variety of reasons, which includes medicinal uses such as anti-diabetic, minimizing blood sugar as well as cholesterol level, anti-microbial, cancer preventing ingredient, and so forth, and culinary uses consisting of soups, curries, coloring agents, young plants consumed just like a vegetable; roasted grains as substitute of coffee in Africa, regulating pests in stored-grain preservation and in scent manufacturing sectors, and so forth.

Acharya et al. (2006) and Zandi et al. (2015) recorded that steroidal sapogenin (4-hydroxyisoleucine), alkaloids, flavonoids, complex carbohydrates, and essential amino acids (diosgenin, tigogenin, and yamogenin) are the most significant phytochemicals found in fenugreek seeds and leaves. Numerous polyphenolic flavonoids, including trigonelline, saponins, quercetin, and phytic acid are present in fenugreek seeds (Roberts, 2011). Additionally, the seeds contain of mucilaginous fiber, lysine, and several uncommon substances that have been found to be effective for restorative actions; prevent cholesterol concentration and lower blood sugar levels. This crop is a valuable resource and a business opportunity that serves the worldwide nutraceuticals, pharmaceuticals, and culinary sectors due to the abundance of wide variety of phytochemicals with therapeutic value (Basu et al., 2017; Sindhu et al., 2017). According to Basu et al. (2014) and Solorio-Sanchez et al. (2014), this spice crop has been utilized just like an herbal medicine by important prehistoric civilizations in the Europe and Asia which includes both of the Chinese and the Indus valley.

1.2 ORIGIN

There are conflicting assertions regarding the origin of fenugreek (Figure 1.2). Vavilov (1926) proposed that it is the native of Mediterranean region, whereas another report (Candolle, 1964) stated that it was Asian in origin. Also, according to Hardman (1969) and Fazli (1978), fenugreek is thought to have originated in Asia. Candolle (1964); Fazli (1978); and Hardman (1969) discovered that fenugreek emerges wild across the Indian states of Punjab and Kashmir, the Persian and Mesopotamian deserts, Asia Minor, and a few nations in Southern Europe, including Italy, Greece, and Spain. On the contrary, according to Dangi et al. (2004), it really originated in Turkey. In addition, fenugreek is said to have been grown as a multi-use crop across several regions of northern Africa, western and southern Asia, Australia, Northern and Southern America, and Europe (Acharya et al., 2006). An additional study reveals that this herb is indigenous to Northern Africa and India (Fuller and Stephens, 2015). According to a source, fenugreek originated in West Africa and is currently grown in nations throughout Asia, including Turkey, Egypt, China, India, and Pakistan (Rashid et al., 2018). This herb is reportedly utilized in the medical, nutraceutical, and pharmaceutical industries. Its seeds and leaves were employed in the culinary and pharmaceutical industries by various countries throughout the world. Such controversies over the fenugreek's origins show that its exact location is still up for debate.

1.3 TAXONOMY

Trigonella L. comes under the *Trifolieae* (Fabaceae) (subtribe), along with the *Trifolium* L., *Medicago* Inc., and *Melilotus* L. According to Dangi et al. (2016), this subtribe is monophyletic belonging to the "*Vicioid clade*," and the taxa of this group have trifoliated leaves with stipules that are attached to the stem but aren't entirely covering it. Within this subtribe, *Trigonella* is the genus with the largest number of taxa and the greatest economic significance. It contains annual or perennial fragrant herbs and may be distinguished by its steady anthers, diadelphous stamens, canular or campanulate sepals with three (small) and two (big) identical lobes, and an ovary with many ovules (Dangi et al., 2016). According to Townsend and Guest (1974), taxa are extensively spread over the arid zones surrounding the Western Asia, Europe, Northern and Southern Africa, Northern America, Mediterranean region, and Southern Australia. According to Petropoulos (2002), Linnaeus identified 260 species, whereas other writers recognized around 62 and 128 species (Rouk and Mengesha, 1963; Fazli, 1967). The specific count of species recognized within *Trigonella* relies upon authors, approved synonymies, and unresolved names.

According to the most current analysis provided by Small (1987), there are presently 62 species in the genus (Dangi et al., 2016). The Plant List featured 98 approved taxa, 97 names that had not yet been resolved, 3 names that had been incorrectly applied, and 27 acceptable synonyms. Among the numerous species of *Trigonella* with commercial potential, *T. foenum-graecum* L. is the sole known species that is commonly cultivated and the most well-known due to its countless benefits and qualities (Basu et al., 2006; Petropoulos, 2002). Following is the commonly found fenugreek species which are also known to exist today: *T. anguina, T. arabica, T. cariensis*, *T. arcuata*, *T. caerulea*, *T. corniculata*, *T. rigida*, *T. suavissima*, *T. torulosa*, *T. hamosa*, *T. cretica*, *T. spinosa*, *T. occulta*, *T. polycerata* (Acharya et al., 2008). Additionally, other authors have suggested a variety of applications for various fenugreek species, although some are currently synonymous or belonging to different genus. For instance, *T. monspeliaca* is synonymous to *Medicago monspeliaca* (L.) Trautv (Hidvegi et al., 1984), *T. polycerata* is synonymous to *Medicago orthoceras* (Kar. & Kir) Trautv (Hardman, 1969; Hardman and Fazli, 1972; Mehra et al., 1996), and *T. radiata* Boiss used for *Medicago radiata* L. (Allen and Allen, 1981), and so forth.

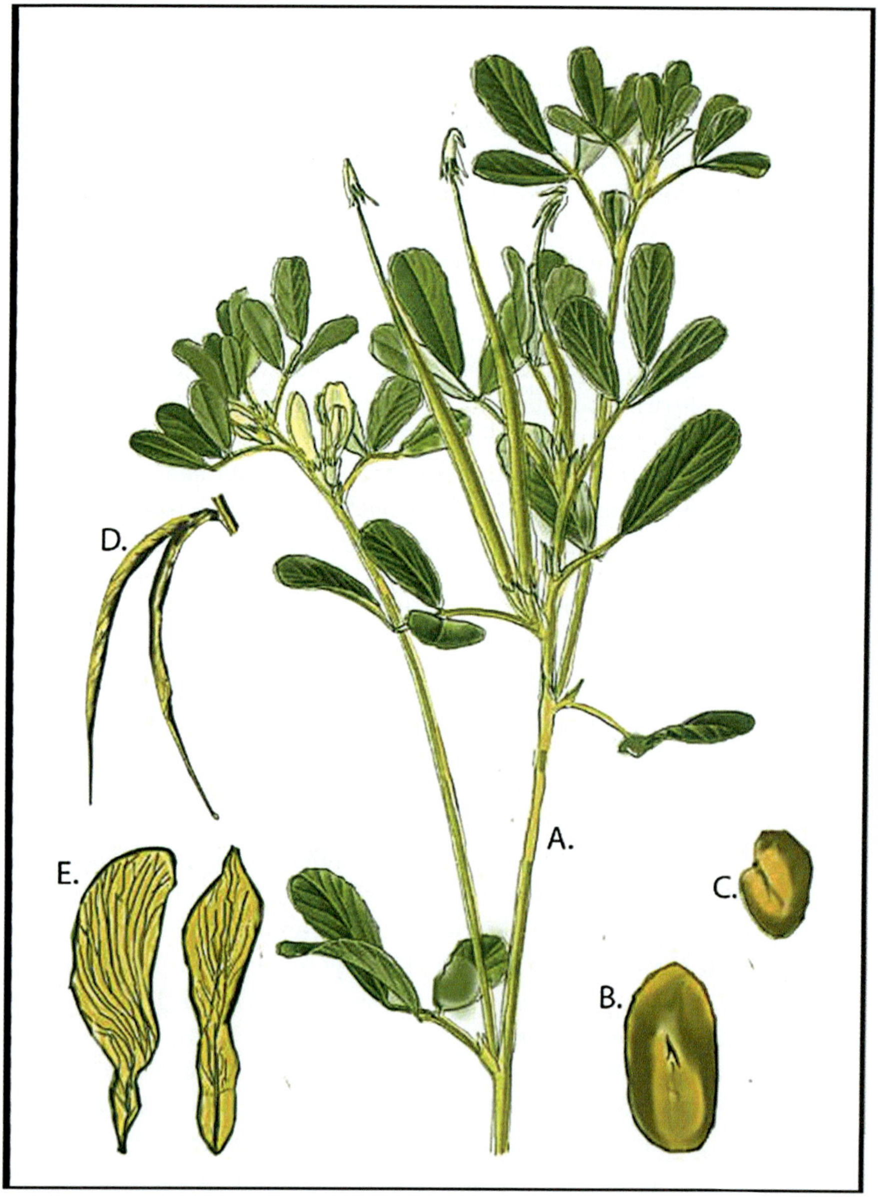

FIGURE 1.1 Fenugreek plant: (A) stem with compound trifoliate leaflets; (B) and (C) seeds; (D) pods; and (E) papilionaceous flower.

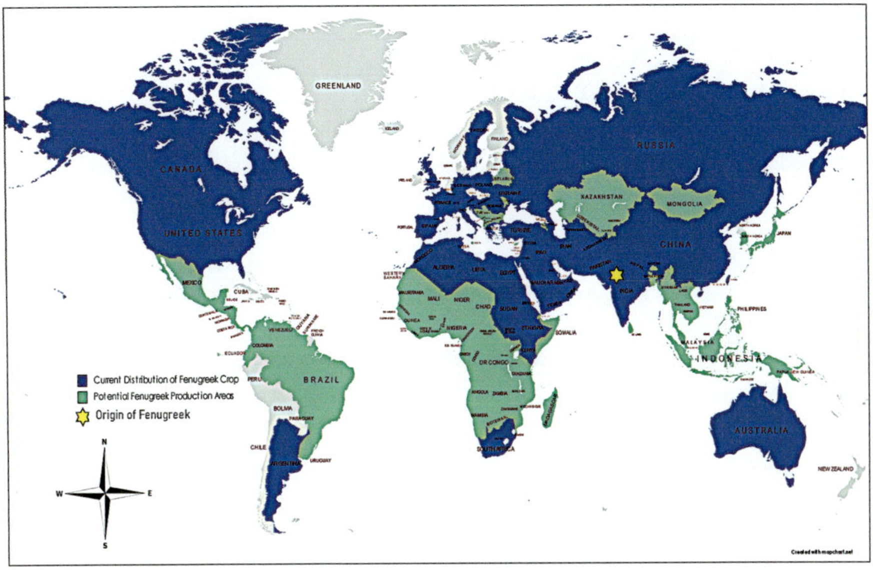

FIGURE 1.2 Map displaying the origin, current, and potential distribution of fenugreek.

1.4 DISTRIBUTION

Seeking novel and alternative crops to withstand challenges to agricultural output brought on by global warming a changing climate, fenugreek is currently being familiarized under innovative crop introduction programs in various countries throughout Africa, America, Asia, and the EU with proper agroclimatic regimes (Basu et al., 2006). As a result, the Old-World nations tend to have higher agricultural production for fenugreek than the New World (Acharya et al., 2006; Mehra et al., 1996). These territories, regions, locations, and nations are referred to as the traditional fenugreek growing countries. It is, however, dispersed all over the world. Overall dispersal of fenugreek is illustrated in Table 1.1 and Figure 1.2. India is the world's leading grower of fenugreek, and Asia is regarded as being among the top producing continents worldwide. Fenugreek is produced in large quantities throughout Asia, with considerable contributions coming from Iran, China, and Pakistan. According to Qadir et al. (2017), Africa is the continent that produces the second-most fenugreek after Asia. The cultivation of fenugreek has also been documented in regions of UK, Russia, West Asia, Mediterranean Europe, USA, and Australia.

TABLE 1.1 Overall Distribution of Fenugreek

Major Areas	Nations
Southern Asia	India, Pakistan, Nepal
Central Asia	Afghanistan, Azerbaijan, Turkmenistan
Middle East	Turkey, Syria, Iraq, Israel, Qatar, Kuwait, Saudi Arabia, Iran, Lebanon, Bahrain, United Arab Emirates, Oman, Yemen, Jordan
Far East	Taiwan, China
European Countries	Italy, Germany, Russia, Romania, Hungary, Switzerland, United Kingdom (UK), Portugal, Greece, Slovenia, Croatia, Spain, Poland, Sweden, Ukraine, France
Northern Africa	Egypt, Algeria, Tunisia, Libya, Morocco, Sudan
Horn of Africa	Eritrea, Ethiopia
Sub-Saharan Africa	South Africa, Kenya
North America	United States (US), Canada
South America	Argentina
Oceania	Australia

This plant contains 1, 6, 1, and 5 species across the lands of Australia, Asia, Africa, and Europe, respectively, according to Oncina et al. (2002). Table 1.2 displays the main *Trigonella* species and their geographical distribution. According to Acharya et al. (2006), fenugreek is also grown in several pockets of south Asia, Europe, Australia northern Africa, and America. From Israel, *Trigonella* species were spread to Iraq, Turkey, Syria, Iran, Morocco Spain. These species are also most prevalent in Turkey. *T. foenum-graecum* cultivars with big seeds are common in the Mediterranean region, whilst those with small seeds are more common in the eastward (Malhotra, 2011). According to Jain et al. (2013), this plant is extensively cultivated in China, Turkey, Iran, Russia, Greece, Ukraine, Nepal, Egypt, Iran, Bangladesh, Spain, France, Pakistan, Morocco, and India.

TABLE 1.2 Important Species of Trigonella and its Distribution

Trigonella Species	Distribution
T. foenum-graecum L.	Caucasus, Central Asia, Eastern Europe
T. Arabica Delile	North-Eastern Africa, Syria to North-Eastern Egypt
T. caerulea (L.)	Central, Western, and Southern Europe; North Africa
T. caerulea (L.)	Eastern Mediterranean continent; South-Eastern Europe
T. stellata Forsk	Middle East, Egypt, North Africa, Arabia, Morocco, Iran, Tunisia, Lebanon, Algeria, Canary Islands, W. Asia, Israel, Kuwait, Iraq
T. corniculata (L.)	Mediterranean continents

India, Pakistan, and China are the countries with the peak number of fenugreek varieties, followed by African (northern) nations (Sudan, Tunisia, Egypt, Algeria, Libya, Morocco), nations Horn of Africa (Eritrea, Ethiopia), African (sub-Saharan) countries (South Africa Kenya), Middle Eastern countries (Oman, Turkey, Jordan, Iraq, Yemen, Israel, Syria, Iran), and European nations (Germany, Azerbaijan, Poland, Switzerland, Italy, Spain, France, Portugal, United Kingdom, Austria, Romania, Slovenia, Greece, Hungary). Of late, other regions have also expressed a keen fascination in fenugreek cultivation, most notably the countries of N. America (including the US and Canada), S. America (Argentina), and Oceania (mostly the Australian islands). Canada has taken an active role in promoting fenugreek production in nations that have not traditionally grown fenugreek, and the "Tristar" cultivar was launched as the first one in North America (Acharya et al., 2007). The US has also ensued intently after Mexico has just recently begun to express some curiosity in the cultivation of the crop. The American and Canadian prairies, as well as dry, semiarid portions of Mexico and the United States, are appropriate locations for the cultivation of fenugreek (Acharya et al., 2007; Basu et al., 2014).

Australia has also expressed a keen attention in fenugreek cultivation and has gradually emerged into a significant fenugreek supplying nation with a number of novel cultivars under development. Fenugreek has traditionally been produced in the Mediterranean region because of its milder temperature, which now extends throughout the Middle East and North African agro-climatic zones (Acharya et al., 2006; Basu et al., 2014). One of the major growing regions is found in China, South Asia (Nepal, India, and Pakistan), and the Horn of Africa, which is another expansion of the fenugreek-producing zone. The fenugreek supplying zone has recently expanded from Australia through the enormous continents of Eurasia and Americas to Africa. Traditional fenugreek growing countries include Taiwan, Turkmenistan, and Afghanistan in Far Eastern Asia, as well as Afghanistan and Turkmenistan in Central Asia.

It is believed that climatic and agro-ecological conditions have a significant impact on fenugreek cultivation (Koocheki et al., 2004; May, 1975). Most often, fenugreek cultivation takes place in nearly every region with fertile land and an ideal climatic condition. Figure 1.2 shows the regions throughout the world across the major continents that Basu et al. (2019) indicate for future prospective fenugreek production areas based on acceptable, parallel, or nearly equivalent agro-climatic conditions.

The semiarid crop's high level of resistance to harsh conditions is well demonstrated (Snehlata and Payal, 2012). It is noted that the fenugreek has a high heterogeneity throughout different coordinates, regardless of desert (non-arable) locations. Tropical and subtropical areas like India and Ethiopia are often recognized for their species richness. The countries of China, Canada, Turkey, Tunisia, Iran, and Pakistan have quite significant genetic variability. Among which, a Mediterranean environment is good for cross, particularly forage cultivation, and extends over Turkey, the northern portion of Tunisia and Iran. The best regions of China for fenugreek growing are probably the central and southern regions. Even though it is often said in the literature that, fenugreek is an herbaceous plant that tolerates frost and extremely cold temperatures, Canada is only given close to 4.95% of the necessary frequency compared to other countries. Unquestionably, India produces the maximum quantity of fenugreek in the total world's production, and because of its unique climatic situations, it also has the greatest genetic diversity in other regions for the multifunctional crop (used as a medicine, an aromatic flavoring, or even as a fodder).

1.4.1 INDIAN SCENARIO

In India, the majority of the country's productivity is assembled in Haryana, Madhya Pradesh, Gujarat West Bengal, and Rajasthan (Figure 1.3). Table 1.3 shows the leading fenugreek growing states of India with higher productivity. Rajasthan occupies largest cropped area as well as production of fenugreek among all states. According to annual report of Directorate of Arecanut and Spices Development (DASD, 2021) 2,03,360 tons of fenugreek seed were produced from 1,33,229 hectares of land and the crop's current productivity is 1526 kg/ha. During 2020–2021, India exported 38,300 tons of fenugreek seed valued at Rs. 24,642 lakhs. United States, Yemen, United Arab Emirates, United Kingdom, Nepal, Bangladesh, Saudi Arabia, Sri Lanka, Germany, and Egypt are the main importers of Indian fenugreek.

1.4.2 CULTIVATED VARIETIES

Selection of a variety is mostly based on how well it adapts to the land and climate, and ideally, it should be resistant to or tolerable of the local diseases and pests. Several varieties have been introduced for growing to various locations. Table 1.4 displays the names of some of the significant cultivated varieties that are suggested for various states.

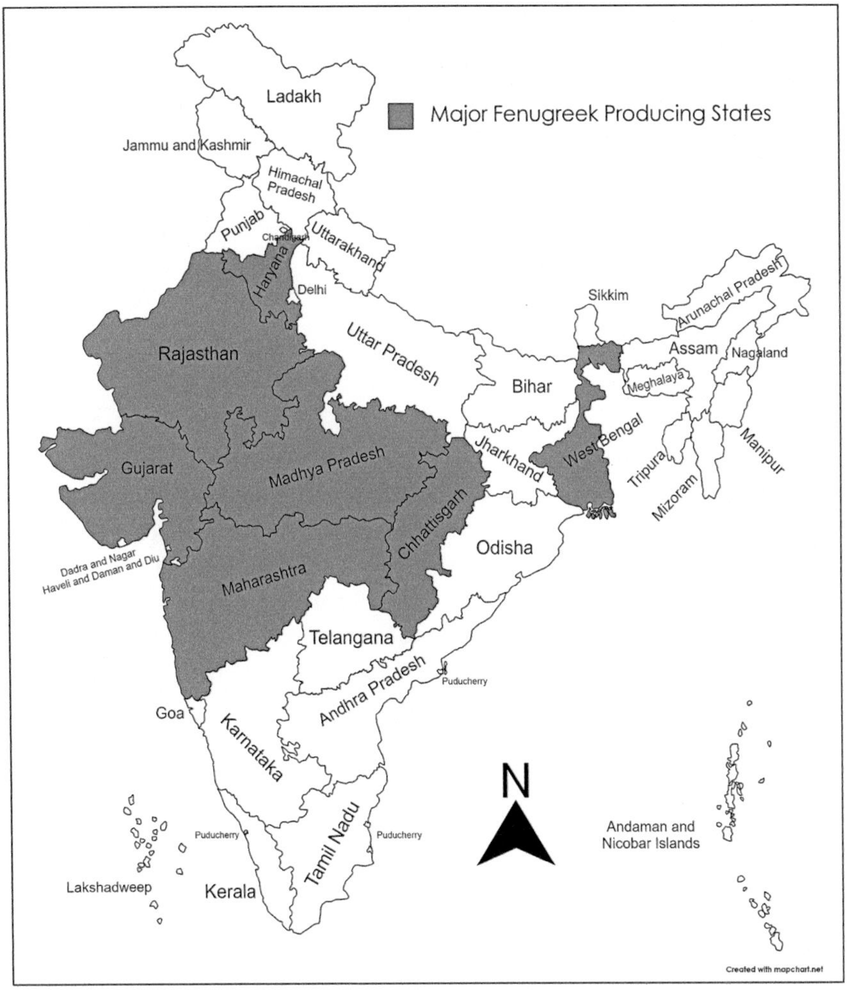

FIGURE 1.3 Map showing leading fenugreek growing states of India.

TABLE 1.3 Major Fenugreek Producing States with Districts

States	Highest Productivity
Rajasthan	Churu, Sikar, Bikaner, Pratapgarh, Jodhpur
Gujarat	Dahod, Banaskatha, Mehasana, Surendranagar, Patan
Madhya Pradesh	Ujjain, Ratlam, Guna, Neemuch, Shajapur, Khargone, Rajgarh, Dhar, Sehore, Betul, Mandsaure
Chhattisgarh	–
Haryana	–
Maharashtra	–

TABLE 1.4 Some Important Cultivated Varieties Recommended for Different States in India

States	Varieties
Rajasthan	Ajmer Methi 1 (AFg-1), Ajmer Methi 2 (AFg-2), Ajmer Methi 3 (AFg-3), Ajmer Methi 4 (AFg-4), Ajmer Methi 5 (AFg-5)
Gujarat	GM-1
Bihar	Rajandra Kranti
Haryana	Hisar Mukta, Hisar Sonali
Uttar Pradesh and Uttaranchal	Pusa Kasuri

1.5 CONCLUSION

Throughout the Indian subcontinent, fenugreek, a conventional fragrant and medicinal crop, has been produced over decades. This crop is produced not only in South Asia but also in certain parts of China, Northern Africa, the Middle East, Mediterranean Europe, Southeast Asia, Canada, Australia, the USA and also in Argentina. India is the world's highest grower of fenugreek, however because of extensive domestic usage, it doesn't account for a significant portion of global fenugreek trade. In dry soil (rain fed) regions with minimal scope for irrigation, fenugreek has excellent growing potential. However, the crop thrives with irrigation, making it appropriate for sustainable agriculture because of its water-saving qualities, which also lowers the expenses of cultivation for farmers. Fenugreek is a crop that is mostly grown on dry terrain, making it an appropriate crop for minimal input agricultural systems in impoverished and emerging countries in Latin America, Asia, and Africa. With proper climatic situations, dry or semi-arid countries or pockets within these continents can offer as possible new sites for fenugreek cultivation. According to reports, climate change and global warming will have a considerable influence on agricultural production all over the world. Therefore, it is crucial to actively seek out opportunities for introducing novel climate-resilient plants such as fenugreek over tropical, subtropical, and sub-temperate agro-climatic zones. By growing novel non-traditional or non-conventional crops in agriculturally neglected or under-explored regions along with the traditional or conventional crops of that area, this will boost agricultural productivity.

Additionally, being an annual legume, it can fix nitrogen from the atmosphere to the soil. Therefore, it is a crop that could potentially be used to reclaim marginal or unsuitable lands for agriculture in order to increase the yield of crops and turn inappropriate lands back into efficient fields for poor as well as small farmers who might not possess accessibility to field for farming.

Additionally, fenugreek may be introduced to the fast-growing cycles of native crops to naturally improve the nitrogen level of the soil. Additionally, it may be included into organic farming techniques. Fenugreek is a versatile plant that may be produced for food, spices, medicine, and even as an agricultural commodity in the pharmaceuticals, nutraceutical, and culinary sectors due to the availability of numerous significant botanical compounds inside its leaves as well as seeds. Therefore, this herb is regarded as a chemurgic plant because of its uses in industry and suitability for use in the little-input farming system of emerging along with impoverished countries.

Because of this, it has tremendous global requirements from the related nutraceutical, pharmaceutical, and nutritious food sectors. Our planet depicts many variations of agricultural ecosystems throughout the globe that are suited for the growing of fenugreek in dry, semi-arid, and arid climates. The crop has the ability to be produced throughout wider regions since it is a chemurgic crop having important financial and commercial significance in the country, even if it is now only planted in a small quantity in prospective fenugreek producing locations, including Central and South America, Africa, and South-eastern Asia. In addition, fenugreek's genetic makeup, metabolic processes, and unique adaptations enable it to function as a proactive fixer of nitrogen with little demand for water and enable the plant to flourish in semi-arid climates with low-input production practices. Therefore, it can be considered a water efficient, energy coherent and climate resilient multi-use herb for underdeveloped, agriculturally reliant nations surrounding the world with semi-arid climates. It can also be readily introduced into localized agricultural cycles in various geographical regions for naturally restoring the earth's fertility.

KEYWORDS

- **cultivars**
- **fenugreek**
- **geographical distribution**
- **medicinal herb**
- **Mediterranean region**
- ***Trigonella* species**

REFERENCES

Acharya, S. N., Blade, S., Mir, Z., & Moyer, J. S. (2007). Tristar fenugreek. *Canadian Journal of Plant Science*, *87*, 901–903.

Acharya, S. N., Thomas, J. E., & Basu, S. K. (2006). Fenugreek: An "old world" crop for the "new world." *Biodiversity*, *7*(3–4), 27–30.

Acharya, S. N., Thomas, J. E., & Basu, S. K. (2008). Fenugreek, an alternative crop for semiarid regions of North America. *Crop Science*, *48*(3), 841–853.

Ahmad, A., Alghamdi, S. S., Mahmood, K., & Afzal, M. (2016). Fenugreek, a multipurpose crop: Potentialities and improvements. *Saudi Journal of Biological Sciences*, *23*(2), 300–310.

Allen, O. N., & Allen, E. K. (1981). *The Leguminosae: A Source Book of Characteristics, Uses, and Nodulation*. University of Wisconsin Press.

Baliga, M., Palatty, P. L., Adnan, M., Naik, T. S., Kamble, P. S., George, T., & D'souza, J. J. (2017). Anti-diabetic effects of leaves of *Trigonella foenum-graecum* L. (Fenugreek): Leads from preclinical studies. *Journal of Food Chemistry and Nanotechnology*, *3*(2), 67–71.

Basu, A., Basu, S. K., Kumar, A., Sharma, M., Chalghoumi, R., Hedi, A., Solorio-Sánchez, F., Balogun, M. O., Hafez, E. E., & Cetzal-Ix, W. (2014). Fenugreek (*Trigonella foenum-graecum* L.), a potential new crop for Latin America. *American Journal of Social Issues & Humanities*, *4*(3), 148–162.

Basu, S. K., Acharya, S. N., & Thomas, J. E. (2006). A report on powdery mildew infestations caused by *Erysiphe polygoni* DC in North American grown fenugreek. *Journal of Mycopathological Research*, *44*(2), 253–256.

Basu, S. K., Acharya, S. N., & Thomas, J. E. (2008). Application of phosphate fertilizer and harvest management for improving fenugreek (*Trigonella foenum-graecum* L.) seed and forage yield in a dark brown soil zone of Canada. *Current Applied Science and Technology*, *8*(1), 1–7.

Basu, S. K., Acharya, S. N., Bandara, M. S., Friebel, D., & Thomas, J. E. (2009). Effects of genotype and environment on seed and forage yield in fenugreek (*Trigonella foenum-graecum* L.) grown in western Canada. *Australian Journal of Crop Science*, *3*(6), 305–314.

Basu, S. K., Zandi, P., & Cetzal-Ix, W. (2017). Opportunities for fenugreek (*Trigonella foenum-graecum* L.) as a chemurgic crop in the emergent global nutraceutical and functional food industries. *International Journal of Agricultural Sciences*, *8*(1), 9–13.

Basu, S. K., Zandi, P., & Cetzal-Ix, W. (2019). Fenugreek (*Trigonella foenum-graecum* L.): Distribution, genetic diversity, and potential to serve as an industrial crop for the global pharmaceutical, nutraceutical, and functional food industries. In R. B. Singh, R. R. Watson, & T. Takahashi (Eds.), *The Role of Functional Food Security in Global Health* (pp. 471–497). Academic Press.

Candolle, A. de. (1964). *Origin of Cultivated Plants*. Hafner Press.

Dangi, R. S., Lagu, M. D., Choudhary, L. B., Ranjekar, P. K., & Gupta, V. S. (2004). Assessment of genetic diversity in *Trigonella foenum-graecum* and *Trigonella caerulea* using ISSR and RAPD markers. *BMC Plant Biology*, *4*(1), 1–11.

Dangi, R., Tamhankar, S., Choudhary, R. K., & Rao, S. (2016). Molecular phylogenetics and systematics of *Trigonella L.* (Fabaceae) based on nuclear ribosomal ITS and chloroplast *trn L* intron sequences. *Genetic Resources and Crop Evolution*, *63*, 79–96.

Directorate of Arecanut and Spices Development (DASD). (2021). *Annual Report 2020–21*. Ministry of Agriculture and Farmers Welfare, Government of India. Calicut, Kerala.

Fazli, F. R. Y. (1967). *Studies in Steroid-Yielding Plants of the Genus Trigonella* (Doctoral dissertation, University of Nottingham).

Fazli, F. R. Y. (1978). The spice fenugreek (*Trigonella foenum-graecum*), its commercial variety of seed as a source of diosgenin. *Tropical Science*, *10*, 66–78.

Fuller, S., & Stephens, J. M. (2015). Diosgenin, 4-hydroxyisoleucine, and fiber from fenugreek: Mechanisms of action and potential effects on metabolic syndrome. *Advances in Nutrition*, *6*(2), 189–197.

Hardman, R. (1969). Pharmaceutical products from plant steroids. In *Steroids* (pp. 196–222).

Hardman, R., & Fazli, F. R. Y. (1972). Methods of screening the genus *Trigonella* for steroidal sapogenins. *Planta Medica*, *21*(2), 131–138.

Hidvegi, M., Lasztity, R., Bekes, F., & El Kady, A. (1984). Contributions to the nutritional characterization of fenugreek (*Trigonella foenum-graecum* L.). *Acta Alimentaria*, *13*, 315–324.

Jain, A., Singh, B., Solanki, R. K., Saxena, S. N., & Kakani, R. K. (2013). Genetic variability and character association in fenugreek (*Trigonella foenum-graecum* L.). *International Journal of Seed Spices*, *2*, 22–28.

Koocheki, A., Nassiri Mahallati, M., Zarea Fizabadi, A., & Jahanbin, G. (2004). Diversity of cropping systems in Iran. *Agronomy and Horticulture*, *63*, 70–83.

Malhotra, S. K. (2011). Fenugreek (*Trigonella foenum-graecum* L.). In *Genetic Resources, Chromosome Engineering, and Crop Improvement* (Chapter 4).

May, R. M. (1975). Pattern of species abundance and diversity. In M. L. Cody & J. M. Diamond (Eds.), *Ecology and Evolution of Communities* (pp. 80–120). Belknap Press (Harvard University).

Mehra, P., Yadav, R., & Kamal, R. (1996). Influence of nicotinic acid on production of trigonelline from *Trigonella polycerata* tissue culture. *Indian Journal of Experimental Biology*, *34*(11), 1147–1149.

Oncina, R., Del Río, J. A., Gomez, P., & Ortuno, A. (2002). Effect of ethylene on diosgenin accumulation in callus cultures of *Trigonella foenum-graecum* L. *Food Chemistry*, *76*(4), 475–479.

Petropoulos, G. A. (2002). *Fenugreek: The Genus Trigonella*. CRC Press.

Qadir, A., Ali, N., Jan, S. A., Rabbani, M. A., Khurshid, H., Nouman, A., & Ullah, F. (2017). Characterization of agromorphological variation in exotic fenugreek (*Trigonella foenum-graecum* L.) germplasm. *Journal of Biodiversity and Environmental Sciences*, *10*(3), 71–79.

Rashid, F., Hussain, S., & Ahmed, Z. (2018). Extraction, purification, and characterization of galactomannan from fenugreek for industrial utilization. *Carbohydrate Polymers*, *180*, 88–95.

Roberts, K. T. (2011). The potential of fenugreek (*Trigonella foenum-graecum*) as a functional food and nutraceutical and its effects on glycemia and lipidemia. *Journal of Medicinal Food*, *14*(12), 1485–1489.

Rouk, H. F., & Mengesha, H. (1963). Fenugreek (*Trigonella foenum-graecum* L.): Its relationships, geography, and economic importance. Imperial Ethiopian College of Agriculture and Mechanical Arts.

Sindhu, A., Tehlan, S. K., & Chaudhury, A. (2017). Analysis of genetic diversity among medicinal therapist *Trigonella foenum-graecum* L. genotypes through RAPD and SSR markers. *Acta Physiologiae Plantarum*, *39*, 1–15.

Small, E. (1987). A taxonomic study of the "medicagoid" *Trigonella* (Leguminosae). *Canadian Journal of Botany*, *65*(6), 1199–1211.

Snehlata, H. S., & Payal, D. R. (2012). Fenugreek (*Trigonella foenum-graecum* L.): An overview. *International Journal of Current Pharmaceutical Review and Research, 2*(4), 169–187.

Solorio-Sánchez, F., Solorio-Sánchez, B., Basu, S. K., Casanova-Lugo, F., Sarabia-Salgado, L., Ku-Vera, J., Aguilar-Pérez, C., Ramírez-Avilés, L., Noguera-Savelli, E., Cetzal-Ix, W., & Infante-Cruz, Á. (2014). Opportunities to grow annual forage legume fenugreek (*Trigonella foenum-graecum* L.) under Mexican silvopastoral system. *American Journal of Social Issues & Humanities, 4*(2), 86–95.

Townsend, C. C., & Guest, E. (1974). *Trigonella L. Flora of Iraq, 3*, 87–112.

Vavilov, N. I. (1926). *Studies On the Origin of Cultivated Plants*. Leningrad.

Wani, S. A., & Kumar, P. (2018). Fenugreek: A review on its nutraceutical properties and utilization in various food products. *Journal of the Saudi Society of Agricultural Sciences, 17*(2), 97–106.

Zandi, P., Basu, S. K., Khatibani, L. B., Balogun, M. O., Aremu, M. O., Sharma, M., Kumar, A., Sengupta, R., Li, X., Li, Y., & Tashi, S. (2015). Fenugreek (*Trigonella foenum-graecum* L.) seed: A review of physiological and biochemical properties and their genetic improvement. *Acta Physiologiae Plantarum, 37*, 1–14.

Zandi, P., Shirani Rad, A. H., Daneshian, J., & Bazrkar-Khatibani, L. (2011). Agronomic and morphologic analysis of fenugreek (*Trigonella foenum-graecum* L.) under nitrogen fertilizer and plant density via factor analysis. *African Journal of Agricultural Research, 6*(5), 1134–1140.

CHAPTER 2

Role of Mineral Nutrients on Growth, Yield, and Active Constituents of Fenugreek

S. VENUGOPAL,[1] A. LAXMAN KUMAR,[2] S. BHARATH KUMAR,[2] B. MARUTHI,[3] M. AKANKSHA,[4] A. C. JNANESHA,[1] and RANJITH KUMAR[1]

[1]*CSIR–Central Institute of Medicinal and Aromatic Plants, Research Center, Boduppal, Hyderabad, Telangana, India*

[2]*Sri Konda Laxman Telangana State Horticultural University, Mulugu, Siddipet, Telangana, India*

[3]*Bidhan Chandra Krishi Viswavidyalaya, Mohanpur, Nadia, West Bengal, India*

[4]*Osmania University, Department of Botany, Hyderabad, Telangana, India*

ABSTRACT

Fenugreek (*Trigonella foenum-graecum* L.), an annual herbaceous plant in the Fabaceae family, is indigenous to the Mediterranean region, North Africa, and western Asia. It is cultivated for its young shoots, leaves, and seeds, which are rich in dietary fiber, protein, and various bioactive compounds. In addition to its widespread use as a spice in many cuisines, fenugreek seeds are also used in traditional medicine to treat numerous ailments. The growth, yield, and quality of fenugreek are heavily influenced by the availability of mineral nutrients. Macronutrients such as nitrogen (N), phosphorus (P), potassium (K), calcium (Ca), magnesium (Mg), and sulfur (S) are crucial for growth. Nitrogen is essential for amino acid and protein production, phosphorus supports reproductive organ growth and root development, and potassium plays a vital role in photosynthesis and enzyme activation. Calcium

Fenugreek: Improvement in Yield Potential Through Plant Growth Regulators (PGRs) and Depolymerized Polysaccharides. Tariq Ahmad Dar, Bilal Ahmad, Moinuddin, and Khalid Rehman Hakeem (Eds.)

ensures cell wall stability, magnesium contributes to chlorophyll synthesis, and sulfur affects protein synthesis and the production of active components. Micronutrients, including iron (Fe), manganese (Mn), zinc (Zn), copper (Cu), boron (B), molybdenum (Mo), and chlorine (Cl), are equally significant. Iron aids chlorophyll synthesis and electron transfer, manganese supports enzyme activation and nitrogen metabolism, and zinc promotes stress tolerance and protein synthesis. Copper is vital for lignin synthesis and enzyme activity, boron aids in cell division and seed development, molybdenum is necessary for nitrogen fixation, and chlorine regulates stomatal movement and osmotic balance. Thus, the availability of macronutrients and micronutrients greatly impacts the quantity and quality of fenugreek production.

2.1 INTRODUCTION

Fenugreek (*Trigonella foenum-graecum* L.) is an annual herb grows up to 60 cm in height that belongs to the Fabaceae family (Pal and Mukherjee, 2020) and native to western Asia, North Africa, and the Mediterranean region (Acharya et al., 2008). It has trifoliate leaves with an oblong or ovate shape and small white or yellow flowers. The plant produces pods that contain 10–20 small, angular, and brownish-yellow seeds (Basha et al., 2020; Ilyina et al., 2017; Sharangi et al., 2018). Around the world, fenugreek is grown in several areas, including Asia, Africa, Europe, and North America (Acharya et al., 2006; Basu et al., 2019). The plant needs moderate temperatures and plenty of sunlight for optimum growth, and it does best in well-drained soils with a pH range of 6 to 7. It is especially suited to climates that are dry and semi-dry (Hordofa and Tolossa, 2020; Nair and Groot, 2021). India is one of the world's top producers of fenugreek, with Pakistan, China, and Ethiopia coming in second and third (Kumar, 2019; Zandi et al., 2017).

Fenugreek has a wide range of applications in numerous industries. Fenugreek seeds and leaves are used in food preparation as a spice and flavoring, especially in Indian, Middle Eastern, and Mediterranean cuisines (Kumar, 2019; Mamatha and Panyam, 2021; Sinha et al., 2007). The seeds are also used to make fenugreek oil, valued for its flavor and aroma in the food industry (Balasubramanian et al., 2016). Fenugreek also has a long history of use in traditional medicine. It is known for its antioxidant, anti-inflammatory, and anti-diabetic properties, along with other potential health benefits (Bafadam et al., 2021; Narapogu et al., 2021; Nathiya et al., 2014). Traditional medical practices like Ayurveda and Unani frequently use fenugreek seeds and leaves to treat various illnesses, such as digestive problems, respiratory issues, and hormonal imbalances (Salman and Qadeer, 2021). Due to its bioactive components, fenugreek has recently drawn interest from the pharmaceutical and nutraceutical industries. The seeds contain compounds with potential therapeutic effects, including saponins, alkaloids, flavonoids, and phenolic compounds (Nagulapalli et al., 2017; Visuvanathan et al., 2022). The anti-cancer, anti-inflammatory, and cholesterol-lowering properties of these bioactive substances have also been investigated (Shahrajabian et al., 2021; Waliat et al., 2023). Mineral nutrients are crucial for plant growth and development, as they play essential roles in metabolic processes and serve as building blocks for various cellular components. Plants obtain mineral nutrients primarily from external sources, with soil serving as the primary source (Bhatla et al., 2018; Maathuis and Diatloff, 2013). The soil's capacity to retain mineral nutrients is influenced by the parent material, weathering processes, organic matter content, and fertilization techniques (Anda et al., 2015; Anderson, 1988). The availability of mineral nutrients is significantly affected by the soil's parent material, as concentrations of essential elements vary across different types of rocks and minerals. For example, granitic parent materials typically contain higher levels of potassium and phosphorus, while basaltic parent materials are often rich in magnesium (Anda et al., 2015; Hamdan and Bumham, 1996; Moore et al., 2022). Ultimately, the composition of the parent material influences the nutrient content and availability in the soil. Weathering processes, such as the physical and chemical breakdown of rocks and minerals, make mineral nutrients available. Physical weathering, such as freeze-thaw cycles or erosion, reduces larger particles to smaller ones, increasing surface area for nutrient release. During chemical weathering, interactions between minerals and water release soluble nutrients. These weathering processes gradually add nutrients to the soil's nutrient pool, enabling plants to absorb them (Lavelle and Spain, 2002; Mavris et al., 2010; Ribeiro et al., 2020; Voroney, 2007). Organic matter is crucial for soil mineral nutrient availability. Organic materials, including compost, animal manures, and plant residues, aid nutrient cycling and serve as a source of slow-release nutrients. As organic matter decomposes, soil microbes release mineral nutrients in forms accessible to plants (Kumar et al., 2019; Shaji et al., 2021). Additionally, organic matter enhances soil structure, water-holding capacity, and nutrient retention, improving overall nutrient availability for plants (Atkinson, 2018; Tadesse et al., 2013; Wang et al., 2016). Both organic and inorganic fertilization techniques are essential for enhancing soil mineral nutrient availability. Inorganic fertilizers provide readily available essential nutrients that plants can absorb, while organic fertilizers decompose gradually, supplying a steady source of mineral nutrients. Fertilizer selection and application rates should be based on crop nutrient requirements and soil nutrient status (Diriba-Shiferaw, 2016; Hofman and Van Cleemput, 2004; Mazid and Khan, 2015).

For the cultivation of fenugreek mineral nutrients are crucial because they are essential for the growth, development, and general productivity of the plant. To support a variety of physiological processes, increase yield, and encourage the synthesis of active components in fenugreek, it is imperative that an adequate and balanced supply of mineral nutrients be provided (Chaichi et al., 2015; Zhang et al., 2021). For the cultivation of fenugreek, macronutrients like nitrogen, phosphorus, and potassium are essential. Nitrogen plays a role in plant vigor, leaf development, and vegetative growth. It significantly contributes to raising fenugreek's yield (Amirnia et al., 2018). The production of seeds, flowers, and roots depend on phosphorus, which also affects fenugreek's overall yield (Dadrasan et al., 2015; Panghal et al., 2015). Potassium is crucial for controlling several physiological processes, such as the activation of enzymes, water absorption, and nutrient transport. It aids in the synthesis of active ingredients, yield, and quality in fenugreek (Kinji and Rahdari, 2015). Additionally important macronutrients that are essential for fenugreek cultivation include calcium, magnesium, and sulfur. The formation of cell walls, the integrity of membranes, and overall plant development all depend on calcium. It aids in lowering diseases and enhancing the qualities of fenugreek (Gupta et al., 2017; Pandey, 2015). Magnesium is a crucial component of chlorophyll, and enough of it encourages photosynthesis, enzyme activity, and the accumulation of active ingredients in fenugreek (Dobrikova et al., 2022). Protein synthesis, enzyme activation, and general plant metabolism all depend on sulfur. It affects the fenugreek plant's development, yield, and quality (Maathuis and Diatloff, 2013).

For the cultivation of fenugreek, trace amounts of micronutrients like iron, zinc, manganese, copper, boron, molybdenum, and chlorine are required. In fenugreek, iron is essential for the synthesis of chlorophyll, enzyme activity, and accumulation of active constituents (Hofman and Van Cleemput, 2004). To improve fenugreek productivity and quality, zinc and manganese are involved in growth, yield, and the synthesis of active constituents (Shahrajabian et al., 2022). Additionally, various physiological processes, growth, and the accumulation of active constituents in fenugreek are influenced by copper, boron, molybdenum, and chlorine (Maathuis and Diatloff, 2013).

In conclusion, mineral nutrients are crucial for fenugreek cultivation because they are crucial for growth, yield, and the synthesis of active constituents. Maximizing fenugreek productivity, improving the quality of its yield, and maximizing the therapeutic and nutritional benefits of this priceless herb all depend on maintaining the proper balance and providing adequate nutrition of macronutrients and micronutrients.

2.2 SOURCES AND AVAILABILITY OF MINERAL NUTRIENTS: OPTIMIZING NUTRIENT SUPPLY FOR SUCCESSFUL FENUGREEK CULTIVATION

For fenugreek cultivation to support growth and development, an adequate supply of mineral nutrients is necessary. These nutrients can be obtained from a variety of sources, and various factors can affect how readily available they are in the soil. The sources and accessibility of mineral nutrients (Figure 2.1) for use in fenugreek cultivation will be covered in this section.

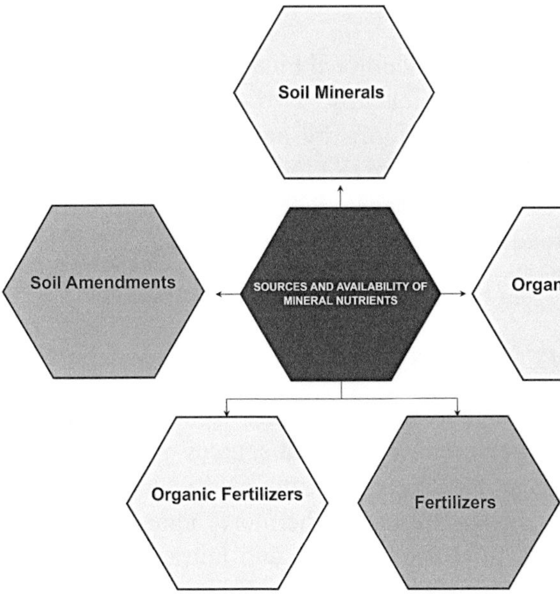

FIGURE 2.1 Sources and availability of mineral nutrients.

2.2.1 *SOIL MINERALS*

For fenugreek plants, the soil is their main source of mineral nutrients. Rocks and other parent materials that make up soil minerals have varying concentrations of the essential elements. The amount and accessibility of nutrients in the soil are influenced by the composition of these minerals (Karamina and Karamina, 2020). These minerals are broken down by weathering processes, which gradually release mineral nutrients for plant uptake into the soil solution.

2.2.2 ORGANIC MATTER

For the cultivation of fenugreek, organic matter is essential for maintaining soil fertility and nutrient availability. Essential mineral nutrients can be found in organic materials like crop residues, animal manures, and compost. Organic matter breaks down after being incorporated into the soil, releasing mineral nutrients in plant-accessible forms (Lima et al., 2009). Additionally enhancing soil structure and water-holding ability, organic matter encourages nutrient retention while minimizing nutrient leaching.

2.2.3 FERTILIZERS

In order to meet the needs for additional mineral nutrients during fenugreek cultivation, fertilizers are frequently used. The essential nutrients are provided in readily accessible forms by inorganic fertilizers like nitrogen (N), phosphorus (P), and potassium (K) fertilizers. To ensure proper nutrient supply during various growth stages of fenugreek, these fertilizers can be applied through broadcasting, side-dressing, or fertigation methods (Lima et al., 2009).

2.2.4 ORGANIC FERTILIZERS

For the cultivation of fenugreek, organic fertilizers such as green manure, compost, and animal manures are excellent sources of mineral nutrients. As they break down, these fertilizers gradually release nutrients, supplying a steady stream of minerals over time. Furthermore, organic fertilizers enhance the health of the soil, microbial activity, and nutrient cycling, promoting overall plant growth and nutrient availability (Lima et al., 2009).

2.2.5 SOIL AMENDMENTS

To change the pH of the soil and increase nutrient availability, soil amendments like lime and gypsum are used. To improve nutrient uptake and raise the pH of acidic soils, lime is frequently applied. Gypsum is employed to lessen the negative effects of soil salinity and enhance soil structure (Karamina and

Karamina, 2020). For fenugreek plants, proper soil amendment techniques help maximize the availability of minerals.

When selecting and applying mineral nutrients for the cultivation of fenugreek, it is crucial to consider variables like soil type, nutrient requirements, and plant growth stage. To evaluate nutrient status and inform fertilizer recommendations, soil testing is a useful tool. Regular soil testing identifies nutrient imbalances or deficiencies, allowing for targeted fertilizer application to satisfy the nutrient needs of fenugreek plants.

2.3 UNVEILING THE ROLE OF MINERAL NUTRIENTS IN FENUGREEK: EFFECTS ON GROWTH, YIELD, AND ACTIVE COMPOUNDS

2.3.1 MACRONUTRIENTS

Plants need macronutrients, which are essential elements, in relatively high quantities for growth and development (Figure 2.2).

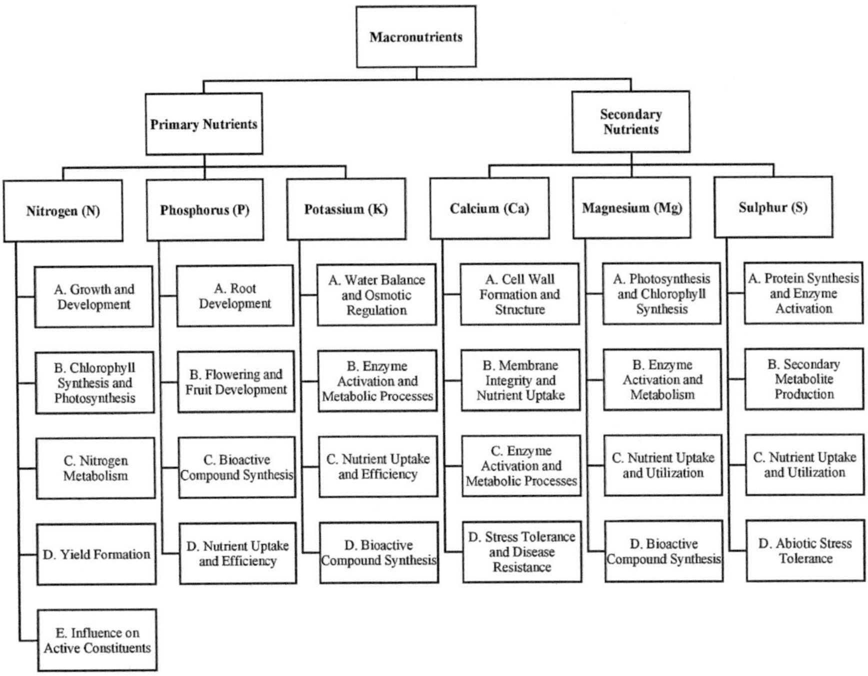

FIGURE 2.2 List of macro-nutrients and their important functions.

2.3.1.1 NITROGEN (N)

The growth, yield, and active ingredients of fenugreek depend heavily on nitrogen, a macronutrient that is essential for plant growth (Mavi et al., 2000). In order to examine how nitrogen affects fenugreek's various properties, such as growth, yield, and active ingredients, this chapter reviews the literature. The results underline how important nitrogen is for promoting fenugreek's vegetative growth, chlorophyll synthesis, photosynthesis, nitrogen metabolism, and yield development. Also discussed are how nitrogen affects the buildup of bioactive substances like alkaloids, flavonoids, and saponins. To maximize fenugreek production and improve its therapeutic and nutritive properties, good nitrogen management techniques are stressed which include:

1. **Growth and Development:** Numerous studies have shown that nitrogen has a favorable impact on fenugreek growth. According to Bisht et al. (2022) fenugreek plants' shoot length, root length, and biomass accumulation were all significantly increased when nitrogen levels were raised. Like this, Ibrahim and Helaly (2017) found that nitrogen fertilization made fenugreek plants grow taller and have more leaf surface.

2. **Chlorophyll Synthesis and Photosynthesis:** Nitrogen is essential for the synthesis of chlorophyll, which affects the effectiveness of photosynthetic processes (Imsande, 1998). According to Ohyama (2010) nitrogen application significantly increased the chlorophyll content in fenugreek leaves, resulting in higher carbon assimilation and improved photosynthetic rates. In turn, this encourages fenugreek's growth and yield.

3. **Nitrogen Metabolism:** Amino acids, proteins, and other nitrogenous compounds all contain nitrogen as a fundamental building block (Ohyama, 2010). For effective nitrogen metabolism and protein synthesis in fenugreek, there must be a sufficient supply of nitrogen. According to Sauvaire et al. (1984) nitrogen fertilization increased the amount of protein in fenugreek seeds, demonstrating the critical role of nitrogen in protein synthesis.

4. **Yield Formation:** Numerous studies have shown that nitrogen has a favorable effect on fenugreek yield. According to Tunçtürk et al. (2011) fenugreek seed yield increased noticeably as nitrogen levels

rose. Nitrogen encourages the growth of reproductive organs like flowers and pods, which increases seed set and overall yield.

5. **Influence on Active Constituents:** The fenugreek plant's ability to accumulate bioactive compounds can be influenced by nitrogen availability. Alkaloids, flavonoids, and saponins in fenugreek plants can have different contents and compositions depending on nitrogen fertilization, according to Qaisi et al. (2022) Increased levels of these active ingredients have been linked to higher nitrogen levels, enhancing the fenugreek's therapeutic and dietary benefits.

2.3.1.2 PHOSPHORUS (P)

The growth, yield, and active ingredients of fenugreek depend heavily on the macronutrient phosphorus. The impact of phosphorus on fenugreek is investigated in this chapter through a review of the literature with an emphasis on its effects on growth, yield, and active ingredients. The research shows how important phosphorus is for fenugreek's bioactive compound synthesis, flowering, fruiting, and root development. In order to increase fenugreek production and maximize its medicinal and nutritional benefits, best practices for phosphorus management are stressed as:

1. **Root Development:** In terms of root architecture and development, phosphorus is essential (Dinkelaker et al., 1995). When there is enough phosphorus available, fenugreek develops a strong root system that is capable of efficiently absorbing nutrients and water. According to Jabborova et al. (2021) the application of phosphorus significantly increased the volume, surface area, and length of the roots in fenugreek plants.

2. **Flowering and Fruit Development:** According to Dashora et al. (2019) phosphorus is essential for the initiation and development of flowers and fruits in fenugreek. According to Meena et al. (2014) fertilizing with phosphorus significantly increased the number of flowers and pods, which in turn increased fenugreek yield and seed set.

3. **Bioactive Compound Synthesis:** The production of bioactive substances in fenugreek is influenced by the phosphorus availability. The main chemical components of fenugreek with medicinal properties include alkaloids, flavonoids, and saponins (Meena et al., 2014).

4. **Nutrient Uptake and Efficiency:** For fenugreek plants to effectively absorb and use other nutrients, phosphorus is necessary. According to Baligar et al. (2017) adequate phosphorus levels maximize nutrient absorption and translocation within the plant, increasing overall nutrient use efficiency. In turn, this encourages balanced growth and supports the best possible fenugreek yield and quality.

2.3.1.3 POTASSIUM (K)

The growth, yield, and active ingredients of fenugreek depend critically on potassium, an essential macronutrient. The impact of potassium on fenugreek is explored in this chapter through a review of the literature, with an emphasis on how it affects growth, yield, and active ingredients. The research underlines the importance of potassium in controlling fenugreek's bioactive compound synthesis, enzyme activation, nutrient translocation, and water balance. To increase fenugreek production and maximize its medicinal and nutritional properties, ideal potassium management practices are stressed as:

1. **Water Balance and Osmotic Regulation:** According to Hassan et al. (2017) potassium is essential for keeping the water balance and osmotic regulation in plants. Fenugreek plants with adequate potassium levels use water more effectively, can withstand drought better, and are less susceptible to water stress.
2. **Enzyme Activation and Metabolic Processes:** Numerous enzymes that are necessary for various plant metabolic processes are activated by potassium. According to Hasanuzzaman et al. (2018) it is essential for photosynthesis, protein synthesis, carbohydrate metabolism, and nutrient translocation. These metabolic processes in fenugreek are enhanced by an adequate potassium supply, which improves growth and yield.
3. **Nutrient Uptake and Efficiency:** The uptake and utilization of nutrients by fenugreek plants are significantly influenced by potassium. It enhances the plant's ability to absorb and move other crucial nutrients like nitrogen and phosphorus. Improved growth, yield, and quality of fenugreek are the results of optimal nutrient use efficiency, which is optimized by adequate potassium levels. According to Bhunia et al. (2006) potassium supplementation improved fenugreek yield attributes and increased the efficiency of nitrogen and phosphorus uptake.

4. **Bioactive Compound Synthesis:** The content and composition of fenugreek's alkaloids, flavonoids, and saponins were positively influenced by potassium application, potentially increasing its therapeutic value (Hasanuzzaman et al., 2018).

2.3.1.4 CALCIUM (Ca)

For fenugreek to grow, yield, and contain active ingredients, calcium, an essential macronutrient, is required. The impact of calcium on fenugreek is investigated in this chapter through a review of the literature with an emphasis on growth, yield, and active ingredients. The results demonstrate the importance of calcium in fenugreek's cell wall formation, membrane integrity, enzyme activation, and production of bioactive compounds. To increase fenugreek production and maximize its therapeutic and nutritive properties, best practices for calcium management are stressed as:

1. **Cell Wall Formation and Structure:** In fenugreek plants, calcium is crucial for cell wall formation and maintenance. According to Demarty et al. (1984) adequate calcium levels contribute to the strength and integrity of cell walls, supporting the structure and promoting plant growth. According to Izydorczyk et al. (2005) calcium application increased the thickness and strength of the cell walls in fenugreek plants.

2. **Membrane Integrity and Nutrient Uptake:** According to Tuna et al. (2007) calcium is essential for maintaining the stability and integrity of plant membranes. It aids in controlling the passage of molecules such as water, nutrients, and other substances across cell membranes. The growth and yield of fenugreek plants are improved overall when calcium levels are sufficient (Loneragan and Snowball, 1969). According to Dehkordi et al. (2021) calcium application improved yield attributes and increased nutrient uptake efficiency in fenugreek.

3. **Enzyme Activation and Metabolic Processes:** For many enzymes involved in crucial metabolic processes in fenugreek, calcium serves as a co-factor. It is essential for the activation of the enzymes needed for hormone regulation, protein synthesis, and photosynthesis (Lautner and Fromm, 2010). These metabolic processes are enhanced

by an adequate calcium supply, which improves fenugreek's growth, yield, and quality.

4. **Stress Tolerance and Disease Resistance:** In plants, including fenugreek, calcium is known to improve disease resistance and stress tolerance. It enhances plant resilience and lessens the damaging effects of abiotic stresses like salinity and drought (Reddy, 2001). The plant's defense mechanisms are strengthened by calcium, which also increases the plant's resistance to various pathogens and diseases. In fenugreek plants showed that calcium application improved drought tolerance and decreased disease incidence (Sadak, 2016).

2.3.1.5 MAGNESIUM (Mg)

For fenugreek to grow, yield, and contain active ingredients, it is crucial that it receive adequate amounts of magnesium. The impact of magnesium on fenugreek is examined in this chapter by reviewing the prior research and focusing on how it affects growth, yield, and active ingredients. The research underlines the importance of magnesium for fenugreek's bioactive compound synthesis, enzyme activation, nutrient uptake, and photosynthesis. To increase fenugreek production and maximize its therapeutic and nutritive properties, best practices for managing magnesium are stressed as:

1. **Photosynthesis and Chlorophyll Synthesis:** Magnesium is crucial for photosynthesis in plants because it is a crucial part of the chlorophyll molecule (Mengel and Kirkby, 1987). Maintaining optimal chlorophyll content and activity is essential for promoting effective photosynthetic processes in fenugreek, which depends on adequate magnesium levels. According to Swamy et al. (2014) applying magnesium to fenugreek plants significantly increased their chlorophyll content and photosynthetic rate.

2. **Enzyme Activation and Metabolism:** Numerous enzymes involved in crucial metabolic processes in fenugreek use magnesium as a cofactor. It is essential for the activation of the enzymes involved in hormone regulation, protein synthesis, and carbohydrate metabolism (Farhat et al., 2014). These metabolic processes are enhanced by an adequate magnesium supply, which helps to improve fenugreek's growth, yield, and quality. According to Swamy et al. (2014)

fenugreek's enzyme activities and metabolic pathways were positively influenced by magnesium supplementation.

3. **Nutrient Uptake and Utilization:** Magnesium is important for fenugreek plants' nutrient uptake and utilization. It makes it easier for the plant to absorb and move vital nutrients like potassium, phosphorus, and nitrogen around. According to Guo et al. (2016) adequate magnesium levels maximize the yield potential of fenugreek while also promoting balanced growth and optimizing nutrient use efficiency.
4. **Bioactive Compound Synthesis:** The synthesis of fenugreek's bioactive compounds is influenced by magnesium availability, which contributes to the herb's therapeutic properties. According to Mengel and Kirkby (1987) magnesium application increased the amount and make-up of bioactive substances like alkaloids, flavonoids, and saponins in fenugreek, potentially increasing its therapeutic value.

2.3.1.6 SULFUR (S)

Sulfur is a crucial macronutrient that is important for fenugreek growth, yield, and active ingredients. The impact of sulfur on fenugreek is explored in this chapter through a review of the literature, with an emphasis on how it affects growth, yield, and active ingredients. The results emphasize the importance of sulfur in fenugreek plant development overall, including protein synthesis, enzymatic activities, secondary metabolite production, and production of secondary metabolites. The best sulfur management techniques are stressed in order to increase fenugreek production and maximize its therapeutic and dietary benefits as:

1. **Protein Synthesis and Enzyme Activation:** Proteins are made up of amino acids, which are essential for their formation (Thakur et al., 2015). Fenugreek plants grow and develop more effectively when there is enough sulfur in the environment, which is necessary for protein synthesis. The activation of various enzymes involved in metabolic processes is another function of sulfur. According to Lal et al. (2015) adding sulfur to fenugreek increased its protein content and enzymatic activity.

2. **Secondary Metabolite Production:** The production of secondary metabolites in fenugreek, which contribute to its therapeutic properties, is influenced by the availability of sulfur. Glycosylates, which have antioxidant and anticancer properties, are sulfur-containing compounds that are formed when sulfur is present (Bernhoft, 2010)
3. **Nutrient Uptake and Utilization:** Sulfur is essential for fenugreek plants' nutrient uptake and utilization. It affects how other crucial nutrients, like nitrogen and phosphorus, are absorbed and transported throughout the plant. Fenugreek grows more quickly, yields more, and is of higher quality when Sulfur levels are adequate (Mitoo and Rahman, 2016).
4. **Abiotic Stress Tolerance:** Sulfur plays a role in the synthesis of substances that help fenugreek tolerate abiotic stress. As a result, plants are better able to withstand environmental stresses like drought, salinity, and heavy metal toxicity (Samanta et al., 2020).

2.3.2 MICRONUTRIENTS

Micronutrients are vital elements that plants require in tiny amounts for appropriate growth, development, and overall health. Micronutrients, although needed in smaller amounts than macronutrients, serve critical roles in a variety of physiological and biochemical processes within plants. Here are some of the most important micronutrients and their roles in plant growth (Figure 2.3).

2.3.2.1 IRON (Fe)

Iron is a necessary micronutrient for the growth, yield, and active components of fenugreek (Boghdady, 2020). This chapter evaluates the available literature to investigate the effects of iron on fenugreek, concentrating on growth, yield, and active components. The findings show the importance of iron in chlorophyll synthesis, enzyme activity, nutrient absorption, and secondary metabolite formation in fenugreek. To increase fenugreek production and maximize its medicinal and nutritional benefits, best iron management practices are stressed as:

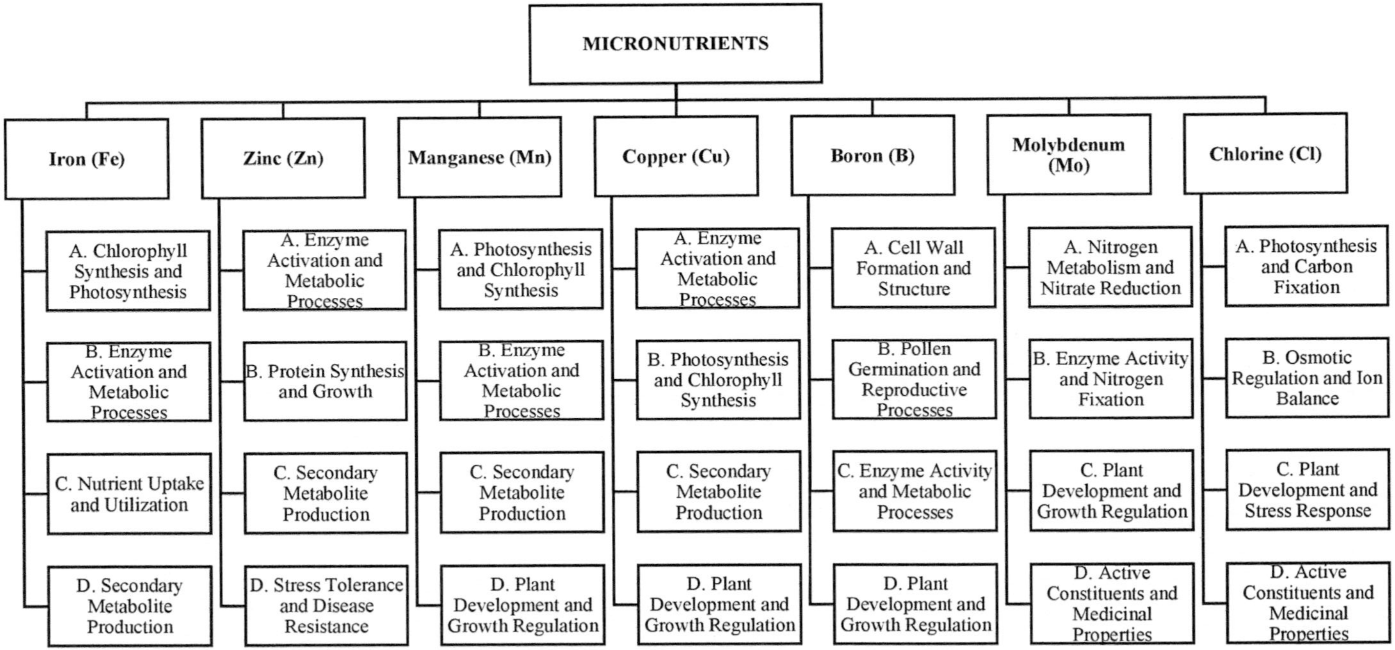

FIGURE 2.3 List of micro-nutrients and their important functions.

1. **Chlorophyll Synthesis and Photosynthesis:** Iron is an essential component of chlorophyll, the pigment responsible for photosynthesis in plants (Rout and Sahoo, 2015). Adequate iron levels are required for chlorophyll synthesis, which directly influences the effectiveness of photosynthetic processes in fenugreek. According to Sadak (2016), iron application significantly increased chlorophyll content and the photosynthetic rate in fenugreek plants.
2. **Enzyme Activation and Metabolic Processes:** Iron is essential for the activation of enzymes involved in various metabolic processes in fenugreek. It plays a vital role in electron transport, respiration, and nitrogen metabolism enzymes (Siddiqui et al., 2022). A sufficient iron supply enhances enzymatic activity, promoting improved growth and development in fenugreek. According to Kausar et al. (2017), iron supplementation improved enzyme activity and metabolic pathways in fenugreek.
3. **Nutrient Uptake and Utilization:** Iron is necessary for fenugreek plants to absorb and utilize various nutrients (Aishwath et al., 2018). It influences the absorption and transport of nutrients such as nitrogen, phosphorus, and potassium. Adequate iron levels enhance nutrient uptake efficiency, promoting balanced development and increasing fenugreek yield potential. According to Sadak (2016), iron supplementation improved nutrient uptake and yield qualities in fenugreek.
4. **Secondary Metabolite Production:** Iron availability impacts secondary metabolite formation in fenugreek, which contributes to its therapeutic effects. Iron is involved in the formation of bioactive chemicals with antioxidant and pharmacological effects, such as alkaloids and flavonoids.

2.3.2.2 ZINC (Zn)

Zinc is an important element for the growth, production, and active components of fenugreek (Hegab, 2018). This chapter examines the available literature to investigate the effects of zinc on fenugreek, concentrating on growth, yield, and active components. The findings emphasize the importance of zinc in fenugreek enzyme activation, protein synthesis, secondary metabolite generation, and overall plant development. To increase fenugreek production and promote its medicinal and nutritional benefits, optimal zinc management methods are emphasized as:

1. **Enzyme Activation and Metabolic Processes:** Zinc is vital for the activation of enzymes that regulate a variety of metabolic processes in fenugreek. It is required for the proper activity of many enzymes, including those involved in glucose metabolism, photosynthesis, and DNA synthesis (Frassinetti et al., 2006). Adequate zinc levels boost enzyme activity, resulting in enhanced fenugreek growth and development. Mabrouk et al. (2019) demonstrated that zinc application significantly increased enzyme activities and metabolic pathways in fenugreek plants.
2. **Protein Synthesis and Growth:** Zinc is needed for protein synthesis and is crucial for plant growth and development. It is required to produce proteins, enzymes, and hormones, all of which are necessary for cell division, tissue differentiation, and overall plant growth. Zinc deficiency can cause retarded growth and yield reduction in fenugreek (Jakhar et al., 2013). Noohpisheh et al. (2021) reported that zinc supplementation improved protein content and growth parameters in fenugreek plants.
3. **Secondary Metabolite Production:** Zinc availability regulates secondary metabolite formation in fenugreek, which contributes to its therapeutic effects. Zinc is involved in the production of bioactive chemicals with antioxidant and pharmacological features, such as alkaloids and flavonoids. A sufficient zinc level can boost the concentration of these beneficial chemicals in fenugreek.
4. **Stress Tolerance and Disease Resistance:** Zinc improves stress tolerance and disease resistance in fenugreek plants (Parwez et al., 2021). It aids in the activation of defense mechanisms and antioxidant systems, which aid in the adaptation of plants to abiotic and biotic stressors. Plants that are deficient in zinc are more vulnerable to disease and environmental stress. Parwez et al. (2021) reported that zinc supplementation improved stress tolerance and disease resistance in fenugreek.

2.3.2.3 MANGANESE (Mn)

Manganese is an important element in the growth, yield, and active components of fenugreek (Alkhlefawi, 2021). This chapter examines the available literature to investigate the effects of manganese on fenugreek, focusing on growth, yield, and active components. The findings emphasize the importance of manganese in fenugreek photosynthesis, enzyme activation, secondary metabolite production, and general plant development. To increase fenugreek production and maximize its medicinal and nutritional benefits, optimal manganese management practices are emphasized as:

1. **Photosynthesis and Chlorophyll Synthesis:** Manganese is required for photosynthesis because it is involved in the water-splitting complex of Photosystem II (Aishwath et al., 2018). It engages in light-dependent processes, allowing light energy to be converted into chemical energy. Hegab (2018) demonstrated that manganese supplementation significantly increased chlorophyll content and photosynthetic rate in fenugreek leaves.
2. **Enzyme Activation and Metabolic Processes:** Manganese is essential for the activation of enzymes involved in a variety of metabolic processes in fenugreek (Aishwath et al., 2018). It is necessary for glucose metabolism, antioxidant defines, nitrogen metabolism, and enzyme activity. Manganese levels that are adequate boost these enzymatic processes, resulting in enhanced growth, nutrient uptake, and general plant health. Baquer et al. (2011) reported that manganese application positively influenced enzyme activities and metabolic pathways in fenugreek plants.
3. **Secondary Metabolite Production:** Manganese supply regulates secondary metabolite formation in fenugreek, which lends to its therapeutic effects. Manganese plays a role in the generation of bioactive molecules with antioxidant and pharmacological effects, such as alkaloids and flavonoids. Adequate manganese amounts may improve fenugreek's bioactive chemical content. Hegab (2018) demonstrated that manganese supplementation increased the content of bioactive compounds and antioxidant activity in fenugreek seeds.
4. **Plant Development and Growth Regulation:** Manganese is involved in several aspects of fenugreek plant development and growth regulation. It has an impact on cell division, the development of roots, and plant morphology in general. A lack of manganese may result in stunted growth, lower yield, and aberrant plant development (McCauley et al., 2009). Baquer et al. (2011) reported that manganese supplementation improved growth parameters, yield attributes, and seed quality in fenugreek.

2.3.2.4 COPPER (Cu)

Copper is a key micronutrient in the growth, yield, and active components of fenugreek. This chapter investigates the available literature to investigate the effects of copper on fenugreek, focusing on growth, yield, and its active ingredients. The findings emphasize the importance of copper in fenugreek enzyme activation, photosynthesis, secondary metabolite generation, and overall plant development. For higher fenugreek production and to enhance its medicinal and nutritional benefits, the best copper management methods are urged as:

1. **Enzyme Activation and Metabolic Processes:** Copper is necessary for the activation of enzymes involved in an array of metabolic activities in fenugreek (Rehman et al., 2009). It serves as a cofactor for many enzymes, including those involved in glucose metabolism, lignin production, and antioxidant defense (Mir et al., 2009). Sufficient levels of copper increase enzyme activity, resulting in enhanced fenugreek growth and development.

2. **Photosynthesis and Chlorophyll Synthesis:** Copper is required for photosynthesis because it aids in electron transport within the photosystem II complex (Droppa and Horváth, 1990). It enhances electron transport, which is essential for the creation of ATP and carbohydrates (Huskisson et al., 2007). Copper shortage can impair photosynthetic efficiency in fenugreek by reducing chlorophyll in the production process.

3. **Secondary Metabolite Production:** Copper availability impacts secondary metabolite formation in fenugreek, which contributes to its therapeutic effects. Copper plays a role in the formation of bioactive chemicals that have antioxidant and medicinal properties, such as alkaloids and phenolic compounds. Adequate copper levels can increase the concentration of these bioactive chemicals in fenugreek.

4. **Plant Development and Growth Regulation:** Copper has an impact on multiple aspects of fenugreek plant development and growth control. It is important in cell wall manufacture, the development of roots, and plant morphology in general. A deficiency of copper can cause reduced growth, lower production, and unusual plant development. Aishwath et al. (2018) reported that copper application improved growth parameters, yield attributes, and seed quality in fenugreek.

2.3.2.5 BORON (B)

Boron is a key component for the growth, yield, and active ingredients of fenugreek (Boghdady, 2017). This chapter explores the available literature to investigate the effects of boron on fenugreek, focusing on growth, yield, and its active ingredients. Boron plays a role in cell wall manufacture, pollen germination, enzyme activity, and general plant development in fenugreek, according to the research. To improve fenugreek production and maximize its medicinal and nutritional benefits, optimal boron management practices are emphasized as:

1. **Cell Wall Formation and Structure:** In fenugreek plants, boron has a role in cell wall development and maintenance (Pandey, 2018). It is vital for cross-linking pectic polysaccharides, which are necessary for the integrity and strength of cell walls. Sufficient boron levels contribute to improved cell wall construction, which leads to increased growth and development in fenugreek.
2. **Pollen Germination and Reproductive Processes:** Boron is essential for pollen germination and fertilization in fenugreek (Seleiman et al., 2020). It is required for pollen tube wall synthesis and the proper functioning of enzymes involved in pollen tube growth. Adequate boron levels promote pollen germination and pollen tube extension, ensuring successful fertilization and seed development in fenugreek. According to Uwakiem (2021), boron management improved pollen germination and fruit set in fenugreek.
3. **Enzyme Activity and Metabolic Processes:** Boron is needed for the activation of many enzymes involved in fenugreek metabolic processes (Pawar and Laware, 2018). It impacts glucose metabolism, nitrogen metabolism, and phenolic compound production enzyme activity. Adequate boron levels can promote these enzyme reactions, which result in improved growth, nutrient uptake, and general plant health.
4. **Plant Development and Growth Regulation:** Boron influences a variety of plant development and growth regulation in fenugreek. It has an impact on root formation, flowering, and plant morphology in general. Boron shortage in fenugreek can cause stunted growth, limited flower production, and poor yield. Beyzi et al. (2019) reported that boron application improved growth parameters, flowering, and seed yield in fenugreek.

2.3.2.6 MOLYBDENUM (Mo)

Molybdenum is an important micronutrient in the growth, yield, and active ingredients of fenugreek. This chapter examines the available literature to investigate the effects of molybdenum on fenugreek, focusing on growth, yield, and its active ingredients. The findings emphasize the vital role of molybdenum in fenugreek nitrogen metabolism, enzyme activity, and overall plant development. To improve fenugreek production and increase its medicinal and nutritional benefits, optimal molybdenum management practices are emphasized as:

1. **Nitrogen Metabolism and Nitrate Reduction:** In fenugreek, molybdenum is required for nitrogen metabolism (Manohar et al., 2017). Its cofactors the enzyme nitrate reductase, which transforms nitrate to ammonium during nitrogen absorption (Beevers and Hageman, 1969). A sufficient dose of molybdenum ensures efficient nitrate reduction, resulting in enhanced nitrogen absorption and utilization in fenugreek. According to Shrivastava et al. (2016) molybdenum supplementation boosted nitrate reductase activity and nitrogen content in fenugreek plants.
2. **Enzyme Activity and Nitrogen Fixation:** Molybdenum is essential for nitrogenase activity, an enzyme involved in biological nitrogen fixation (Threatt and Rees, 2023). Nitrogen fixation is the process through which fenugreek converts ambient nitrogen into a useful form for plant growth (Raut, 2020). Sufficient molybdenum levels are required for optimum nitrogen fixation in fenugreek, which improves plant growth, production, and nitrogen availability. According to Shrivastava et al. (2016) molybdenum treatment enhanced nitrogen fixation capacity and biomass output in fenugreek.
3. **Plant Development and Growth Regulation:** Molybdenum has an impact on various aspects of plant development and growth regulation in fenugreek. It influences root growth, leaf expansion, and plant morphology (Roychoudhury and Chakraborty, 2022). In fenugreek, inadequate amounts of molybdenum can cause stunted growth, decreased leaf size, and poor yield. According to Shrivastava et al. (2016) molybdenum supplementation increased growth attributes, leaf area, and seed yield in fenugreek.
4. **Active Constituents and Medicinal Properties:** The availability of molybdenum influences the synthesis of active compounds in fenugreek which promotes therapeutic qualities. Molybdenum helps

in the production of several bioactive chemicals, including alkaloids and flavonoids, which have antioxidant and medicinal properties. Optimal molybdenum levels can increase the concentration of these active ingredients in fenugreek (Gendy, 2013).

2.3.2.7 CHLORINE (Cl)

Chlorine is a crucial micronutrient for the growth, yield, and active ingredients of fenugreek (Prajapati, 2019). This chapter analyses the available literature to investigate the effects of chlorine on fenugreek, focusing on growth, yield, and its active ingredients. The findings indicate the significant role of chlorine in fenugreek photosynthesis, osmotic control, and overall plant development. To improve fenugreek production and maximize its medicinal and nutritional benefits, optimal chlorine management techniques are highlighted as:

1. **Photosynthesis and Carbon Fixation:** Chlorine is involved in photosynthesis, the process by which plants convert light energy into chemical energy (Arnon, 1959). It plays a role in maintaining the structure and function of chloroplasts, where photosynthesis occurs (Colmenero-Flores et al., 2019). Adequate chlorine levels contribute to efficient carbon fixation and energy production, leading to improved growth and yield in fenugreek.
2. **Osmotic Regulation and Ion Balance:** Chlorine plays a crucial role in osmotic regulation and ion balance in fenugreek plants (Abbas et al., 2021). It helps maintain proper water balance by regulating osmotic potential (Franco-Navarro et al., 2016). Chlorine ions also contribute to maintaining electrical neutrality by balancing other essential ions within plant tissues (Dainty, 1962). Adequate chlorine levels ensure optimal osmotic regulation, ion transport, and overall plant health in fenugreek.
3. **Plant Development and Stress Response:** Chlorine influences various aspects of plant development and stress response in fenugreek (Hussein and Aqlan, 2012). It plays a role in root development, leaf expansion, and overall plant morphology. Chlorine deficiency can lead to reduced growth, chlorosis, and increased susceptibility to stressors (Hajiboland, 2012).
4. **Active Constituents and Medicinal Properties:** Chlorine availability can affect the synthesis of active constituents in fenugreek,

which contribute to its medicinal properties. Chlorine is involved in the biosynthesis of various bioactive compounds, including alkaloids and flavonoids, which have antioxidant and therapeutic effects. Optimal chlorine levels can enhance the content of these active constituents in fenugreek (Hussein and Aqlan, 2012).

2.4 CONCLUSION

Mineral nutrients play a crucial role in the growth, yield, and active constituents of fenugreek. Optimal nutrient management practices are essential for maximizing fenugreek cultivation and harnessing its full potential in terms of medicinal and nutritional properties. Further research is warranted to explore specific nutrient requirements, interactions, and their effects on fenugreek cultivation, ensuring sustainable production and enhancing its value as a therapeutic and functional food crop.

KEYWORDS

- **active components**
- **chlorophyll synthesis**
- **enzyme activity**
- **fenugreek**
- **iron management practices**
- **micronutrient**
- **nutrient absorption**
- **photosynthesis**
- **photosynthetic rate**
- **secondary metabolite formation**

REFERENCES

Abbas, S., Javed, M. T., Ali, Q., Azeem, M., & Ali, S. (2021). Nutrient deficiency stress and relation with plant growth and development. In *Engineering Tolerance in Crop Plants Against Abiotic Stress* (pp. 239–262). CRC Press.

Acharya, S. N., Thomas, J. E., & Basu, S. K. (2006). Fenugreek: An "old world" crop for the "new world." *Biodiversity, 7*(3, 4), 27–30.

Acharya, S. N., Thomas, J. E., & Basu, S. K. (2008). Fenugreek, an alternative crop for semiarid regions of North America. *Crop Science, 48*(3), 841–853.

Aishwath, O. P., Meena, R. D., & Dubey, P. N. (2018). Role of micronutrients on morphophysiological and nutritional parameters including disease suppression of fenugreek. *International Journal of Seed Spices, 8*, 24–31.

Alkhlefawi, A. M. K. (2022). Impact of nano-manganese and Siapton on the growth of fenugreek. *Al-Qadisiyah Journal for Agriculture Sciences, 12*(2), 49–57.

Amirnia, R., Ghiyasi, M., Sadighfard, S., & Mogaddam, S. S. (2018). Effect of some macronutrient's foliar application on the quantitative and qualitative traits of fenugreek (*Trigonella foenum-graecum* L.). *Gaziosmanpaşa Üniversitesi Ziraat Fakültesi Dergisi, 35*(3), 305–318.

Anda, M., Suryani, E., & Subardja, D. (2015). Strategy to reduce fertilizer application in volcanic paddy soils: Nutrient reserves approach from parent materials. *Soil and Tillage Research, 150*, 10–20.

Anderson, D. W. (1988). The effect of parent material and soil development on nutrient cycling in temperate ecosystems. *Biogeochemistry, 5*, 71–97.

Arnon, D. I. (1959). Conversion of light into chemical energy in photosynthesis. *Nature, 184*, 10–21.

Atkinson, C. J. (2018). How good is the evidence that soil-applied biochar improves water-holding capacity? *Soil Use and Management, 34*(2), 177–186.

Bafadam, S., Mahmoudabady, M., Niazmand, S., Rezaee, S. A., & Soukhtanloo, M. (2021). Cardioprotective effects of fenugreek (*Trigonella foenum-graecum*) seed extract in streptozotocin-induced diabetic rats. *Journal of Cardiovascular and Thoracic Research, 13*(1), 28.

Baghbani-Arani, A., Modarres-Sanavy, S. A. M., Mashhadi-Akbar-Boojar, M., & Mokhtassi-Bidgoli, A. (2017). Towards improving the agronomic performance, chlorophyll fluorescence parameters, and pigments in fenugreek using zeolite and vermicompost under deficit water stress. *Industrial Crops and Products, 109*, 346–357.

Balasubramanian, S., Roselin, P., Singh, K. K., Zachariah, J., & Saxena, S. N. (2016). Postharvest processing and benefits of black pepper, coriander, cinnamon, fenugreek, and turmeric spices. *Critical Reviews in Food Science and Nutrition, 56*(10), 1585–1607.

Baligar, V. C., Fageria, N. K., & He, Z. L. (2001). Nutrient use efficiency in plants. *Communications in Soil Science and Plant Analysis, 32*(7, 8), 921–950.

Baquer, N. Z., Kumar, P., Taha, A., Kale, R. K., Cowsik, S. M., & McLean, P. (2011). Metabolic and molecular action of *Trigonella foenum-graecum* (fenugreek) and trace metals in experimental diabetic tissues. *Journal of Biosciences, 36*, 383–396.

Basha, S. K. M., & Reddy, P. S. K. (2020). Ethnobotanical trees of Sri Lankamalleswara Wildlife Sanctuary, Eastern Ghats, Andhra Pradesh. In *Medicinal Plants: Biodiversity, Sustainable Utilization and Conservation* (pp. 317–373).

Basu, S. K., Zandi, P., & Cetzal-Ix, W. (2019). Fenugreek (*Trigonella foenum-graecum* L.): Distribution, genetic diversity, and potential to serve as an industrial crop for the global pharmaceutical, nutraceutical, and functional food industries. In *The Role of Functional Food Security in Global Health* (pp. 471–497). Academic Press.

Beevers, L., & Hageman, R. H. (1969). Nitrate reduction in higher plants. *Annual Review of Plant Physiology, 20*(1), 495–522.

Bernhoft, A. J. A. (2010). A brief review on bioactive compounds in plants. In *Bioactive Compounds in Plants—Benefits and Risks for Man and Animals* (pp. 11–17). The Norwegian Academy of Science and Letters.

Beyzi, E., Güneş, A., Arslan, M., & Şatana, A. (2019). Effects of foliar boron treatments on yield and yield components of fenugreek (*Trigonella foenum-graecum* L.): Detection by PCA analysis. *Communications in Soil Science and Plant Analysis, 50*(16), 2023–2032.

Bhatla, S. C., Lal, M., Kathpalia, R., & Bhatla, S. C. (2008). Plant mineral nutrition. In *Plant Physiology, Development and Metabolism* (pp. 37–81).

Bhunia, S. R., Chauhan, R. P. S., Yadav, B. S., & Bhati, A. S. (2006). Effect of phosphorus, irrigation, and *Rhizobium* on productivity, water use, and nutrient uptake in fenugreek (*Trigonella foenum-graecum*). *Indian Journal of Agronomy, 51*(3), 239–241.

Bisht, S., Singh, S., Singh, M., & Sharma, J. G. (2022). Augmentative role of *Piriformospora indica* fungus and plant growth promoting bacteria in mitigating salinity stress in *Trigonella foenum-graecum*. *Journal of Applied Biology and Biotechnology, 10*(1), 85–94.

Boghdady, M. (2017). Role of micronutrients in improving yield and quality of seeds in fenugreek plants (*Trigonella foenum-graecum* L.). *Egyptian Journal of Agronomy, 39*(3), 267–277.

Chaichi, M. R., Dadresan, M., Hosseini, M. B., Pourbabaie, A., Yazdani, D., & Zandvakili, O. R. (2015). Effect of biofertilizers on the growth, productivity, and nutrient absorption of fenugreek (*Trigonella foenum-graecum* L.). *International Journal of Agriculture Innovations and Research, 3*(5), 2319–1473.

Colmenero-Flores, J. M., Franco-Navarro, J. D., Cubero-Font, P., Peinado-Torrubia, P., & Rosales, M. A. (2019). Chloride as a beneficial macronutrient in higher plants: New roles and regulation. *International Journal of Molecular Sciences, 20*(19), 4686.

Dadrasan, M., Chaichi, M. R., Pourbabaee, A. A., Yazdani, D., & Keshavarz-Afshar, R. (2015). Deficit irrigation and biological fertilizer influence on yield and trigonelline production of fenugreek. *Industrial Crops and Products, 77*, 156–162.

Dainty, J. (1962). Ion transport and electrical potentials in plant cells. *Annual Review of Plant Physiology, 13*(1), 379–402.

Dashora, L. N., Mundra, S. L., Choudhary, J., & Choudhary, P. (2019). Effect of inorganic and organic sources of fertilization on productivity of fenugreek (*Trigonella foenum-graecum* L.) under agro-climatic conditions of Southern Rajasthan. *Journal of Pharmacognosy and Phytochemistry, 8*(4), 1886–1888.

Dehkordi, R. A., Roghani, S. R., Mafakheri, S., & Asghari, B. (2021). Effect of biostimulants on morpho-physiological traits of various ecotypes of fenugreek (*Trigonella foenum-graecum* L.) under water deficit stress. *Scientia Horticulturae, 283*, 110077.

Demarty, M., Morvan, C., & Thellier, M. (1984). Calcium and the cell wall. *Plant, Cell & Environment, 7*, 441–448.

Dinkelaker, B., Hengeler, C., & Marschner, H. (1995). Distribution and function of proteoid roots and other root clusters. *Botanica Acta, 108*(3), 183–200.

Diriba-Shiferaw, G. (2016). Review of management strategies of constraints in garlic (*Allium sativum* L.) production. *The Journal of Agricultural Sciences, 11*(3), 186–207.

Dobrikova, A., Apostolova, E., Adamakis, I. D. S., Hanć, A., Sperdouli, I., & Moustakas, M. (2022). Combined impact of excess zinc and cadmium on elemental uptake, leaf anatomy and pigments, antioxidant capacity, and function of photosynthetic apparatus in clary sage (*Salvia sclarea* L.). *Plants, 11*(18), 2407.

Droppa, M., & Horváth, G. (1990). The role of copper in photosynthesis. *Critical Reviews in Plant Sciences, 9*(2), 111–123.

Farhat, N., Elkhouni, A., Zorrig, W., Smaoui, A., Abdelly, C., & Rabhi, M. (2016). Effects of magnesium deficiency on photosynthesis and carbohydrate partitioning. *Acta Physiologiae Plantarum, 38*(6), 145.

Fischer, W. W., Hemp, J., & Johnson, J. E. (2015). Manganese and the evolution of photosynthesis. *Origins of Life and Evolution of Biospheres, 45*, 351–357.

Franco-Navarro, J. D., Brumós, J., Rosales, M. A., Cubero-Font, P., Talón, M., & Colmenero-Flores, J. M. (2016). Chloride regulates leaf cell size and water relations in tobacco plants. *Journal of Experimental Botany, 67*(3), 873–891.

Frassinetti, S., Bronzetti, G. L., Caltavuturo, L., Cini, M., & Della Croce, C. (2006). The role of zinc in life: A review. *Journal of Environmental Pathology, Toxicology and Oncology, 25*(3), 597–610.

Gendy, A. S. H. (2013). Growth, yield and chemical constituents of fenugreek as influenced by *Rhizobium* inoculation and molybdenum foliar spray. *Middle East Journal of Agricultural Research, 2*(3), 84–92.

Guo, W., Nazim, H., Liang, Z., & Yang, D. (2016). Magnesium deficiency in plants: An urgent problem. *The Crop Journal, 4*(2), 83–91.

Gupta, N., Debnath, S., Sharma, S., Sharma, P., & Purohit, J. (2017). Role of nutrients in controlling the plant diseases in sustainable agriculture. In *Agriculturally Important Microbes for Sustainable Agriculture: Volume 2: Applications in Crop Production and Protection* (pp. 217–262).

Hajiboland, R. (2012). Effect of micronutrient deficiencies on plants stress responses. In *Abiotic Stress Responses in Plants: Metabolism, Productivity, and Sustainability* (pp. 283–329). Springer.

Hamdan, J., & Bumham, C. P. (1996). The contribution of nutrients from parent material in three deeply weathered soils of Peninsular Malaysia. *Geoderma, 74*(3, 4), 219–233.

Hasanuzzaman, M., Bhuyan, M. B., Nahar, K., Hossain, M. S., Mahmud, J. A., Hossen, M. S., & Fujita, M. (2018). Potassium: A vital regulator of plant responses and tolerance to abiotic stresses. *Agronomy, 8*(3), 31.

Hassan, M. U., Aamer, M., Chattha, M. U., Ullah, M. A., Sulaman, S., Nawaz, M., & Guoqin, H. (2017). The role of potassium in plants under drought stress: A mini-review. *Journal of Basic & Applied Sciences, 13*, 268–271.

Hegab, R. (2018). Effect of foliar application of micronutrients and compost tea on chemical composition and the productivity of fenugreek plants under Baloza-North Sinai conditions. *Journal of Soil Sciences and Agricultural Engineering, 9*(9), 403–410.

Hofman, G., & Van Cleemput, O. (2004). Soil and plant nitrogen. *International Fertilizer Industry Association*, 1–41.

Hordofa, T. S., & Tolossa, T. T. (2020). Cultivation and postharvest handling practices affecting yield and quality of major spices crops in Ethiopia: A review. *Cogent Food & Agriculture, 6*(1), 1788896.

Huskisson, E., Maggini, S., & Ruf, M. (2007). The role of vitamins and minerals in energy metabolism and well-being. *Journal of International Medical Research, 35*(3), 277–289.

Hussein, E. A., & Aqlan, E. M. (2012). Effect of mannitol and sodium chloride on some total secondary metabolites of fenugreek calli cultured *in vitro*. *Plant Tissue Culture and Biotechnology, 21*(1), 35–43.

Ibrahim, F. R., & Helaly, A. A. E. (2017). Growth and productivity response to nitrogen, potassium and humic acid of fenugreek (*Trigonella foenum-graecum* L.) plant. *Middle East Journal of Agricultural Research*, *6*(4), 1526–1535.

Ilyina, A., Ramos-González, R., Segura-Ceniceros, E. P., Vargas-Segura, A., Martínez-Hernández, J. L., Zaynullin, R., & Korotina, T. (2017). Alimentary and medicinal plants in functional nutrition. *Universidad Autónoma de Coahuila*, 314.

Imsande, J. (1998). Iron, sulfur, and chlorophyll deficiencies: A need for an integrative approach in plant physiology. *Physiologia Plantarum*, *103*(1), 139–144.

Izydorczyk, M., Cui, S. W., & Wang, Q. (2005). Polysaccharide gums: Structures, functional properties, and applications. In S. W. Cui (Ed.), *Food Carbohydrates: Chemistry, Physical Properties, and Applications* (pp. 293–299). CRC Press.

Jabborova, D., Annapurna, K., Choudhary, R., Bhowmik, S. N., Desouky, S. E., Selim, S., & Elkelish, A. (2021). Interactive impact of biochar and arbuscular mycorrhizal on root morphology, physiological properties of fenugreek (*Trigonella foenum-graecum* L.) and soil enzymatic activities. *Agronomy*, *11*(11), 2341.

Jakhar, R. K., Yadav, B. L., & Choudhary, M. R. (2013). Irrigation water quality and zinc on growth and yield of fenugreek (*Trigonella foenum-graecum* L.). *Journal of Spices and Aromatic Crops*, *22*(2), 170.

Karamina, H., & Fikrinda, W. (2020). Soil amendment impact to soil organic matter and physical properties on the three soil types after second corn cultivation. *AIMS Agriculture and Food*, *5*(1), 150–169.

Kausar, S., Faizan, S., & Haneef, I. (2017). Nitrogen level affects growth and reactive oxygen scavenging of fenugreek irrigated with wastewater. *Tropical Plant Research*, *4*(2), 210–224.

Kinji, F. R., & Rahdari, P. (2012). Shortage on (magnesium and calcium) dry weight, fresh weight, root and shoot length, leaf relative water content (RWC), chlorophyll content, and malondialdehyde activity in fenugreek (*Trigonella foenum-graecum*). *Greener Journal of Agricultural Sciences*, *2*(7).

Kumar, A. (2019). *Fenugreek*. Sankalp Publication.

Kumar, R., Kumar, R., & Prakash, O. (2019). Chapter 5: The impact of chemical fertilizers on our environment and ecosystem. In Chief Ed (Ed.), *Agriculture and Fertilization* (pp. 69–78).

Lal, G., Singh, B., Mehta, R. S., Singh, R., & Maheria, S. P. (2015). Performance of fenugreek (*Trigonella foenum-graecum* L.) as influenced by sulfur and zinc. *International Journal of Seed Spices*, *5*(1), 29–33.

Lautner, S., & Fromm, J. (2010). Calcium-dependent physiological processes in trees. *Plant Biology*, *12*(2), 268–274.

Lavelle, P., & Spain, A. (2002). *Soil Ecology*. Springer Science & Business Media.

Lima, D. L., Santos, S. M., Scherer, H. W., Schneider, R. J., Duarte, A. C., Santos, E. B., & Esteves, V. I. (2009). Effects of organic and inorganic amendments on soil organic matter properties. *Geoderma*, *150*(1, 2), 38–45.

Loneragan, J. F., & Snowball, K. (1969). Calcium requirements of plants. *Australian Journal of Agricultural Research*, *20*(3), 465–478.

Maathuis, F. J., & Diatloff, E. (2013). Roles and functions of plant mineral nutrients. In F. J. Maathuis (Ed.), *Plant Mineral Nutrients: Methods and Protocols* (pp. 1–21). Springer Science & Business Media.

Mabrouk, B., Kâab, S. B., Rezgui, M., Majdoub, N., da Silva, J. T., & Kaab, L. B. B. (2019). Salicylic acid alleviates arsenic and zinc toxicity in the process of reserve mobilization in germinating fenugreek (*Trigonella foenum-graecum* L.) seeds. *South African Journal of Botany*, *124*, 235–243.

Mamatha, N. C., & Panyam, K. R. (2021). Agronomic practices in fenugreek. In A. Sharma, D. Singh, & J. Prasad (Eds.), *Fenugreek: Biology and Applications* (pp. 83–97). Springer Nature.

Manohar, M. S., Shivran, A. C., & Choudhary, A. (2017). Effect of sulfur and molybdenum on yield attributes, yield, and economics of fenugreek. *Chemical Science Review and Letters*, *7*(21), 94–97.

Mavi, D., Lal, S., Singh, K. S. B. A., & Singh, N. (2000). Response of fenugreek (*Trigonella foenum-graecum* L.) to seed rate, nitrogen, and phosphorus fertilizer. *Haryana Journal of Horticultural Sciences*, *29*, 244–246.

Mavris, C., Egli, M., Plötze, M., Blum, J. D., Mirabella, A., Giaccai, D., & Haeberli, W. (2010). Initial stages of weathering and soil formation in the Morteratsch proglacial area (Upper Engadine, Switzerland). *Geoderma*, *155*(3, 4), 359–371.

Mazid, M., & Khan, T. A. (2015). Future of bio-fertilizers in Indian agriculture: An overview. *International Journal of Agricultural and Food Research*, *3*(3).

McCauley, A., Jones, C., & Jacobsen, J. (2009). Plant nutrient functions and deficiency and toxicity symptoms. *Nutrient Management Module, 9*, 1–16.

Meena, S. S., Mehta, R. S., Bairwa, M., & Meena, R. D. (2014). Productivity and profitability of fenugreek (*Trigonella foenum-graecum* L.) as influenced by bio-fertilizers and plant growth regulators. *Legume Research-An International Journal, 37*(6), 646–650.

Mengel, K., & Kirkby, E. A. (1987). *Principles of Plant Nutrition*. International Potash Institute.

Mir, A. R., Pichtel, J., & Hayat, S. (2021). Copper: Uptake, toxicity and tolerance in plants and management of Cu-contaminated soil. *Biometals, 34*(4), 737–759.

Mitoo, S., & Rahman, M. (2016). Effect of phosphorus and sulfur fertilization on vegetative growth and seed yield of fenugreek (*Trigonella foenum-graecum* L.) (Doctoral dissertation, Department of Agricultural Botany, Sher-E-Bangla Agricultural University Dhaka-1207).

Moore, J. A., Kimsey, M. J., Garrison-Johnston, M., Shaw, T. M., Mika, P., & Poolakkal, J. (2022). Geologic soil parent material influence on forest surface soil chemical characteristics in the inland northwest, USA. *Forests, 13*(9), 1363.

Nagulapalli Venkata, K. C., Swaroop, A., Bagchi, D., & Bishayee, A. (2017). A small plant with big benefits: Fenugreek (*Trigonella foenum-graecum* Linn.) for disease prevention and health promotion. *Molecular Nutrition and Food Research, 61*(6), 1600950.

Nair, M. B., & Groot, M. J. (2021). Medicinal plants for home herbal gardens, Institutional gardens, and animal health. *Natural Livestock Farming India*, 59–60.

Narapogu, V., Swathi, C., Polsani, M. R., & Premendran, J. (2021). Evaluation of anti-diabetic and anti-inflammatory activities of Fenugreek (*Trigonella foenum-graecum*) seed extracts on Albino Wistar Rats. *National Journal of Physiology, Pharmacy and Pharmacology, 11*(11), 1277–1282.

Nathiya, S., Durga, M., & Devasena, T. (2014). Therapeutic role of *Trigonella foenum-graecum* [fenugreek] – A review. *International Journal of Pharmaceutical Sciences Review and Research, 27*(2), 74–80.

Noohpisheh, Z., Amiri, H., Mohammadi, A., & Farhadi, S. (2021). Effect of the foliar application of zinc oxide nanoparticles on some biochemical and physiological parameters of *Trigonella foenum-graecum* under salinity stress. *Plant Biosystems-An International Journal Dealing with all Aspects of Plant Biology, 155*(2), 267–280.

Ohyama, T. (2010). Nitrogen as a major essential element of plants. In *Nitrogen Assimilation in Plants* (pp. 1–17).

Pal, D., & Mukherjee, S. (2020). Fenugreek (Trigonella foenum) seeds in health and nutrition. In *Nuts and Seeds in Health and Disease Prevention* (pp. 161–170). Academic Press.

Pandey, N. (2018). Role of plant nutrients in plant growth and physiology. In *Plant Nutrients and Abiotic Stress Tolerance* (pp. 51–93).

Pandey, R. (2015). Mineral nutrition of plants. In *Plant Biology and Biotechnology: Volume I: Plant Diversity, Organization, Function and Improvement* (pp. 499–538).

Panghal, V. P. S., Duhan, D. S., & Lal, M. (2015). Influence of biofertilizers along with inorganic source of nitrogen and phosphorus on growth, seed yield and quality of fenugreek (*Trigonella foenum-graecum* L.). *Research on Crops, 16*(2), 288–292.

Parwez, R., Nabi, A., Mukarram, M., Aftab, T., Khan, M. M. A., & Naeem, M. (2021). Various mitigation approaches applied to confer abiotic stress tolerance in fenugreek (*Trigonella foenum-graecum* L.): A review. In *Fenugreek: Biology and Applications* (pp. 137–185).

Pawar, V. A., & Laware, S. L. (2018). Seed priming: A critical review. *International Journal of Scientific Research in Biological Sciences, 5*, 94–101.

Prajapati, J. (2019). The essential nutrient element for the crop. *International Journal of All Research Writings, 1*(1), 17–22.

Qaisi, A. M., Al Tawaha, A. R., Imran, & Al-Rifaee, M. D. (2022). Effects of chitosan and biochar-mended soil on growth, yield, and yield components and mineral composition of fenugreek. *Gesunde Pflanzen*, 1–12.

Raut, R. P. (2020). Isolation and identification of Rhizobium from root nodules of fenugreek plant collected from village Vangaon and to study its effect on soil fertility and plant growth in research interventions and advancements in plant sciences. In *Bhumi Publishing* (pp. 53–62).

Reddy, A. S. (2001). Calcium: Silver bullet in signaling. *Plant Science, 160*(3), 381–404.

Rehman, M., Liu, L., Wang, Q., Saleem, M. H., Bashir, S., Ullah, S., & Peng, D. (2019). Copper environmental toxicology, recent advances, and future outlook: A review. *Environmental Science and Pollution Research, 26*, 18003–18016.

Ribeiro, I. D. A., Volpiano, C. G., Vargas, L. K., Granada, C. E., Lisboa, B. B., & Passaglia, L. M. P. (2020). Use of mineral weathering bacteria to enhance nutrient availability in crops: A review. *Frontiers in Plant Science, 11*, 590774.

Rout, G. R., & Sahoo, S. (2015). Role of iron in plant growth and metabolism. *Reviews in Agricultural Science, 3*, 1–24.

Roychoudhury, A., & Chakraborty, S. (2022). Cobalt and molybdenum: Deficiency, toxicity, and nutritional role in plant growth and development. In *Plant Nutrition and Food Security in the Era of Climate Change* (pp. 255–270). Academic Press.

Sadak, M. S. (2016). Mitigation of drought stress on fenugreek plant by foliar application of trehalose. *International Journal of Chemistry Technology Research, 9*, 147–155.

Salman, M. T., & Qadeer, F. (2021). Pharmacological actions and therapeutic potential of *Trigonella foenum-graecum* L. In *Fenugreek: Biology and Applications* (pp. 523–537).

Samanta, S., Singh, A., & Roychoudhury, A. (2020). Involvement of sulfur in the regulation of abiotic stress tolerance in plants. In *Protective Chemical Agents in the Amelioration of Plant Abiotic Stress: Biochemical and Molecular Perspectives* (pp. 437–466).

Sauvaire, Y., Girardon, P., Baccou, J. C., & Ristérucci, A. M. (1984). Changes in growth, protein, free amino acids of developing seeds and pods of fenugreek. *Phytochemistry, 23*(3), 479–486.

Seleiman, M. F., Almutairi, K. F., Alotaibi, M., Shami, A., Alhammad, B. A., & Battaglia, M. L. (2020). Nano-fertilization as an emerging fertilization technique: Why can modern agriculture benefit from its use? *Plants, 10*(1), 2.

Shahrajabian, M. H., Sun, W., & Cheng, Q. (2022). Foliar application of nutrients on medicinal and aromatic plants: The sustainable approaches for higher and better production. *Beni-Suef University Journal of Basic and Applied Sciences, 11*(1), 1–10.

Shahrajabian, M. H., Sun, W., Shen, H., & Cheng, Q. (2021). A mini-review of galactomannas and diosgenin in fenugreek. *Pharmacognosy Communications, 11*(1), 26–30.

Shaji, H., Chandran, V., & Mathew, L. (2021). Organic fertilizers as a route to controlled release of nutrients. In *Controlled Release Fertilizers for Sustainable Agriculture* (pp. 231–245). Academic Press.

Sharangi, A. B., Bhutia, P. H., Raj, A. C., & Sreenivas, M. (2018). Underexploited spice crops: Present status, agrotechnology, and future research directions. In *CRC Press* (pp. 230–248).

Shiurkar, N. L., & Barcchiya, J. (2016). Effect of different levels of phosphorus and biofertilizers on yield and quality of fenugreek (*Trigonella foenum-graecum* L.). *Advances in Life Sciences, 5*(21), 10059–10063.

Shrivastava, M., Ahanger, M. A., & Agarwal, R. M. (2016). Improved growth of *Trigonella foenum-graecum* L. with potassium supplementation involves physiological and biochemical implications. *Journal of Functional and Environmental Botany, 6*(2), 84–101.

Siddiqui, H., Singh, P., Arif, Y., Sami, F., Naaz, R., & Hayat, S. (2022). Role of micronutrients in providing abiotic stress tolerance. In *Microbial Biofertilizers and Micronutrient Availability: The Role of Zinc in Agriculture and Human Health* (pp. 115–136).

Sinha, S., Gupta, A. K., & Bhatt, K. (2007). Uptake and translocation of metals in fenugreek grown on soil amended with tannery sludge: Involvement of antioxidants. *Ecotoxicology and Environmental Safety, 67*(2), 267–277.

Swamy, K. N., Vardhini, B. V., Ramakrishna, B., Anuradha, S., Siddulu, N., & Rao, S. S. R. (2014). Role of 28-homobrassinolide on growth biochemical parameters of *Trigonella foenum-graecum* L. plants subjected to lead toxicity. *International Journal of Multidisciplinary and Current Research, 2*, 317–321.

Tadesse, T., Dechassa, N., Bayu, W., & Gebeyehu, S. (2013). Effects of farmyard manure and inorganic fertilizer application on soil physico-chemical properties and nutrient balance in rain-fed lowland rice ecosystem. *American Journal of Plant Sciences, 4*(2), 8.

Thakur, N., Singh, A., Ravi, K. T., & Sarkar, S. (2021). Effect of Sulfur on growth and development in sunflower. *The Pharma Innovation Journal, 10*(11), 133–138.

Threatt, S. D., & Rees, D. C. (2023). Biological nitrogen fixation in theory, practice, and reality: A perspective on the molybdenum nitrogenase system. *FEBS Letters, 597*(1), 45–58.

Tuna, A. L., Kaya, C., Ashraf, M., Altunlu, H., Yokas, I., & Yagmur, B. (2007). The effects of calcium sulfate on growth, membrane stability, and nutrient uptake of tomato plants grown under salt stress. *Environmental and Experimental Botany, 59*(2), 173–178.

Tunçtürk, R., Celen, A. E., & Tunçtürk, M. (2011). The effects of nitrogen and Sulfur fertilizers on the yield and quality of fenugreek (*Trigonella foenum-graecum* L.). *Turkish Journal of Field Crops, 16*(1), 69–75.

Uwakiem, M. K. (2021). Growth and productivity of red globe grapevines as affected by spraying of fenugreek seed sprout, nano-boron, and moringa extract. *Hortscience Journal of Suez Canal University, 10*(1), 49–61.

Visuvanathan, T., Than, L. T. L., Stanslas, J., Chew, S. Y., & Vellasamy, S. (2022). Revisiting *Trigonella foenum-graecum* L.: Pharmacology and therapeutic potentialities. *Plants, 11*(11), 1450.

Voroney, R. P. (2007). The soil habitat. In *Soil Microbiology, Ecology and Biochemistry* (pp. 25–49). Academic Press.

Vuolo, M. M., Lima, V. S., & Junior, M. R. M. (2019). Phenolic compounds: Structure, classification, and antioxidant power. In *Bioactive Compounds* (pp. 33–50). Woodhead Publishing.

Waliat, S., Arshad, M. S., Hanif, H., Ejaz, A., Khalid, W., Kauser, S., & Al-Farga, A. (2023). A review on bioactive compounds in sprouts: Extraction techniques, food applications, and health functionality. *International Journal of Food Properties, 26*(1), 647–665.

Wang, Y., Zhang, L., Yang, H., Yan, G., Xu, Z., Chen, C., & Zhang, D. (2016). Biochar nutrient availability rather than its water holding capacity governs the growth of both C3 and C4 plants. *Journal of Soils and Sediments, 16*, 801–810.

Zandi, P., Basu, S. K., Cetzal-Ix, W., Kordrostami, M., Chalaras, S. K., & Khatibai, L. B. (2017). Fenugreek (*Trigonella foenum-graecum* L.): An important medicinal and aromatic crop. In *Active Ingredients from Aromatic and Medicinal Plants* (pp. 207–224).

Zhang, Y., Xiao, Z., Ager, E., Kong, L., & Tan, L. (2021). Nutritional quality and health benefits of microgreens, a crop of modern agriculture. *Journal of Future Foods, 1*(1), 58–66.

CHAPTER 3

Characterization of Bioactive Compounds (BACs) in Fenugreek

PRIYA SINGLA,[1] PIVERJEET KAUR DHILLON,[2] and MADHU[3]

[1]*Food and Nutrition Department, Desh Bhagat University, Mandi Gobindgarh, District Fatehgarh Sahib, Punjab, India*

[2]*Home Science Department, Krishi Vigyan Kendra, Tarn Taran, Guru Angad Dev Veterinary and Animal Sciences University, Ludhiana, Punjab, India*

[3]*Home Science Department, Kashi Naresh Government PG College, Gyanpur, Bhadohi, Uttar Pradesh, India*

ABSTRACT

An herb/legume, fenugreek comes under family Fabaceae with scientific name *Trigonella foenum-graecum* and cultivated annually and globally and included as a spice in human dietaries. It is highly perishable for storage, due to its high respiration rate. Apart from its good nutritional composition, it is gaining consideration throughout the globe, because of its non-nutritive components, that is, bioactive compounds (BACs) which are potent antioxidants. High humidity and temperature alter its bioactive composition. BACs in fenugreek include both volatile and non-volatile compounds *viz.* alkaloids, flavonoids, polyphenols, steroids, stolone, and so forth. The number of compounds can be dependent of using specific extraction techniques like ultra HPLC (High Performance Liquid Chromatography), HPLC, and so on, since the laboratory studies reported five to seven BACs in different procedures. These compounds have numerous uses ranging from food aesthetics to medical applications for flavor enhancement and treatment of fatal ordeals, correspondingly. Besides, the multiple component utilization

Fenugreek: Improvement in Yield Potential Through Plant Growth Regulators (PGRs) and Depolymerized Polysaccharides. Tariq Ahmad Dar, Bilal Ahmad, Moinuddin, and Khalid Rehman Hakeem (Eds.)

of this food (in form of leaves, seeds, and oil) draw attention to its charisma in the diverse industrial sectors. Methanolic extract of leaves contains ferulic acid, catechin, gallic acid, caffeic acid, coumaric acid, chlorogenic acid, ellagic acid, naringenin, quercetin which are important polyphenolic compounds. This crop exhibits antioxidant, antimicrobial, anticarcinogenic, anti-anorexia, hypocholesterolemic, and hepatoprotective effects. It also acts as gastric stimulant. Conversely, a strong odor and bitter taste are attributed to volatile oils and alkaloid compounds present in it but can be removed prior to its consumption. In various countries, fenugreek leaves are included in daily diets as well as manufacturing of value-added food products, for instance, it is supplemented to wheat and maize flour during bread preparation, and fenugreek seeds are the main ingredient in preparation of maple syrup, vanilla, and butterscotch flavors, because it possesses a main alkaloid called "trigonelline." Total phenol content (TPC) might be augmented using different methods of processing such as roasting, germination, and soaking. Studies reported the TPC as 54.4, 80.8 and 48.5 mg GAE/g in the seed flours obtained after treatment with above-mentioned three methods, in that order, whereas raw seed flour contained 45.4 mg GAE/g of sample. Hence, the total antioxidant potential of processed seed flours can also be enhanced from 13.9 up to 55.8%, due to decline in phytic acid content. Moreover, content of flavonoids and phenols can be increased via pressure cooking. Fenugreek is known as a medicinal plant since >4000 years in countries like India, China, North America, and some regions of Australia. Fenugreek extract has a dose and time-dependent recuperative effect during cancer as documented in previous clinical investigations because it suppressed the metastasis due to its BAC (flavone and isoflavone) content. Similarly, antioxidant quercetin decreases absorption of glucose among intestine and increment in secretion of insulin from beta cells of pancreases through inhibiting glucosidase and 11-β-hydroxysteroid dehydrogenate type 1 enzymes, thereby, acts as an anti-diabetic agent. It is important to tap fenugreek's potential by targeting its bioactive compounds, in order to provide a cost-effective substitute for serious medical challenges such as cancer, diabetes, and heart diseases.

3.1 INTRODUCTION

Fenugreek is a dicotyledonous and annual plant. The scientific name of fenugreek is *Trigonella foenum-graecum* L., and it belongs to the family Leguminosae (Fabaceae) and the subfamily Papilionaceae. This plant has roots with nodules, golden seeds of yellow color, a branched stem, trifoliate leaves, and white-colored triangular flowers (Acharya et al., 2008). The height, length, and width of raw fenugreek seeds have been reported as 1.77 mm, 0.41 cm, and 0.24 cm, respectively. The volumetric density of the seeds ranges between 683 and 692 kilograms per cubic meter, with an average bulk density of 687.62 kg/m^3 (Adhikari and Rai, 2021). Fenugreek originates from Western Asia and Southeastern Europe (Khoja et al., 2021), but it has gained attention due to increasing worldwide production over the past decade. For instance, fenugreek production in India rose from 116,000 to 247,000 tons from 2011 to 2016 (Navarro Del Hierro et al., 2020).

Fresh and dried fenugreek leaves and seeds are widely recognized components of the human diet. The seed contains a yellow-colored, rigid embryo in the center, enclosed by an endosperm—a horny, white, and partially transparent layer (Wani and Kumar, 2018). These seeds can be consumed raw, sprouted, or cooked and are generally used as flavoring agents in curry powders, tea blends, or as dietary supplements (Yao et al., 2020). The bitter flavor of fenugreek seeds is due to the presence of components such as coumarin, diosgenin, saponin, and steroidal compounds. Furthermore, several antinutrients like alkaloids, cyanogen, oxalates, enzyme and protease inhibitors, phytate, saponin, tannin, and hemagglutinins (lectin) are present in fenugreek seeds and can hinder the metabolism (digestion, absorption, and utilization) of both macro- and micro-nutrients (such as carbohydrates, proteins, iron, zinc, and calcium) by inhibiting enzymatic activity (digestion enzymes like trypsin, pepsin, pancreatin, and α-amylase). However, extracts of fenugreek exhibit advanced nutraceutical properties (Singh et al., 2020).

Fenugreek comprises approximately 58% carbohydrates and 25% dietary fiber. Protein and fat content range from 23% to 26% and 6% to 7%, respectively. Iron and ascorbic acid content are available at 33 mg and 12 mg per 100 g of fenugreek, respectively (Ojha et al., 2018; Wani and Kumar, 2018). Regarding the nutritive value of fenugreek seeds, they contain about 28% mucilage, 22% protein, and 5% bitter, aromatic oil (Khare, 2004). Fenugreek contains alkaloids such as choline and trigonelline, and proteins such as albumin, histidine, lecithin, and globulin (Khare, 2004; Mathur and Choudhry, 2009; Raju et al., 2001). A 100 g portion of fenugreek endosperm contains 44 g of protein (Madhava Naidu et al., 2011). Fenugreek also has abundant amounts of B-complex vitamins, especially choline, pyridoxine, and niacin (Sharma, 1986). Besides its macronutrient composition, fenugreek seeds are a good source of bioactive components and phytochemicals, including alkaloids, carotenoids, saponins, tocopherols, phenolic compounds, and phytosterols. They are also rich in amino acids, vitamins, essential fatty acids, and minerals (Aasim et al., 2018; Hernández-Ledesma, 2019). Beyond these properties and composition, interest in fenugreek seeds has centered primarily on fenugreek extract, which is rich in bioactive compounds for nutraceutical applications (Navarro Del Hierro et al., 2020).

3.2 BIOACTIVE COMPOUNDS IN FENUGREEK

Bioactive compounds are essential and non-essential part of our food, and it is having an impact on our heath (Bahubali and Bornare, 2021). The major bioactive compounds like polyphenols, especially, isovitexin, and rhaponticin are present in seeds (He et al., 2015). There are various chemical constituents such as triamcinolone, osapogenin, 4-hydroxyisoleucine and galactomannas are available in seed part. Other important bioactive components like vitexin, trigonelline, orientin, isovitexin, and isoorientin were analyzed using "high performance liquid chromatography technique." Simultaneously, "ultra-high-performance liquid chromatography-electrospray ionization-tandem mass spectrometry" technique was used to estimate trigonelline, sarsasapogenin, pinitol, isovitexin, isoorientin, and 4-hydroxyisoleucine under MRM mode (multiple reactions monitoring; Singh et al., 2020).

Among the above bioactive components, extracts of fenugreek seed rich in saponin are gaining attention. Saponin are considered as a vast class of glycosidic components, having molecular weight of 1223.3 g/mol. It is composed of water repellent steroid aglycone and triterpenoid, named as sapogenin. These are linked with ≥1 hydrophile sugar elements through an ester or ether glycosidic attachment (Güçlü-Üstündağ, and Mazza, 2007). Fenugreek mainly contains steroid saponins. Saponin and their hydrolyzed extract acquire from seeds of fenugreek were obtained as 31.13 and 2.01 g/100 g (Navarro Del Hierro et al., 2020). The composition of chemical compounds of fenugreek (cotyledons, husk, and seeds) revealed that saponin content was present in higher amount in 100 g sample of endosperm (4.6 g; Wani and Kumar, 2018).

Fenugreek also contains various other chemical compounds, including steroid saponins (also known as steroidal sapogenins). Among the oily part of embryo, diosgenin was found. The two furostanol ($C_{27}H_{46}O_2$) glycosides are a diosgenin precursor (having opened F ring), present in fenugreek recognized as hederagin ($C_{30}H_{48}O_4$) glycosides. In stem part of plant, alkaloids for example, nicotinic acid, trigocoumarin, trigonelline, and trimethyl coumarin are existing. Likewise, leafy part consists of 7 saponins called graecunins, considered as diosgenin glycosides (Wani and Kumar, 2018). Commercially extracted seeds composed off diosgenin (0.1 to 0.9%; Bahubali and Bornare, 2021).

Table 3.1 describes the chemical components of bioactive compounds present in fenugreek.

TABLE 3.1 Bioactive Compounds of Fenugreek and Their Chemical Components

Sl. No.	Bioactive Compounds	Chemical Components
1.	Alkaloids	Betain, carpaine, choline, gentianine, neurin, trigonelline, trimethylamine
2.	Amino acid	Arginine, cystine, histidine, 4-hydroxy isoleucine, isoleucine, leucine, lysine, tryptophan, tyrosine
3.	Coumarins	Trimethylcoumarin, trigocumarin, methyl coumarin
4.	Fatty acids and lipids	Triacylglycerides, stearic acid, phosphatidylinositol, phosphatidylethanolamine, phosphatidylcholine, palmitic acid, oleic acid, monoacylglycerols, linoleic acid, free fatty acids, diacylglycerols, cycloartenol, campesterol, beta-sitosterol, alpha-linolenic acid.
5.	Fibers	Triacylglycerols, phosphatidylcholine, neutral detergent fiber lipids, gum
6.	Flavonoids	Vitexin, vecenin-1, tricine, saponaretin, rutin, quercetin, orientin, naringenin, irilone, isovitexin, isoorientin, luteolin, lilin, kaempferol, glucopyranoside, daidzein, calycosin, apigenin
7.	Steroid saponins	Yamogenin, yuccagenin, tigogenin, saponaretin, sarsasapogenin, smilagenin, neogitogenin, gitogenin, diosgenin.
8.	Saponins	Trigonoesides, trigofoenosides A-G, fenugrin B, graecunins, glycoside, foenugracin, hederagin, fenugreekine
9.	Other components	Ascorbic acid, biotin, folic acid, gum, nicotine acid, riboflavin, thiamine, vitamin A

Source: Kousalya et al. (2019); Srivastava et al. (2020); Wani and Kumar (2018).

3.2.1 BIOACTIVE COMPOUND WISE HEALTH BENEFITS OF FENUGREEK SEEDS

1. **Galactomannas:** It has the capacity to lower the low-density lipoprotein (LDL) level. Thus, effective during hypercholesterolemic conditions. Moreover, it also helps in improving blood pressure, fibrinolysis, and lipid (Hannan et al., 2007).
2. **4-Hydroxyisoleucine:** It has a pivotal role in ameliorating glucose level especially in patients of diabetes mellitus (types 1 and 2) by maintaining homeostasis (Hamden et al., 2010).
3. **Osapogenin:** It has an anti-breast cancer activity because of its constant molecular kinetics technique, lower bounding energy computation and high dockage score (Rampogu et al., 2018).

4. **Triamcinolone (Alkaloid):** Fenugreek seeds contain Triamcinolone (alkaloid), acts as a probiotic (Bahubali and Bornare, 2021; Majeed et al., 2018).

The seeds are mostly used for spice production or medicinal purposes (Türker and İŞleroğlu, 2023). Fenugreek is an essential constituent of 19th century and patent medication for postmenopause and dysmenorrhea. In Ayurvedic texts, the medicinal value of fenugreek seeds is quoted as a powerful aphrodisiac herb. In Egypt, it was consumed during and after delivery to simplify the birth of child, increase lactation, to relieve menstrual cramps, and other stomachache. In China, it is used for treating edema and pain of abdomen (Wani and Kumar, 2018). Fenugreek helps in detoxification of the body, because of its diaphoretic properties. It has lymphatic cleansing activity by which it removes dormant cells, immobilized or proteins and hazardous wastes from the body. Figure 3.1 describes chemical structure for some bioactive compounds.

3.3 ANTIOXIDANT POTENTIAL

Fenugreek husk is consisted of higher total polyphenols. According to DPPH method (concentration of 200 μg), antioxidant action of extract of seeds, husk, and endosperm expressed as around 56, 64, and 72%, respectively (Madhava Naidu et al., 2011). Its oil showed a powerful antioxidant action in terms of ABTS and DPPH with values 161 and 173 μg/ml, respectively. The value of 173 mg/g DPPH activity depicts a great antioxidant action in fenugreek oil because of the presence of phytol and palmitic acid analyzed using Gas chromatography-mass spectrometry technique. Moreover, total flavonoid and total phenolic content in oil of seeds were found as 14.4 mg quercitin and 39.0 mg gallic acid equivalent per g ram of sample, correspondingly. High flavonoid and phenolic content reflect higher antioxidant activity of the plants. Resultantly, oil of seeds might be efficacious in case of urinary infections, sexual disorder, inflammation, cancer, and asthma. The above values are according to degree of 50% inhibition. Furthermore, other antioxidants present in oil also showed antiinflammatory action (Akbari et al., 2019; Arora and Pandey-Rai, 2014; Benzidia et al., 2019).

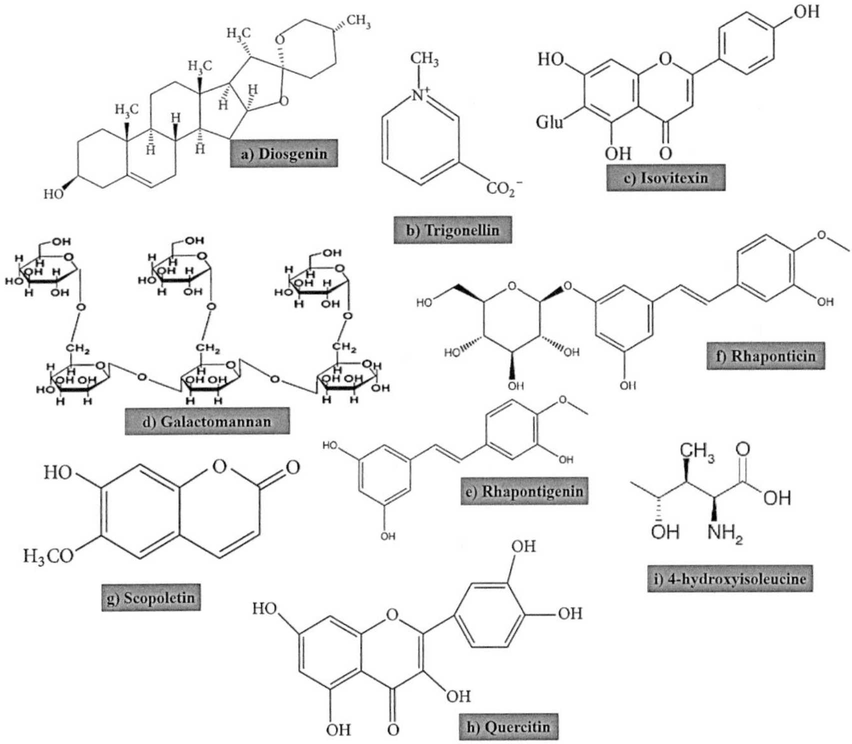

FIGURE 3.1 Chemical structures of some bioactive compounds of fenugreek.

In addition, antioxidant content determined using ABTS was stronger in comparison to DPPH as values for ABTS is lesser and for DPPH is higher. It means lower value in fenugreek oil shows high antioxidant capability. Because ABTS has much sensitiveness to identify antioxidant action by making chemical kinetics faster, therefore, resulting to high activity of antioxidants. Moreover, both antioxidants consist of proton radicals, it decreased as it exposes to proton ROS (radical scavengers; Floegel et al., 2011; Lee et al., 2015).

3.4 EXTRACTION METHODS

Extraction is a unit operation conducted to separate bioactive compounds from plant materials (Chouhan et al., 2019). Extraction conditions have a great impact on the extraction yield as well as activity of the bioactive compounds. Important parameters for an extraction process are generally defined as ration of sample to solvent, temperature, and solvent form (Zuorro et al., 2016). Fenugreek leaves have a great potential for obtaining biomaterials, and simple extraction methods can be used for this purpose. For extraction of bio-actives, several ordinary to modern methods are available such as enzyme, microwave, and ultrasound assisted, Soxhlet, reflux, maceration, percolation, pressurized liquid, super critical fluid, and pulsed electric field. With the exception of these, few techniques based on water can be used for extracting bioactive components (Bahubali and Bornare, 2021). Further, the suitable extraction methods for obtaining diverse compounds are discussed in subsections.

3.4.1 EXTRACTION OF BIOACTIVE COMPOUNDS

3.4.1.1 SAPONIN EXTRACTION

Fenugreek seeds were ground using a Retsch mill with a 3-unit knife for one minute at 10,000 rounds per minute (rpm). The ground seeds were then sieved with a vertical sifter to obtain a particle size range of 100 to 250 micrometers. For extraction, samples were mixed with methanol in a 1:10 ratio (w/v) in a glass vial, and sonication was performed using an ultrasound transducer at 20 kilohertz with the direct sonication method for approximately 15 minutes. The mixture was then centrifuged for 15 minutes at a relative centrifugal force (RCF) of 3400×g.

The clear supernatant was vortexed by adding hexane for one minute to remove fats. After centrifuging for 10 minutes at an RCF of 2688×g, the upper hexane layer was discarded, and the lower methanol extraction layer was collected. The methanol layer was evaporated using a rotary evaporator. Milli-Q purified water was then added to the dried residue in a 1:20 ratio (sample, w/v) in centrifuge tubes. After the residue dissolved, a mixture of n-butanol and water (2:1, v/v) was added to the homogeneous solution. The tubes were vortexed for one minute and centrifuged for 10 minutes at an RCF of 2688×g, and the upper layer was collected. This procedure was repeated once with the lower layer.

Both collected upper layers were dried under vacuum, and the yield of extraction was calculated. The remaining lower layer of fenugreek extract was stored at –20°C for further analysis (Navarro Del Hierro et al., 2020).

3.4.1.2 SAPOGENIN EXTRACTION

A 2 M hydrochloric acid solution was added to the saponin extract in a 50:1 (acid, V: sample, W) ratio in glass vials, then heated for one hour at a maintained temperature of 100°C using a block heater. The mixture was then cooled on ice for five minutes. Ethanol (ethyl acetate) was added to the mixture in a 1:1 (V/V) ratio, vortexed for one minute, and centrifuged at 3400 x g for five minutes, after which the upper layer was collected. The extraction procedure with ethyl acetate was repeated on the bottom layer. Finally, the two collected upper layers were dried using a nitrogen stream, and the extraction yield was calculated. The remaining bottom layer of the extract was stored at –20°C for further analysis (Navarro Del Hierro et al., 2020).

3.4.1.3 PREPARATION OF EXTRACT

Fenugreek leaves with <10% moisture was taken, crushed with hand and sifting was done for removal of soil and dust particles at pore size of 630 micrometer. Now the samples were placed at 25°C for extraction process. Sample was taken in beaker with lid and solution of ethanol + water (0 to 100%) was added in it in the ratio of 10 to 50 gram per liter. Then, lid was closed and kept on a magnetized stirrer for shaking about two hours at 400 round per minutes and temperature must be between 25 and 60°C. Afterwards, the shaken sample was centrifuged for 5 minutes at an RCF of 4637 g. Filtration was performed using coarse filtration paper, and the filtered extract was used for further analyses (Türker and İŞleroğlu, 2023).

1. **Total Saponins:** For estimation, several reagents were prepared. Vanillin reagent was prepared by dissolving 0.8 g of vanillin in ethanol to a final volume of 100 ml. Sulfuric acid solution was prepared by adding distilled water to 72 ml of sulfuric acid to make a final volume of 100 ml. A prepared sample extract (0.2 ml) was taken in test tubes, followed by the addition of 0.35 ml vanillin solution and 0.8 ml distilled water. Then, 1.25 ml sulfuric acid was added to each test tube, which was then capped. After gentle mixing, the caps were removed, and the test tubes were placed in a water bath at 60°C. After 10 minutes in the incubator, the test tubes were immediately transferred to ice for cooling. Finally, the absorbance of the solution was measured at a wavelength of 544 nanometers using

a spectrophotometer. The results were presented as milligrams of diosgenin per gram of dried sample. This method was described by Akbari et al. (2019), with some modifications made by Türker and İşleroğlu (2023).

2. **ABTS (2,2'-Azino-Bis (3-Ethylbenzothiazoline-6-Sulfonic Acid):** It is modified method (Pająk et al., 2019). For estimation, some reagents were prepared. (i) ABTS and potassium persulfate solution was prepared by mixing equal volume of 2.45 millimolar solution of potassium persulfate and 7 millimolar stock solution of ABTS; and (ii) preparation of solution buffer using 20 millimolar of $C_2H_3NaO_2$ (sodium acetate) with maintenance of 4.5 pH.

   Reagent A was now incubated in the dark conditions for about 16 h at room temperature. Afterwards, dilution of one ml of reagent A with reagent B (54 ml) was done. The value of absorbance was fixed at 0.700 at wavelength of 734 nanometer for this dilution and considered as reagent C.

   Then, sample extract of 150 ml was taken and reagent C (2850 microliter) was added in it. After incubation of half an hour in dark conditions and at room temperature, its absorbance was taken. The values were expressed as mg/g and equivalent to Trolox (Türker and İŞleroğlu, 2023).

3.4.1.4 MICROWAVE-ASSISTED AND METHANOL-BASED EXTRACTION METHODS

Samples of fenugreek leaves were cleaned to remove dust, and impurities present in it using tap water and kept on tissue paper to remove extra water. Now, 50 g of leaves were kept for drying in an oven and used variety of powers (30 to 100 Watt) to operate and drying leaves. Reading of moisture loss was noted at different time intervals until weight gets stable. A microwave-based extraction system was used to extract bioactive compounds from leaves. Here, rotary system and condenser was used to liquefy the moisture/aqueous vapor comprising bioactive components in flask and stored at a maintained temperature of 4°C for analysis of different antioxidant activities and other analysis.

In addition, other extraction method was also performed. Five gram of leaf sample was taken and blended with 80% solution of methanol. Homogenization or shaking was done for total of 45 minutes at 8000 rounds per minute. The upper liquid layer from residual sample was separated and used for further analysis of various components such as DPPH (2,2-diphenyl-1-picrylhydrazyl), total flavonoids and phenols (Khan et al., 2022; Terrence Justine et al., 2019).

1. **DPPH (2,2-Diphenyl-1-Picrylhydrazyl) Activity:** A solution of DPPH was prepared by taking weighed amount of DPPH (4 g) and methanol was added to make the volume till 100 ml. Now, methanol and microwave extracted fenugreek sample of 0.5 milliliter was taken separately and added in prepared DPPH solution with vigorous and continues shaking so that it mixes properly. Later, the above solution was immediately kept at darkest place for overnight. Measurement of absorbance of solution was taken using spectrophotometer at wavelength of 515 nanometer and results were presented in percentage (Barkat et al., 2018; Curi et al., 2019; Khan et al., 2022).
2. **Total Flavonoid Content:** A sample extract (one ml) prepared using methanol and microwave extraction method was taken separately and added in 3 ml solution of methanol comprising CH_3CO_2K (potassium acetate) and $AlCl_3$ (aluminum chloride). Further, dilution was done with addition of 5.6 ml distilled water in above solution of sample extract and kept for 30 minutes at room temperature. Finally, measurement of absorbance of solution was taken using spectrophotometer at wavelength of 415 nanometer and values were calculated and presented in equivalents of quercetin (used as standard). The reference value was taken from prepared standard curve (Ghafar et al., 2017; Khan et al., 2022).
3. **Total Phenol Content:** A reagent solution was prepared by mixing 1 ml of Folin-Ciocalteu reagent with distilled water to a final volume of 10 ml. Then, 0.1 ml of extract (prepared using methanol and the microwave extraction method) was taken separately, and 0.75 ml of the prepared reagent solution was added to it. After incubating the sample mixture for 10 minutes, 0.75 ml of sodium carbonate (Na_2CO_3) solution was added, and the mixture was kept in a dark place for 45 minutes. Finally, the absorbance of the solution was measured using a spectrophotometer at a wavelength of 765 nanometers. A standard curve was prepared using gallic acid at concentrations ranging from 0 to 40 mg/L, and results were calculated and expressed in gallic acid equivalents (mg/g). This method was originally described by

Sun et al. (2005) and Lu et al. (2011), with some modifications made by Khan et al. (2022).

3.4.1.5 *TOTAL TANNINS*

A 0.5-gram ground seed sample was placed in a beaker, and 75 mL of water was added. The mixture was then boiled for 30 minutes, followed by centrifugation for 20 minutes at 2000 rpm. After collecting the clear liquid (supernatant), sodium carbonate (Na_2CO_3) and Folin Denis reagent were added to the extract, and the volume was adjusted to 100 mL with distilled water while gently shaking the solution. Absorbance was measured using a spectrophotometer at a wavelength of 700 nm, precisely 30 minutes after preparation (Saini et al., 2016).

3.4.1.6 *FRAP ANTIOXIDANT (FERRIC REDUCING ABILITY OF PLASMA)*

The method used for determination of FRAP was $K_3[Fe(CN)_6]$–$FeCl_3$ (Potassium Ferricyanide–Ferric Chloride). Some reagents were prepared before estimation: (i) Phosphate buffer (0.2 Molar, 6.6 pH); (ii) potassium ferricyanide $\{K_3[Fe(CN)_6]\}$ amounting one gram was diluted and made volume up to 100 ml with distilled water; (iii) TCA (Tri Chloroacetic Acid) weighing of 10 grams was taken and diluted with distilled water to make volume up to 100 ml and (iv) $FeCl_3$ (Ferric chloride) solution was prepared by dissolving one gram in distilled water and volume raises till 100 ml.

Different concentration of extracts Sample extract was taken of various dilutions and 2.5 ml of reagent A and B was added in it. After incubation of 20 minutes at a temperature of 50C, 2.5 ml of reagent C was added. Then, from this prepared sample solution, took 2.5 ml and in this, distilled water (2.5 ml) and reagent D (0.5 ml) was added. The measurement of absorbance was taken exactly after half an hour using spectrophotometer at wavelength of 700 nanometer (Saini et al., 2016).

3.5 MEDICINAL SIGNIFICANCE

Fenugreek greens and seeds have medicinal applications in human history. Fenugreek is beneficial for digestion (Wani and Kumar, 2018) and is an apt supplement of cereal foods as it possesses high protein, beta-carotene, lysine, dietary fiber, calcium, and iron content (Hooda and Jood, 2003). Moreover, at alkaline conditions (>7 pH), protein of fenugreek is pretty soluble (Meghwal, 2012). Fenugreek has numerous pharmacological properties due to its antibacterial, antidiabetic, antioxidant, antitumor, hypocholesterolemic, hypolipidemic, and hepatoprotective effects and it acts as remedy during lactation, chronic cough, edema, gout, and paralysis. The seeds diminish the CaC_2O_4, calcium oxalate (a compound responsible for kidney stones) amount in the kidneys (Charles et al., 2022). Fenugreek leaves are containing fair amounts of vitamins (folic acid, niacin, riboflavin, thiamine, and ascorbic acid), minerals (calcium, iron, and phosphorus), and important functional components especially, flavonoid, and carotene. The presence of flavonoids occurs mainly in the complex form of glycosides like O and C-glycosidic bonds; and quercetin, vitexin, apigenin, and so on, are the major flavonoids (Nagulapalli Venkata et al., 2017). The afore-mentioned compounds have the capacity to manage infections (chronic cough, diarrhea), and chronic ailments such as diabetes mellitus (type I and type II), rickets and ulcers (Anoopkumar et al., 2020; Aylanc et al., 2020). Fenugreek helps to clear respiratory congestion, eases sore throat. It also relieves asthma, colds, influenza, constipation, pneumonia, sinusitis, laryngitis, and tuberculosis (Wani and Kumar, 2018).

The fenugreek shows antidiabetic effects. Consuming fenugreek could help controlling weight of the body and glycogen in the liver; and have a noteworthy impression on various enzymes involved in carbohydrate metabolism (Srinivasan, 2006). Therefore, the use of fenugreek seeds as a substitute of oral insulin for lowering glucose level of blood (Madar and Stark, 2002). Moreover, fenugreek lowers the serum cholesterol and raises HDL cholesterol by increasing the amount of fecal excretion (Annida and Stanely, 2004). Thus, fenugreek leaves are freshly consumed.

As the chemical composition is concerned, triamcinolone as alkaloid, osapogenin, 4-hydroxyisoleucine and galactomannas as a fiber, are the main components present in fenugreek with medicinal properties (Bahubali and Bornare, 2021). Figure 3.2 presents medicinal significance of fenugreek.

3.6 CULINARY IMPORTANCE

Apart from medicinal implications, numerous culinary usages of fenugreek such as thickening, prebiotic, emulsifying, and so on, are there (Hefnawy and Ramadan, 2011). It is used to enhance sensory characteristics (flavor,

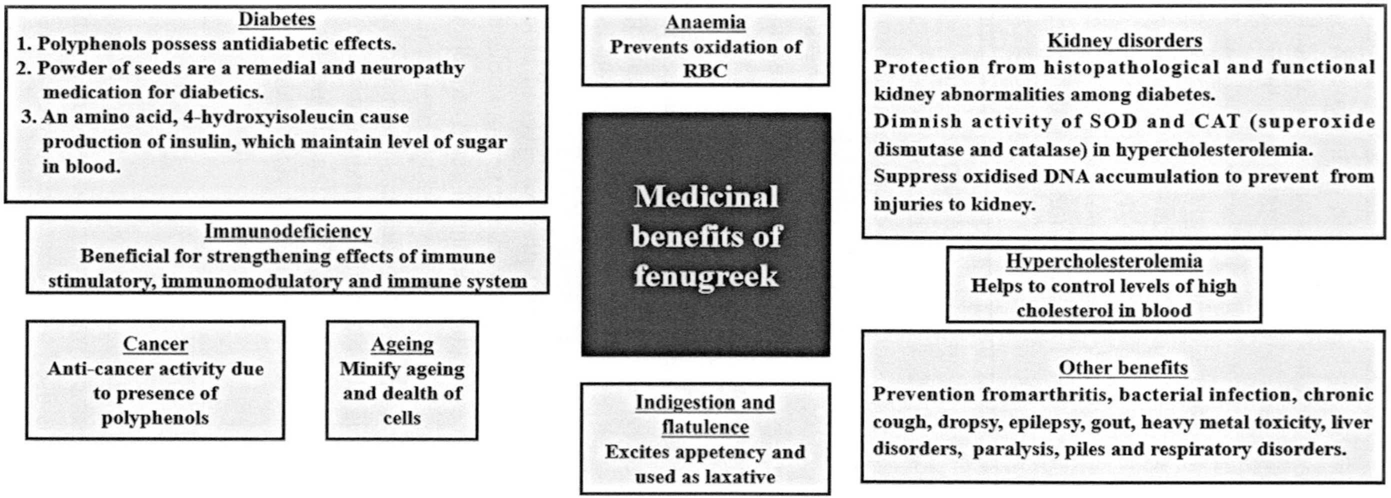

FIGURE 3.2 Medicinal benefits of fenugreek.

Source: Adapted from Srivastava et al., 2020

texture, and color) of food materials. It is renowned for its volatile as well as gum, fiber contents. Fiber (about 25%) present in fenugreek seed changes the texture of food. Because of its high fiber, gum, and protein content, it is utilized for adhesion, stabilization, and emulsification in food preparation.

This herb is being incorporated from ages as a food ingredient throughout the world and it has been a well-liked ingredient in spice mixtures, pickles, and curry powders in various countries such as India, Bangladesh, Pakistan, and so forth. The wood flavor and aroma of fenugreek is responsible for production of imitated molasses (maple syrup). Indian population uses leaves and stems of fenugreek plant for preparation of "*methi ka saag*" and it is commonly used during winter season, and the leaves in dried form and seeds in both raw and roasted form are added in diverse cuisines. Sprouted and cooked seeds are also consumed. In Ethiopia, Egypt, and Africa, fenugreek is added during bread preparation and in Switzerland, it is added for flavoring the cheese. In the United States, fenugreek is added as spice mix for stews and soups (Wani and Kumar, 2018). In India and Nepal, fenugreek is called "*Methi*," and the seeds and leaves are incorporated to flavor the food dishes (Bahubali and Bornare, 2021).

Fenugreek, a functional food ingredient improves appearance, emulsifying, encapsulating, gelling, texture, thickness, and stabilizing abilities due to gum present in it. Moreover, the dietary fiber present in it have the potential to develop nutritious milk, cereal, and beverage-based products. Fenugreek concentrate can be blended with other spice mixtures. Fenugreek flavored candies, dressings, milk shakes, sweets, and soups can also be prepared. It is fortified in flour for baked and extruded preparations such as tortilla, pizza, pasta, noodles, muffins, flat bread, bread, cake, and so forth. Conversely, the seeds taste bitter as saponins are present (Pandey and Awasthi, 2015). Nevertheless, it can be incorporated at acceptable levels. Unwanted fat intake can be prevented through fried snacks if high fiber flour is blended with the basic flour, since it absorbs 8 to 15% less oil during frying of the final product (Kumar and Maliakel, 2008).

Anti-nutrients interfere in the digestion as well as absorption of micronutrients present in foods. These must be removed to get nutritive value of food substances (Thakur et al., 2019). To trim down these factors, different methods of processing such as germination, roasting, soaking, and so on, are effective (Adhikari and Rai, 2021).

3.7 EFFECT OF PROCESSING ON BIOACTIVE COMPOUNDS

Diverse processing methods (i) sun drying; (ii) solar drying (Shrivastava and Kumar, 2016); (iii) microwave drying (Borda-Yepes et al., 2019); (iv) hot air drying (Bishnoi et al., 2016); (v) convective drying; (vi) fluidized bed drying (Jawake et al., 2018a, b); and (vii) freeze drying (Kushwaha and Mustafa, 2012) have been employed to fenugreek components to make it edible. These methods are having great efficiency for drying; however, loss of bioactive compounds is the main concern especially those, are soluble in water. It might be due to more time of processing and sometimes, high cost is factor (Thamkaew et al., 2021).

Roasting breaks down large quantities of N-methylnicotinate (trigonelline) into niacin (nicotinic acid) and associated azines (pyridines). In contrast, soaking of seeds in water (1:5) for 12 hours in darkened conditions, increases the ascorbic acid and phenol content significantly whereas decreased tannin content (Ojha et al., 2018). Similarly, if seeds are germinated at relative humidity of 90% and 25°C temperature for at least 3 days (72 hours), increase in content of tannins, phenolics, and ascorbic acid of germinated, dormant, and imbibed seed powder was found. Ascorbic acid in germinated fenugreek seed powder was higher (0.07 ± 1.27 mg%) than the ascorbic acid content (5.6 mg%) in raw fenugreek seeds. This increment might be because of increase in the galactonolactone dehydrogenase activity of soaked seed (Moriyama and Oba, 2008).

Disparity in variety of crop, geographic conditions, complication in plant composition, extraction time and method, solvents used for extraction are several factors may attribute to the fluctuation found in profile of phenolics (Wijekoon et al., 2011). Total phenol content of germinated (48 hours) fenugreek seeds was also increased by 46% in comparison to raw while the tannin content was gradually decreased (532.1 to 210.1 mcg Gallic acid per gram) with development of sprouts (Pandey and Awasthi, 2015; Shimelis and Rakshit, 2008). This increment might be due to bioaccumulation, biosynthesis, and destruction of solidified polyphenols, especially tannic acid, gallotannins, and ellagitannins (hydrolysable tannins) and the hydroxylation of more glycosylated flavonoids (Randhir et al., 2004). Roasted fenugreek seeds contained 86 mg/g of phenols present in it whereas it was 73 mg/g in raw seed (Manju, Khatkhar, and Parveen, 2015). Thermal treatment employed to plant foods through roasting, cause dehydration of water located within cells, encourage denaturation of protein and triggering various chemical and enzymatic reactions, result in phenols availability (Rizki et al., 2015).

3.7.1 EFFECT OF PROCESSING ON ANTIOXIDANT ACTIVITY AND DPPH

The values of DPPH (2,2-diphenyl-1-picrylhydrazyl) for raw fenugreek seeds were 5.95 mg decimeter/milliliter. Different processing methods such as germination, roasting, and soaking decreased the values to 1.28, 3.38, and 1.54 mg decimeter/milliliter, correspondingly. Antioxidant activity (AA) was 16.95%, 26.98%, 62.28% and 75.66%, in that order (Pandey and Awasthi, 2015). Another study revealed that DPPH activity for soaked, germinated, and dried seed was 23,616 (14%), 1582 (55%) and 27,975 μg/ml (10%), respectively, against standard value of 615 μg gallic acid/ml (90%; Ojha et al., 2018). Roasting destroys bioactive components, and conjugated phenols are released in high amount from the roasted seeds. However, antioxidants (melanoidins) are formed by roasting at intense heat and by using Maillard reaction (Rashid et al., 2015; Wani et al., 2017). Autogenous antioxidant catalyst prevents antioxidant components of the natural plant from oxidation as heat inactivates enzyme. During germination, several biological changes occur, it increases activity of endogenic hydrolytic catalyst which alter an endosperm and modify the chemical action of tocopherol and vitamin C. It is further responsible for amino acid increment (Sharma and Gujral, 2010). Sprouting for at least 3 days (72 hours is an auspicious and efficacious method as well as helpful for reducing the antinutritional content and improving the composition of antioxidants, phytochemicals, and other nutrients (Adhikari and Rai, 2021; Table 3.2).

TABLE 3.2 Effect of Processing on Antioxidant Activity of Fenugreek

Activity	Raw	Soaking	Roasting	Germination	Pressure Cooking/ Boiling
ABTS (μg/ml)***	698	950	72.5%****	848	950.0
DPPH (%)*	13.7	35.8	24.8	48.8	27.7
Flavonoid (mgQE/g)**	4.23	2.11	3.92$	2.10	2.33
FRAP (mg AA/g)*	4.9	5.2	3.4	6.5	5.2
Phenols (mg GAE/g)*	173.4	158.0	172.3	184.3	178.2
Tannin (mg TA/g)*	3.9	4.3	2.6	5.5	5.3

Source: *Saini et al. (2016); **Salam et al. (2023); ***El-Salam et al. (2022); ****Yazicioglu (2022); $Manju et al. (2016).

3.7.2 EFFECT OF PROCESSING ON THE ANTINUTRITIONAL FACTORS

1. **Phytic Acid:** Raw fenugreek seeds contain higher levels of phytates than processed seeds (Atlaw and Kumar, 2018; Singh et al., 2015),

which may be due to the efficiency of phosphorus utilization (Deme et al., 2017). Roasting reduced phytate content by 36.4% to 39.9% when seeds were roasted at 185°C for 15 minutes (Adhikari and Rai, 2021). Similarly, soaking seeds for 12 hours also resulted in an 8.7% to 9.0% reduction in phytate content, although this effect was not statistically significant ($p > 0.05$; Adhikari and Rai, 2021; Hooda and Jood, 2003; Pandey and Awasthi, 2015). Dried, germinated fenugreek seeds (germinated for 72 hours) contained 38.7 mg of phytates per 100 g, showing a reduction of about 55%. During germination, seeds utilize phytates as a source of phosphoric acid (inorganic phosphate), leading to improved nutritional content and palatability (Ojha et al., 2018; Wang et al., 1997). Germination activates enzymatic (phosphatase and phytase) activity, promoting hydrolysis (Mabrouki et al., 2015). Both roasting and germination reduce phytate content, likely due to the activation of phytase enzymes, with further enhancement during germination (Pandey and Awasthi, 2015).

2. **Tannin:** Raw seeds contain 675 mg of tannin per 100 g of sample (Atlaw and Kumar, 2018). Tannin content was reduced to 481 mg, 436 mg, and 225 mg in roasted, soaked, and dried germinated seeds, respectively. An approximate decrease in tannin content (67%) was observed in germinated fenugreek seeds (Maria and Victoria, 2018; Ojha et al., 2018). Tannin content decreased progressively (from 532.1 to 210.1 mcg gallic acid per gram) with sprout development (Shimelis and Rakshit, 2008). This reduction may result from factors such as tannin leaching in water, polyphenol binding with protein or carbohydrate, formation of water-repellent bonds between tannins and enzymes or proteins, and enzymatic hydroxylation and polyphenol oxidase activity (Rao and Deosthale, 1982; Rusydi and Azlan, 2012; Saharan et al., 2002; Shimelis and Rakshit, 2008). In conclusion, germination, soaking, and roasting significantly affect tannin content ($p > 0.05$; Adhikari and Rai, 2021).

3.8 SUMMARY

It is to retrieve that various bioactive compounds extracted from fenugreek can have potential health benefits. Moreover, fenugreek components when processed through suitable methods such as germination results into nutrient availability. Therefore, suitable processing and extraction methods for obtaining bioactive compounds from fenugreek can be recommended for their antioxidant capacity to curb diverse ailments.

3.9 FUTURE RESEARCH PERSPECTIVE

Copious research studies have been conducted with regard to bioactive component extraction of fenugreek. Considering existing research evidences, it is pertinent to mention the significance of encapsulation and nanoencapsulation technologies in view of the health benefits of fenugreek powder from different components such as seed, leaves, and so forth. Therefore, there is a need to intensify the work in this area of interest because of protein, dietary fiber content utilization of fenugreek. Conclusively, extraction methods need to be improved to recover the BACs as maximum at lower cost.

KEYWORDS

- **alkaline conditions**
- **antitumor**
- **ascorbic acid**
- **beta-carotene**
- **calcium oxalate**
- **chronic cough**
- **dietary fiber**
- **flavonoid**
- **minerals**
- **pharmacological properties**
- **vitamins**

REFERENCES

Aasim, M., Baloch, F. S., Nadeem, M. A., Bakhsh, A., Sameeullah, M., & Day, S. (2018). Fenugreek (*Trigonella foenum-graecum L.*): An underutilized edible plant of the modern world. In M. Ozturk, K. R. Hakeem, M. Ashraf, & M. S. A. Ahmad (Eds.), *Global*

Perspectives on Underutilized Crops (pp. 381–408). Springer International Publishing. https://doi.org/10.1007/978-3-319-77776-4_12.

Acharya, S. N., Thomas, J. E., & Basu, S. K. (2008). Fenugreek, an alternative crop for semiarid regions of North America. *Crop Science, 48*(3), 841. https://doi.org/10.2135/cropsci2007.09.0519.

Adhikari, B., & Rai, S. (2021). Processing impact on nutritional, antinutritional, and phytochemical of fenugreek seeds (*Trigonella foenum-graecum L.*). *Himalayan Journal of Science and Technology, 5*(01), 36–44. https://doi.org/10.3126/hijost.v5i01.42131.

Akbari, S., Abdurahman, N. H., Yunus, R. M., Alara, O. R., & Abayomi, O. O. (2019). Extraction, characterization and antioxidant activity of fenugreek (*Trigonella foenum-graecum*) seed oil. *Materials Science for Energy Technologies, 2*(2), 349–355. https://doi.org/10.1016/j.mset.2018.12.001.

Annida, B., & Stanely Mainzen Prince, P. (2004). Supplementation of fenugreek leaves lower lipid profile in streptozotocin-induced diabetic rats. *Journal of Medicinal Food, 7*(2), 153–156. https://doi.org/10.1089/1096620041224201.

Anoopkumar, A. N., Aneesh, E. M., & Sudhikumar, A. V. (2020). Exploring the mode of action of isolated bioactive compounds by induced reactive oxygen species generation in *Aedes aegypti*: A microbes-based double-edged weapon to fight against arboviral diseases. *International Journal of Tropical Insect Science, 40*(3), 573–585. https://doi.org/10.1007/s42690-020-00104-z.

Arora, N., & Pandey-Rai, S. (2014). GC–MS analysis of the essential oil of *Celastrus paniculatus* Willd. seeds and antioxidant, anti-inflammatory study of its various solvent extracts. *Industrial Crops and Products, 61*, 345–351. https://doi.org/10.1016/j.indcrop.2014.07.025.

Atlaw, T. K., & Kumar, J. Y. (2018). Effect of germination on nutritional composition and functional properties of fenugreek (*Trigonella foenum-graecum* Linn) seed. *Food Science and Quality Management, 76*, 18–23.

Aylanc, V., Eskin, B., Zengin, G., Dursun, M., & Cakmak, Y. S. (2020). *In vitro* studies on different extracts of fenugreek (*Trigonella spruneriana* BOISS.): Phytochemical profile, antioxidant activity, and enzyme inhibition potential. *Journal of Food Biochemistry, 44*(11), e13463. https://doi.org/10.1111/jfbc.13463.

Bahubali, M. S. B., & Bornare, D. T. (2021). Study of bioactive components present in oil extracted from fenugreek seed: Review. *IJHAF, 5*(5), 14–17. https://doi.org/10.22161/ijhaf.5.5.2.

Barkat, N., Singh, J., Jayaprakasha, G. K., & Patil, B. S. (2018). Effect of harvest time on the levels of phytochemicals, free radical-scavenging activity, α-amylase inhibition and bile acid-binding capacity of spinach (*Spinacia oleracea*). *Journal of the Science of Food and Agriculture, 98*(9), 3468–3477. https://doi.org/10.1002/jsfa.8862.

Benzidia, B., Barbouchi, M., Hammouch, H., Belahbib, N., Zouarhi, M., Erramli, H., Ait Daoud, N., Badrane, N., & Hajjaji, N. (2019). Chemical composition and antioxidant activity of tannins extract from green rind of *Aloe vera* (L.) Burm. F. *Journal of King Saud University–Science, 31*(4), 1175–1181. https://doi.org/10.1016/j.jksus.2018.05.022.

Bishnoi, S., Sheoran, R., Ray, A., & Sindhu, S. C. (2016). Mathematical modeling of hot air drying of fenugreek leaves (*Trigonella foenum-graecum*) in cabinet dryer. *International Journal of Food Sciences and Nutrition, 5*(3), 170–175.

Borda-Yepes, V. H., Chejne, F., Daza-Olivella, L. V., Alzate-Arbelaez, A. F., Rojano, B. A., & Raghavan, V. G. S. (2019). Effect of microwave and infrared drying over polyphenol content in *Vaccinium meridionale* (Swartz) dry leaves. *Journal of Food Processing Engineering, 42*(1), e12939. https://doi.org/10.1111/jfpe.12939.

Bouyanzer, A., Hammouti, B., Majidi, L., & Haloui, B. (2010). Testing natural fenugreek as an ecofriendly inhibitor for steel corrosion in 1 M HCl. *Portugaliae Electrochimica Acta, 28*(3), 165–172. https://doi.org/10.4152/pea.201003165

Charles, A. C., Dadmohammadi, Y., & Abbaspourrad, A. (2022). Food and cosmetic applications of the avocado seed: A review. *Food Function, 13*(13), 6894–6901. https://doi.org/10.1039/D1FO02438H.

Chouhan, K. B. S., Tandey, R., Sen, K. K., Mehta, R., & Mandal, V. (2019). Extraction of phenolic principles: Value addition through effective sample pretreatment and operational improvement. *Food Measure, 13*(1), 177–186. https://doi.org/10.1007/s11694-018-9931-0.

Curi, P. N., Salgado, D. L., Mendonça, K., Pio, R., Ferreira, J. L. G., & Souza, V. R. de. (2019). Influence of microwave processing on the bioactive compounds, antioxidant activity and sensory acceptance of blackberry jelly. *Food Science and Technology, 39*(2), 386–391. https://doi.org/10.1590/fst.18618.

Deme, T., Haki, G., Retta, N., Woldegiorgis, A., & Geleta, M. (2017). Mineral and anti-nutritional contents of niger seed (*Guizotia abyssinica* (L.f.) Cass.), linseed (*Linum usitatissimum* L.) and sesame (*Sesamum indicum* L.) varieties grown in Ethiopia. *Foods, 6*(4), 27. https://doi.org/10.3390/foods6040027.

El-Salam, S. G. S. A., Rashed, M. M., Ibrahim, N. A., Rahim, E. A., Aly, T. A. A., & Al-Farga, A. (2022). Phytochemical screening and *in-vitro* biological properties of unprocessed and household processed fenugreek (*Trigonella foenum-graecum* Linn.) seeds and leaves [Preprint]. In Review. https://doi.org/10.21203/rs.3.rs-1952713/v1.

Floegel, A., Kim, D. O., Chung, S. J., Koo, S. I., & Chun, O. K. (2011). Comparison of ABTS/DPPH assays to measure antioxidant capacity in popular antioxidant-rich US foods. *Journal of Food Composition and Analysis, 24*(7), 1043–1048. https://doi.org/10.1016/j.jfca.2011.01.008.

Ghafar, F., Tengku Nazrin, T. N. N., Mohd Salleh, M. R., Nor Hadi, N., Ahmad, N., Hamzah, A. A., Mohd Yusof, Z. A., & Azman, I. N. (2017). Total phenolic content and total flavonoid content in *Moringa oleifera* seed. *Scientific Heritage Journal, 1*(1), 23–25. https://doi.org/10.26480/gws.01.2017.23.25.

Ghayur, M. N., Abdalla, M., Khalid, A., Ahmad, S., & Gilani, A. H. (2022). *Trigonella foenum-graecum* methanolic extract on isolated smooth muscles and acetylcholinesterase enzyme: An *in vitro* and mechanistic *in silico* investigation. *BioMed Research International, 2022*, 1–12. https://doi.org/10.1155/2022/4849464.

Güçlü-Üstündağ, Ö., & Mazza, G. (2007). Saponins: Properties, applications, and processing. *Critical Reviews in Food Science and Nutrition, 47*(3), 231–258. https://doi.org/10.1080/10408390600698197.

Hamden, K., Jaouadi, B., Carreau, S., Bejar, S., & Elfeki, A. (2010). Inhibitory effect of fenugreek galactomannan on digestive enzymes related to diabetes, hyperlipidemia, and liver-kidney dysfunctions. *Biotechnology & Bioprocess Engineering, 15*(3), 407–413. https://doi.org/10.1007/s12257-009-3037-9.

Hannan, J. M. A., Ali, L., Rokeya, B., Khaleque, J., Akhter, M., Flatt, P. R., & Abdel-Wahab, Y. H. A. (2007). Soluble dietary fiber fraction of *Trigonella foenum-graecum* (fenugreek) seed improves glucose homeostasis in animal models of type 1 and type 2 diabetes by delaying carbohydrate digestion and absorption, and enhancing insulin action. *British Journal of Nutrition, 97*(3), 514–521. https://doi.org/10.1017/S0007114507657869.

He, Y., Ding, C., Wang, X., Wang, H., & Suo, Y. (2015). Using response surface methodology to optimize countercurrent chromatographic separation of polyphenol compounds from fenugreek (*Trigonella foenum-graecum L.*) seeds. *Journal of Liquid Chromatography & Related Technologies*, *38*(1), 29–35. https://doi.org/10.1080/10826076.2013.864981.

Hefnawy, H. T. M., & Ramadan, M. F. (2011). Physicochemical characteristics of soy protein isolate and fenugreek gum dispersed systems. *Journal of Food Science and Technology, 48*(3), 371–377. https://doi.org/10.1007/s13197-010-0203-1.

Hernández-Ledesma, B. (2019). Quinoa (*Chenopodium Quinoa Willd.*) as a source of bioactive compounds: A review. *Bioactive Compounds in Health and Disease*, *2*(3), 27. https://doi.org/10.31989/bchd.v2i3.556.

Hooda, S., & Jood, S. (2003). Effect of soaking and germination on nutrient and antinutrient contents of fenugreek (*Trigonella foenum graecum* L.). *Journal of Food Biochemistry, 27*(2), 165–176. https://doi.org/10.1111/j.1745-4514.2003.tb00274.x.

Jawake, P. D., Mudgal, V. D., & Rajpurohit, D. (2018a). Fluidized bed drying kinetics of fenugreek (*Trigonella foenum-graecum* L.) leaves. Retrieved from: https://doi.org/10.13140/RG.2.2.30056.52483.

Jawake, P. D., Mudgal, V. D., & Rajpurohit, D. (2018b). Convective drying of fenugreek (*Trigonella foenum-graecum* L.) leaves. Retrieved from: https://doi.org/10.13140/RG.2.2.26701.08165.

Jiratchariyakul, W., & Mahady, G. B. (2013). Overview of botanical status in EU, USA, and Thailand. *Evidence-Based Complementary and Alternative Medicine, 2013*, 1–13. https://doi.org/10.1155/2013/480128.

Khan, M. K. I., Ghauri, Y. M., Alvi, T., Amin, U., Khan, M. I., Nazir, A., Saeed, F., Aadil, R. M., Nadeem, M. T., Babu, I., & Maan, A. A. (2022). Microwave assisted drying and extraction technique; kinetic modelling, energy consumption and influence on antioxidant compounds of fenugreek leaves. *Food Science and Technology, 42*, e56020. https://doi.org/10.1590/fst.56020.

Khare, C. P. (2004). *Indian Herbal Remedies*. Springer Berlin Heidelberg. https://doi.org/10.1007/978-3-642-18659-2.

Khoja, K. K., Aslam, M. F., Sharp, P. A., & Latunde-Dada, G. O. (2021). *In vitro* bioaccessibility and bioavailability of iron from fenugreek, baobab, and moringa. *Food Chemistry*, *335*, 127671. https://doi.org/10.1016/j.foodchem.2020.127671.

Khorshidian, N., Iran, Yousefi Asli, M., Iran, Arab, M., Iran, Adeli Mirzaie, A., & Mortazavian, A. M. (2016). Fenugreek: Potential applications as a functional food and nutraceutical. *Nutraceuticals and Food Science Research, 3*(1), 5–16. https://doi.org/10.18869/acadpub.nfsr.3.1.5.

Kim, J., Kim, C. S., Lee, Y. M., Sohn, E., Jo, K., Shin, S. D., & Kim, J. S. (2013). Scopoletin inhibits rat aldose reductase activity and cataractogenesis in galactose-fed rats. *Evidence-Based Complementary and Alternative Medicine, 2013*, 1–8. https://doi.org/10.1155/2013/787138.

Kolodziejczyk-Czepas, J., & Czepas, J. (2019). Rhaponticin as an anti-inflammatory component of rhubarb: A minireview of the current state of the art and prospects for future research. *Phytochemistry Reviews, 18*(5), 1375–1386. https://doi.org/10.1007/s11101-019-09652-w.

Kousalya, N., Ishwarya, R., Logeshwaran, V., Sabarinath, K., Sandhiya, S., & Arun, P. (2019). A review about bioactive compounds of fenugreek (*Trigonella foenum-graceum*). *International Journal of Engineering Applied Sciences and Technology*, *4*(7), 332–335.

Kumar, I. M. K., & Maliakel, B. P. (2008). Fenugreek dietary fiber: A novel class of functional food ingredient. *Agro Food Industry Hi Tech, 19*(2), 18–21.

Kushwaha, S., & Mustafa, N. (2012). Comparison of cabinet, microwave and freeze-drying methods on physical and nutritional quality of fenugreek leaves. *Research Journal of Agricultural Science, 3*(4), 977–978.

Laila, O., Murtaza, I., Abdin, M. Z., Ahmad, S., Ganai, N. A., & Jehangir, M. (2014). Development and validation of HPTLC method for simultaneous estimation of diosgenin and quercetin in fenugreek seeds (*Trigonella foenum-graecum*). *ISRN Chromatography, 2014*, 1–8. https://doi.org/10.1155/2014/583047.

Lee, K. J., Oh, Y. C., Cho, W. K., & Ma, J. Y. (2015). Antioxidant and anti-inflammatory activity determination of 100 kinds of pure chemical compounds using offline and online screening HPLC assay. *Evidence-Based Complementary and Alternative Medicine, 2015*, 1–13. https://doi.org/10.1155/2015/165457.

Lu, X., Wang, J., Al-Qadiri, H. M., Ross, C. F., Powers, J. R., Tang, J., & Rasco, B. A. (2011). Determination of total phenolic content and antioxidant capacity of onion (*Allium cepa*) and shallot (*Allium oschaninii*) using infrared spectroscopy. *Food Chemistry, 129*(2), 637–644. https://doi.org/10.1016/j.foodchem.2011.04.105.

Mabrouki, S., Omri, B., & Abdouli, H. (2015). Chemical, functional and nutritional characteristics of raw, autoclaved and germinated fenugreek seeds. *Journal of New Sciences, Agriculture and Biotechnology, 16*(1), 541–551.

Madar, Z., & Stark, A. H. (2002). New legume sources as therapeutic agents. *British Journal of Nutrition, 88*(S3), 287–292. https://doi.org/10.1079/BJN2002719.

Madhava Naidu, M., Shyamala, B. N., Pura Naik, J., Sulochanamma, G., & Srinivas, P. (2011). Chemical composition and antioxidant activity of the husk and endosperm of fenugreek seeds. *LWT–Food Science and Technology, 44*(2), 451–456. https://doi.org/10.1016/j.lwt.2010.08.013.

Majeed, M., Nagabhushanam, K., Arumugam, S., Majeed, S., & Ali, F. (2018). *Bacillus coagulans* MTCC 5856 for the management of major depression with irritable bowel syndrome: A randomised, double-blind, placebo-controlled, multi-center, pilot clinical study. *Food & Nutrition Research, 62*(0). https://doi.org/10.29219/fnr.v62.1218.

Manju, Khatkhar, B. S., & Parveen. (2015). Effect of germination and roasting on nutritive composition and anti-nutrients in fenugreek seeds. *International Journal of Science Technology & Engineering, 3*(06), 31–35.

Maria, M. F., & Victoria, A. T. (2018). Changes in nutritional, functional and pasting properties of raw and germinated seeds of white sesame (*Sesamum indicum* L.) grown in Nigeria. *Acta Scientific Nutritional Health, 2*(11), 07–15.

Mathur, P., & Choudhry, M. (2009). Consumption pattern of fenugreek seeds in Rajasthani families. *Journal of Human Ecology, 25*(1), 9–12. https://doi.org/10.1080/09709274.2009.11906127.

Meghwal, M. (2012). A review on the functional properties, nutritional content, medicinal utilization and potential application of fenugreek. *Journal of Food Process Technology, 3*(9), 1–10. https://doi.org/10.4172/2157-7110.1000181.

Moriyama, M., & Oba, K. (2008). Comparative study on the vitamin C contents of the food legume seeds. *Journal of Nutritional Science and Vitaminology, 54*(1), 1–6. https://doi.org/10.3177/jnsv.54.1.

Nagulapalli Venkata, K. C., Swaroop, A., Bagchi, D., & Bishayee, A. (2017). A small plant with big benefits: Fenugreek (*Trigonella foenum-graecum* Linn.) for disease prevention and health promotion. *Molecular Nutrition & Food Research, 61*(6), 1600950. https://doi.org/10.1002/mnfr.201600950.

Navarro Del Hierro, J., Reglero, G., & Martin, D. (2020). Chemical characterization and bioaccessibility of bioactive compounds from saponin-rich extracts and their acid-hydrolysates obtained from fenugreek and quinoa. *Foods*, *9*(9), 1159. https://doi.org/10.3390/foods9091159.

Ojha, P., Prajapati, P., & Karki, T. B. (2018). Soaking and germination effect on bioactive components of fenugreek seeds (*Trigonella foenum graecum L.*). *International Food Research Journal*, *25*(2), 690–694.

Pająk, P., Socha, R., Broniek, J., Królikowska, K., & Fortuna, T. (2019). Antioxidant properties, phenolic and mineral composition of germinated chia, golden flax, evening primrose, phacelia and fenugreek. *Food Chemistry, 275*, 69–76. https://doi.org/10.1016/j.foodchem.2018.09.081.

Pandey, H., & Awasthi, P. (2015). Effect of processing techniques on nutritional composition and antioxidant activity of fenugreek (*Trigonella foenum-graecum*) seed flour. *Journal of Food Science and Technology, 52*(2), 1054–1060. https://doi.org/10.1007/s13197-013-1057-0.

Raju, J., Gupta, D., Rao, A. R., Yadava, P. K., & Baquer, N. Z. (2001). *Trigonella foenum graecum* (fenugreek) seed powder improves glucose homeostasis in alloxan diabetic rat tissues by reversing the altered glycolytic, gluconeogenic and lipogenic enzymes. *Molecular and Cellular Biochemistry*, *224*(1/2), 45–51. https://doi.org/10.1023/A:1011974630828.

Rampogu, S., Parameswaran, S., Lemuel, M. R., & Lee, K. W. (2018). Exploring the therapeutic ability of fenugreek against type 2 diabetes and breast cancer employing molecular docking and molecular dynamics simulations. *Evidence-Based Complementary and Alternative Medicine, 2018*, 1943203. https://doi.org/10.1155/2018/1943203.

Randhir, R., Lin, Y. T., & Shetty, K. (2004). Phenolics, their antioxidant and antimicrobial activity in dark germinated fenugreek sprouts in response to peptide and phytochemical elicitors. *Asia Pacific Journal of Clinical Nutrition*, *13*(3), 295–307.

Rao, P. U., & Deosthale, Y. G. (1982). Tannin content of pulses: Varietal differences and effects of germination and cooking. *Journal of the Science of Food and Agriculture*, *33*(10), 1013–1016. https://doi.org/10.1002/jsfa.2740331012.

Rashid, U., Gani, A., Shah, A., Ahmad, M., Baba, W. N., & Masoodi, F. A. (2015). Effect of sand roasting on the antioxidant and antiproliferative activity of barley (*Hordeum vulgare*). *Nutrafoods*, *14*(4), 227–236. https://doi.org/10.1007/s13749-015-0054-5.

Rizki, H., Kzaiber, F., Elharfi, M., Ennahli, S., & Hanine, H. (2015). Effects of roasting temperature and time on the physicochemical properties of sesame (*Sesamum indicum* L.) seeds. *International Journal of Innovation and Applied Studies*, *11*(1), 148–155.

Rusydi, M. R. M., & Azlan, A. (2012). Effect of germination on total phenolic, tannin and phytic acid contents in soybean and peanut. *International Food Research Journal*, *19*(2), 673–677.

Saharan, K., Khetarpaul, N., & Bishnoi, S. (2002). Antinutrients and protein digestibility of fababean and ricebean as affected by soaking, dehulling and germination. *Journal of Food Science and Technology, Mysore*, *39*(4), 418–422.

Saini, P., Dubey, N., Singh, P., & Singh, A. (2016). Effect of processing methods on proximate composition and antioxidant activity of fenugreek (*Trigonella foenum-graecum*) seeds. *International Journal of Agricultural and Food Science, 6*(2), 82–87.

Salam, S. G. A., Rashed, M. M., Ibrahim, N. A., Rahim, E. A. A., Aly, T. A. A., & Al-Farga, A. (2023). Phytochemical screening and *in-vitro* biological properties of unprocessed and household processed fenugreek (*Trigonella foenum-graecum* Linn.) seeds and leaves. *Scientific Reports*, *13*(1), 7032. https://doi.org/10.1038/s41598-023-31888-y.

Sharma, P., & Gujral, H. S. (2010). Antioxidant and polyphenol oxidase activity of germinated barley and its milling fractions. *Food Chemistry*, *120*(3), 673–678. https://doi.org/10.1016/j.foodchem.2009.10.059.

Sharma, R. D. (1986). Effect of fenugreek seeds and leaves on blood glucose and serum insulin responses in human subjects. *Nutrition Research*, *6*(12), 1353–1364. https://doi.org/10.1016/S0271-5317(86)80020-3.

Shimelis, E. A., & Rakshit, S. K. (2008). Influence of natural and controlled fermentations on α-galactosides, antinutrients and protein digestibility of beans (*Phaseolus vulgaris* L.). *International Journal of Food Science & Technology*, *43*(4), 658–665. https://doi.org/10.1111/j.1365-2621.2006.01506.x.

Shrivastava, V., & Kumar, A. (2016). Experimental investigation on the comparison of fenugreek drying in an indirect solar dryer and under open sun. *Heat and Mass Transfer*, *52*(9), 1963–1972. https://doi.org/10.1007/s00231-015-1721-1.

Singh, K., Colaço, M., & Misquith, S. (2015). Effect of germination on the proximate, nutritive and anti-nutrient value of fenugreek (*Trigonella foenum-graecum*). *Research Journal of Chemistry and Environment*, *19*(1), 16–21.

Singh, P., Bajpai, V., Gond, V., Kumar, A., Tadigoppula, N., & Kumar, B. (2020). Determination of bioactive compounds of fenugreek (*Trigonella foenum-graecum*) seeds using LC-MS techniques. In M. Jain & R. Garg (Eds.), *Legume Genomics* (pp. 377–393). Methods in Molecular Biology; Springer US. https://doi.org/10.1007/978-1-0716-0235-5_21.

Srinivasan, K. (2006). Fenugreek (*Trigonella foenum-graecum*): A review of health beneficial physiological effects. *Food Reviews International, 22*(2), 203–224. https://doi.org/10.1080/87559120600586315.

Srivastava, A., Singh, Z., Verma, V., & Choedon, T. (2020). Potential health benefits of fenugreek with multiple pharmacological properties: In N. Mishra (Ed.), *Advances in medical diagnosis, treatment, and care* (pp. 137–153). IGI Global. https://doi.org/10.4018/978-1-7998-2524-1.ch011.

Sun, T., Tang, J., & Powers, J. R. (2005). Effect of pectolytic enzyme preparations on the phenolic composition and antioxidant activity of asparagus juice. *Journal of Agricultural and Food Chemistry, 53*(1), 42–48. https://doi.org/10.1021/jf0491299.

Terrence Justine, V., Mustafa, M., Kankara, S. S., & Go, R. (2019). Effect of drying methods and extraction solvents on phenolic antioxidants and antioxidant activity of *Scurrula ferruginea* (Jack) Danser (Loranthaceae) leaf extracts. *JSM, 48*(7), 1383–1393. https://doi.org/10.17576/jsm-2019-4807-07.

Thakur, A., Sharma, V., & Thakur, A. (2019). An overview of antinutritional factors in foods. *International Journal of Chemical Studies*, *7*(1), 2472–2479.

Thamkaew, G., Sjöholm, I., & Galindo, F. G. (2021). A review of drying methods for improving the quality of dried herbs. *Critical Reviews in Food Science and Nutrition*, *61*(11), 1763–1786. https://doi.org/10.1080/10408398.2020.1765309.

Türker, İ., & İŞleroğlu, H. (2023). Extraction of bioactive compounds from fenugreek leaves by maceration with D-optimal design. *Gıda, 48*(2), 305–316. https://doi.org/10.15237/gida.GD22108.

Wang, N., Lewis, M. J., Brennan, J. G., & Westby, A. (1997). Effect of processing methods on nutrients and anti-nutritional factors in cowpea. *Food Chemistry*, *58*(1, 2), 59–68. https://doi.org/10.1016/S0308-8146(96)00212-9.

Wani, I. A., Hamid, H., Hamdani, A. M., Gani, A., & Ashwar, B. A. (2017). Physico-chemical, rheological and antioxidant properties of sweet chestnut (*Castanea sativa* Mill.) as affected by pan and microwave roasting. *Journal of Advanced Research*, *8*(4), 399–405. https://doi.org/10.1016/j.jare.2017.05.005.

Wani, S. A., & Kumar, P. (2018). Fenugreek: A review on its nutraceutical properties and utilization in various food products. *Journal of the Saudi Society of Agricultural Sciences*, *17*(2), 97–106. https://doi.org/10.1016/j.jssas.2016.01.007.

Wijekoon, M. M. J. O., Bhat, R., & Karim, A. A. (2011). Effect of extraction solvents on the phenolic compounds and antioxidant activities of Bunga kantan (*Etlingera elatior* Jack.) inflorescence. *Journal of Food Composition and Analysis*, *24*(4, 5), 615–619. https://doi.org/10.1016/j.jfca.2010.09.018.

Yao, D., Zhang, B., Zhu, J., Zhang, Q., Hu, Y., Wang, S., Wang, Y., Cao, H., & Xiao, J. (2020). Advances on application of fenugreek seeds as functional foods: Pharmacology, clinical application, products, patents, and market. *Critical Reviews in Food Science and Nutrition*, *60*(14), 2342–2352. https://doi.org/10.1080/10408398.2019.1635567.

Yao, Y., Cheng, X., Wang, L., Wang, S., & Ren, G. (2011). A determination of potential α-glucosidase inhibitors from azuki beans (*Vigna angularis*). *International Journal of Molecular Sciences, 12*(10), 6445–6451. https://doi.org/10.3390/ijms12106445.

Yazicioglu, N. (2022). Sensory analysis of microwave-roasted fenugreek and coffee blend. *Legume Science*, *2022*, e166. https://doi.org/10.1002/leg3.166.

Zuorro, A., Maffei, G., & Lavecchia, R. (2016). Reuse potential of artichoke (*Cynara scolymus* L.) waste for the recovery of phenolic compounds and bioenergy. *Journal of Cleaner Production, 111*, 279–284. https://doi.org/10.1016/j.jclepro.2015.06.011.

CHAPTER 4

Effects of Different Plant Growth Regulators on Growth, Yield, and Active Constituents of Fenugreek

AJMAT JAHAN,[1] M. MASROOR A. KHAN,[1] MOHD. GULFISHAN,[2] KHAN BILAL MUKHTAR AHMED,[3] BILAL AHMAD,[4] and YAWAR SADIQ[1]

[1]*Advanced Plant Physiology Section, Department of Botany, Aligarh Muslim University, Aligarh, Uttar Pradesh, India*

[2]*School of Agricultural Sciences, Glocal University, Saharanpur, Uttar Pradesh, India*

[3]*Centre of Agricultural Education, faculty of Agricultural Sciences, AMU-Aligarh, UP, India*

[4]*Department of Botany, Govt Degree College, KhanSahib, Budgam, affliated with University of Kashmir Hazratbal Srinagar, India*

ABSTRACT

Fenugreek (*Trigonella foenum-graecum* L.) is a versatile, winter-grown annual herbaceous crop in North India. The young plants are used as a vegetable and fodder, while the seeds are primarily utilized as a condiment and in the pharmaceutical industry, particularly in the production of Ayurvedic medicines. Fenugreek is a rich source of minerals, protein, vitamin C, and other nutrients. It is predominantly grown under semi-arid conditions in Rajasthan, Tamil Nadu, Uttar Pradesh, Gujarat, and Madhya Pradesh. Due to its high therapeutic potential, demand for fenugreek seeds and leaves is increasing. Diosgenin, a compound found in fenugreek seeds, is widely used in the manufacture of contraceptive pills, serving as an alternative to

Fenugreek: Improvement in Yield Potential Through Plant Growth Regulators (PGRs) and Depolymerized Polysaccharides. Tariq Ahmad Dar, Bilal Ahmad, Moinuddin, and Khalid Rehman Hakeem (Eds.)

medicinal yam. To meet rising demand, improved cultural practices must be implemented to enhance production.

Plant growth regulators (PGRs) are compounds that, when applied in small doses, can significantly modify plant growth, either stimulating or inhibiting it. These are small chemical molecules that, even at very low concentrations (often less than 10–7 M), influence the physiological responses of plants to environmental stimuli. PGRs are extensively used in biotechnology and agriculture to modify plant growth and development. Numerous spice crops have shown increased yield and quality due to the positive effects of various PGRs. This chapter discusses the role of naphthaleneacetic acid, brassinosteroids, and salicylic acid on fenugreek growth, yield, and active ingredient composition.

4.1 INTRODUCTION

Trigonella foenum-graecum, the biological name for fenugreek, is an annual herbaceous plant also known as "Methi" in popular culture. The primary uses of fenugreek cultivation are for the production of its seeds and leaves. While the seeds offer medicinal properties such as anti-diabetic, anti-cancer, and cholesterol-lowering effects, the leaves are frequently used in food preparation.

Fenugreek is mainly grown in India's semi-arid states of Rajasthan, Tamil Nadu, Gujarat, Uttar Pradesh, and Madhya Pradesh. The leaves contain 1.1% fiber, 1.5% minerals, 4.4% protein, 0.9% fat, 86.1% moisture, and 6% carbohydrates. Other nutrients present in the leaves include riboflavin, carotene, thiamine, niacin, calcium, zinc, iron, phosphorus, and vitamin C (Rao, 2003). Fresh fenugreek leaves contain 220.97 mg of ascorbic acid and 19 mg of β-carotene per 100 g of leaves (Yadav and Sehgal, 1997).

Because of its greater therapeutic potential, fenugreek seeds and leaves are in more demand. Diosgenin, a substance found in fenugreek seeds, is primarily employed in the manufacture of contraceptive tablets. It is utilized in place of medicinal yam. Through the implementation of better cultural practices, production must be raised in order to fulfil the demand. The *Trigonella foenum-graecum* L. plant, also known as fenugreek, is one of the 50 species in the Fabaceae family. The fenugreek plant is regarded as a multi-use plant that is utilized in many nations throughout the world for human food and medication as well as plant reformation and fertilizing for soil and animal feed.

Fenugreek is currently grown worldwide, primarily as a medicinal herb, particularly in China, India, Morocco, Ukraine, Egypt, Ethiopia, Greece,

Canada, and Turkey (Dadrasan et al., 2015). Recent studies have shown that fenugreek possesses anti-diabetic, anticarcinogenic, antioxidant, anti-inflammatory, anti-ulcer, antibacterial, antifungal, and neuroprotective properties (Ouzir et al., 2016). According to Moradi and Moradi (2013), *Trigonella* is used as a vegetable in India and Pakistan, as a spice in Switzerland, Egypt, and Iran, and as a flavoring for cheese in Switzerland, Egypt, and Iran. The bioactive components of fenugreek, such as saponins and sapogenins, have shown potential for managing diabetes (Zhang et al., 2020). Fenugreek reportedly originated in West Africa and is now cultivated in several Asian countries, including Turkey, Egypt, China, India, and Pakistan, and is used across the medical, pharmaceutical, and nutraceutical industries. Fenugreek seeds and leaves are used by various countries worldwide for culinary and medicinal purposes (Rashid et al., 2018).

In Europe, fenugreek is represented by five species, with one species each in Australia and Africa, and six species found across Asian countries. There are about 260 different species of fenugreek, although their origins vary (Petropoulos, 2002, 2002a, b). The most recognized species of fenugreek include *Trigonella rigida*, *Trigonella caerulea*, *Trigonella anguina*, *Trigonella arcuata*, *Trigonella arabica*, *Trigonella cariensis*, *Trigonella hamosa*, *Trigonella suavissima*, *Trigonella orniculata*, *Trigonella torulosa*, *Trigonella cretica*, *Trigonella spinosa*, *Trigonella occulta*, and *Trigonella polycerata*.

Fenugreek seeds have been verified by current pharmacological investigations to contain anti-fungal, anti-bacterial, anti-inflammatory, anti-obesity, anti-ulcer, anti-analgesic, anthelmintic, and allelopathic properties (Yao et al., 2020). Due to the aforementioned advantages, fenugreek is frequently employed in pharmaceutical, food and cosmetic productions. Some regions solely use the plant's dried or fresh leaves, whilst other regions use the seeds as seasonings or blend them with other spices (Zhou et al., 2012). Fenugreek has a variety of industrial applications in addition to being used in traditional medicine to cure edema and burns (Figure 4.1). Additionally, the crop's seed extract is utilized in various nations, particularly India, to stop hair loss (Nayak et al., 2013).

4.2 *TRIGONELLA* AND THE ABIOTIC STRESS

The detrimental impact of non-living forces on the growth of a plant's physiology, and development is referred to as abiotic stress in plants. These includes UV rays, heat stress, temperature stress, salinity stress, nutritional stress, heavy metal stress, and water stress. These abiotic stressors have a variety of effects on plants.

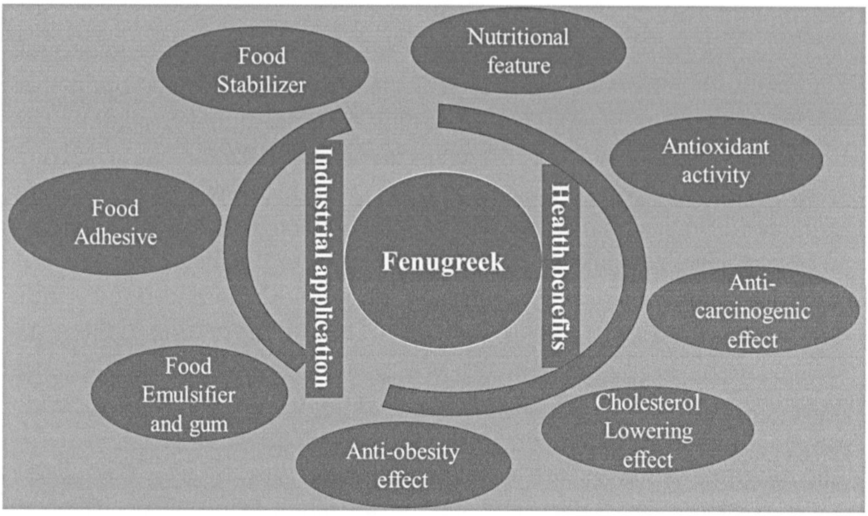

FIGURE 4.1 Various applications of fenugreek.

Due to increasing industrialization and anthropogenic activities, heavy metal (HM) contamination has become a major concern. HM contamination negatively impacts crucial metabolic processes in plants, including seed germination, photosynthesis, nutrient uptake, and water interactions (Ozyigit et al., 2018; Rucińska-Sobkowiak, 2016; Sethy and Ghosh, 2013). This stunts plant growth and development. Some heavy metals are essential for plant metabolism at trace levels but can be harmful when concentrations exceed safe thresholds (Tchounwou et al., 2012), while others are toxic even at low doses due to their non-essential nature. Heavy metals also induce oxidative stress, which leads to excessive production of reactive oxygen species (ROS) such as hydrogen peroxide, hydroxyl radicals, and superoxide ions. This ROS generation compromises the antioxidant defense system, causing cellular damage and disrupting ionic balance. Key biological molecules, including lipids, DNA, and proteins, are damaged as a result. Consequently, plants undergo physiological and molecular changes in response to the abiotic stress caused by high HM concentrations (Panchal et al., 2013). At low concentrations, several heavy metals, such as manganese (Mn), copper (Cu), iron (Fe), cobalt (Co), molybdenum (Mo), chromium (Cr), nickel (Ni), and zinc (Zn), function as essential nutrients for plant growth and development.

A soil with excessive sodium (Na^+) or chloride (Cl^-) ions induces salinity stress, which inhibits plant growth (Munns, 2005). For example, *Trigonella* showed no effect on germination at up to 140 mM NaCl; however, while it could tolerate NaCl concentrations as high as 175 mM, various morphological changes, including shorter stems, fewer leaves, decreased fresh weight, and increased nodule formation, reduced the crop's overall productivity and yield (Abdelmoumen et al., 2009). Soil salinity is a major global issue that impacts plant metabolism and structure, impeding growth and development due to cytotoxicity and nutrient deficiencies caused by an excess of ions like Na^+ and Cl^-. Salinity is also often accompanied by oxidative stress due to the production of reactive oxygen species (ROS; Gao et al., 2008; Maathuis et al., 2014; Negrão et al., 2011). This stress is a significant barrier to agricultural productivity worldwide (Sadeghzadeh-Ahari et al., 2010). Studies have shown that when fenugreek is exposed to salinity stress, its protein content and membrane stability index significantly decrease compared to controls. Salinity stress is also associated with increased membrane lipid peroxidation, malondialdehyde levels, and catalase activity. Notably, higher catalase levels in plants exposed to NaCl indicate an enhanced tolerance and defense against oxidative damage caused by salinity (Koca et al., 2007; Rahnama and Ebrahimzadeh, 2005).

Fenugreek frequently experiences drought or water stress due to unpredictable and inconsistent precipitation patterns. All aspects of growth, photosynthesis, yield factors, protein content, and carbohydrate levels decreased in *Trigonella* plants irrigated with lower water-holding capacity; however, flavonoid and phenolic content increased under drought conditions (Sadak, 2016). Temperature stress data revealed a similar trend. Transmission electron microscopy images from another study showed cellular abnormalities in the chloroplasts, nucleus, and intercellular spaces of heat-stressed fenugreek leaves exposed to 35°C (Osman et al., 2015). Increasing drought conditions reduce the primary absorption rate, which in turn lowers the germination rate (Iqbal et al., 2015). The results demonstrated that plants exposed to drought experience diminished leaf water potential, negatively affecting physiological activities and yield. Biological yield is lower under water stress than in non-stress conditions, and this yield—an essential factor in enhancing production—is significantly impacted by the surrounding environment. Additionally, the harvest index, a key parameter for improving yield, is greatly influenced by environmental conditions (Chauhan et al., 2017; Siddique et al., 2001).

T. foenum graecum occasionally experiences ultraviolet (UV) stress in its native environment. By changing the plant's physiological and biochemical profile, UV rays can cause the plant to have gastronomic and healing characteristics (Kumari and Prasad, 2013, 2014). However, plant growth and development are slowed down when they are exposed for a prolonged length of time above the sub-lethal level.

4.3 *TRIGONELLA* AND PLANT GROWTH REGULATORS (PGRS)

A relatively small amount of an organic substance that is either naturally occurring or synthesized chemically which functions as a chemical messenger in order to regulate intercellular interactions as well as to affect the expansion and growth of plant organs, tissues, and cells is termed as regulator of the growth of plant. It could either hasten or slow down the plant's growth. Natural plant hormones such as cytokinins, auxin, strigolactones, ethylene, gibberellic acids, jasmonic acid, abscisic acid, salicylic acid, and brassinosteroids are examples of plant growth regulators (PGRs). Triacontanol, thidiazuron, and other strong PGRs are made synthetically (Gaspar et al., 1996). The PGRs can change a variety of physiological and biochemical plant processes, which can impact how well plants grow and develop in general (Davies, 2013). Numerous spice crops have benefited from the positive effects of several plant growth regulators, which have been demonstrated to increase yield and quality. Because they operate as a type of chemical catalyst in plants and enhance the plant height, leaves, and branches per plant, pods per plant, pod length, grains per pod, quality and yield of the grain. Growth regulators are used in agriculture and are regarded as chemical biochemical techniques that help plants utilize their physiological and genetic potential to the fullest extent and invest nutrients in an efficient manner. Growth regulators also aid in the understanding of compositional processes and its constituent parts for their impact on plant growth and development.

The application of a plant growth regulator (PGR) to plants can be done in various ways, including foliar sprays (Sajjad et al., 2014), drenches, sprenches, seed soaking, liner dips (Schnelle and Barrett, 2010), and grooming of bulbs, seeds, and tubers, as well as rhizome soaking (Whipker et al., 2005). The most popular and efficient techniques in most plants are foliar spraying and immersion. Additionally, seed treatment has been shown to be highly beneficial for legumes, especially fenugreek. The effectiveness of PGRs in eliciting biological responses depends on the application method, assimilation by plants, treatment duration, PGR concentration, plant species, and environmental factors (Wroblewska and Debicz, 2013). This chapter discusses the effects of brassinosteroids, salicylic acid, and naphthaleneacetic acid (NAA) on *Trigonella*.

4.3.1 BRASSINOSTEROID

Brassinosteroids are a class of plant hormones that regulate various physiological processes, including cell elongation and expansion, flowering, photomorphogenesis, stomatal formation, pollen development, vascular differentiation, seed germination, and senescence (Ashraf et al., 2010; Fariduddin et al., 2014). They also protect plants from abiotic stresses, such as extreme temperatures (Sadura and Janeczko, 2018), salinity (Abbas et al., 2013), herbicides and pesticides (Sharma et al., 2013; Xia et al., 2006), heavy metals (Cao et al., 2005), drought, and light stress (Mahesh et al., 2013). *Trigonella foenum-graecum* treated with 1.0 ppm brassinolide demonstrated a positive response in harvest under both normal and drought conditions (Singh et al., 2018).

In fenugreek, the application of 24-epibrassinolide at 2.0 M and 1.0 M dosages of 28-homobrassinolide has been shown to alleviate cadmium and lead stress (Swamy et al., 2011). Plant growth regulators (PGRs) substantially improved biochemical parameters in fenugreek, including growth and yield, by activating reactive oxygen species (ROS) scavenging through both non-enzymatic and enzymatic antioxidants, thus mitigating oxidative stress. Additionally, brassinosteroids (BRs) enhanced antioxidant enzyme activities in the presence of zinc (Al-Duraid et al., 2019; Rahnama and Ebrahimzadeh, 2005), lead (Radwan and Kokate, 1980), and chromium (Sharma et al., 2013). Under high-temperature stress, 24-epibrassinosteroid at 2 ppm and 5 ppm significantly boosted squalene epoxidase expression, with expression levels rising 20- and 10-fold, respectively, compared to normal conditions. After 24 hours, treatment with 2 ppm of 24-epibrassinosteroid increased sterol methyltransferase expression sixfold relative to 23°C and untreated controls. In contrast, squalene synthase expression was reduced under high temperatures and varied 24-epibrassinosteroid levels compared to normal conditions (Sharma et al., 2011).

Fenugreek varieties such as RMt-305, RMt-1, RMt-361, Hissar, and Sonali were treated at the vegetative stage with different concentrations (0.0, 0.25, 0.50, 0.75, and 1.00 ppm) of brassinolide. Approximately 50% of the treated plants were subjected to water stress, with brassinolide proving most effective at 1.0 ppm in both water-stressed and non-stressed environments (Siddique et al., 2001). Compared to the control, 24-epibrassinolide at various concentrations improved mineral content in *Trigonella* seeds during germination and in the leaves of plants grown from seeds pretreated with 24-epibrassinolide. While 0.1 ppm 24-epibrassinolide elevated magnesium (90 ppm) and potassium (8700 ppm) in germinated seeds to their highest levels, 2.5 M 24-epibrassinolide was the most effective pretreatment for increasing calcium (94 ppm) and iron (300 ppm). Similar results were observed in leaves from plants grown from pretreated seeds, where 2.5 M

24-epibrassinolide increased mineral levels except for magnesium (Munns, 2005).

Growth and metabolic characteristics of *Trigonella foenum-graecum* L. plants exposed to lead toxicity were measured, revealing that 28-homobrassinolide enhanced photosynthetic pigments, which boosted plant biomass and accelerated photosynthesis in fenugreek plants. In plants exposed to lead, biochemical parameters such as carbohydrates, proteins, and nucleic acids were reduced, but low doses of 28-homobrassinolide mitigated the toxicity. These findings demonstrate that 28-homobrassinolide effectively reduced lead toxicity in fenugreek plants (Swamy et al., 2011). When 0.50 ppm brassinosteroid was applied as a foliar spray, seed yield in fenugreek increased by 14.6% compared to the control (Gaspar et al., 1996).

4.3.2 SALICYLIC ACID

Salicylic acid is a hydroxyl phenolic acid synthesized from phenylalanine via the isochorismate pathway in the chloroplast and the phenylpropanoid pathway in the cytoplasm (Meena et al., 2014). It was named after its discovery in the bark and leaves of willow trees (*Salix*). Salicylic acid synthesis increases when plants are exposed to environmental stressors, and it can be produced in the form of free salicylic acid, carboxylated esters, or phenolic glycosides.

According to Hayat et al. (2007) and Hasanuzzaman et al. (2017), salicylic acid plays a role in various physiological mechanisms in plants, including seed germination, seed production, photosynthesis, thermogenesis, respiration, vegetative growth, senescence, and protection against different stresses. Additionally, it enhances photosynthesis under abiotic stress by promoting CO_2 assimilation, carboxylase activity, stomatal conductance, ribulose-1,5-bisphosphate activity, and the photosynthetic apparatus in various plants (Fariduddin et al., 2003; Swamy et al., 2014). Although salicylic acid can delay germination and cause oxidative damage, there are mixed findings on its effect on seed germination in some plants. However, priming *Trigonella* seeds with a salicylic acid concentration of 2800 M significantly improved germination percentage and seedling vigor index (Métraux, 2002). In certain cases, it also enhances seed vigor.

Trigonella seeds treated with 100 mg L^{-1} salicylic acid (SA) and grown under cadmium (Cd) concentrations of 0, 10, 20, and 30 mg L^{-1} showed protection against cadmium toxicity, reducing the adverse effects of cadmium on germination (Dwivedi and Gurjar, 2023). Salicylic acid, a powerful ameliorator, provides defense against heavy metal toxicity, primarily through mechanisms involving reactive oxygen species (ROS) scavenging (Yao et al., 2020). According to Firoz et al. (2019), the application of salicylic acid enhanced the growth of fenugreek plants under stress by mitigating the adverse effects of salinity. This resulted in improved plant height, dry weight, and significantly higher levels of photosynthetic pigments and yield components.

Salicylic acid also positively affected fenugreek germination. Abdelhameed et al. (2021) reported a substantial increase in photosynthetic pigments and the accumulation of phenolics, amino acids, and flavonoids in treated plants. Under water stress, Hamad and Ali (2021) observed enhancements in plumule length, germination percentage, radicle length, dry weights of both plumule and radicle, and the vitality index in fenugreek plants treated with SA. In *Trigonella foenum-graecum* L. plants exposed to varying Cd concentrations (0, 10, 20, 30 mg/L), SA treatment alleviated the negative effects on germination, shoot and root elongation, and seedling dry weight, enhancing these parameters (Dwivedi and Gurjar, 2023). Pour et al. (2012) also observed improved seed germination and seedling growth in fenugreek plants subjected to salt stress and treated with SA compared to untreated stressed plants.

In *Trigonella*, increased arsenic (As) and zinc (Zn) levels boosted the activities of superoxide dismutase (SOD), catalase (CAT), ascorbate peroxidase (APX), and glutathione reductase (GR). Administering 0.2 mM SA reduced the harmful effects of As and Zn, enhancing radicle development and increasing protein, soluble sugar, and free amino acid concentrations in cotyledons and radicles. Hydrolytic enzyme activity was also improved, though cell viability decreased even with SA treatment (Maathuis et al., 2014).

4.3.3 NAPHTHALENEACETIC ACID (NAA)

Auxins are essential for various physiological processes in plants, including cell elongation and division, vascular tissue formation, root initiation, differentiation, flowering, apical dominance, leaf senescence, and abscission of leaves and fruit. Pre-soaking fenugreek plants in a 50 ppm naphthaleneacetic acid (NAA) solution significantly enhanced growth, yield, and alkaloid production (Dadrasan et al., 2015). When leaf and shoot apical meristems were treated with 1 mg/L of 2,4-dichlorophenoxyacetic acid (2,4-D), *in vitro* callus induction occurred, increasing callus fresh and dry weights at 15 and 45 days after culture. In leaf cultures, the highest diosgenin production was observed after 45 days of culture (Reddy and Hore, 2020). Auxin treatments have also been shown to mitigate toxic metal effects on plants up to a certain level (Yuan et al., 2013). Foliar application of 10 μM indole-3-acetic acid (IAA) to *Trigonella* seedlings was effective in reducing oxidative damage caused by cadmium by enhancing antioxidant levels and activating genes for enzymes in the ascorbate-glutathione pathway (Balakrishnan and Arunprasath, 2018).

Callus induction using NAA was found to improve the fenugreek plant's capacity for secondary metabolite production and biomass accumulation. Explant treatment with 10 mg/L of 2,4-D for 1 hour not only promoted rapid callus induction but also increased trigonelline production in the cultured callus (Pour et al., 2012). Meena et al. (2014) reported that foliar spraying with 10 and 20 ppm of NAA enhanced yield, growth parameters, and yield attributes in *Trigonella*. Salicylic acid (100 mg/L) applied as a foliar spray boosted biomass production and physicochemical properties in two varieties, Deli Kabul and Kasuri, of *Trigonella foenum-graecum* L. under saline conditions (Ashraf et al., 2010).

Foliar application of NAA at 50 ppm at intervals of 40-, 60-, and 80-days increased growth, yield attributes, and economic returns in *Trigonella*, resulting in more pods per plant, higher seed and straw yields, increased test weight, and higher net and gross returns compared to the control (Sheikhi et al., 2021). Fenugreek treated with 50 mg/L NAA foliar spray showed improvements in plant height, secondary branches, leaf area index, chlorophyll content, leaf yield, fresh and dry weights, seed antioxidant activity, phenolic compounds, antioxidant capacity, and free radical scavenging (Abdelmoumen et al., 2009). Balakrishnan and Arunprasath (2018) found that NAA enhanced morphological parameters such as shoot and root length, fresh and dry weights, and increased biochemical components, including carotenoids, chlorophyll, starch, sugar, protein, and amino acids compared to untreated plants. Spraying with 20 ppm NAA at 25 and 55 days after sowing significantly boosted fenugreek yield and growth (Godara et al., 2017). Applying 0, 50, and 100 mg/L NAA increased antioxidant activity, yield parameters, and growth parameters (Al-Duraid et al., 2019). Reddy and Hore (2020) observed improvements in plant height, primary and secondary branch numbers, pod numbers per plant, seed numbers per pod, and total yield per plant with foliar applications of 25, 50, and 75 ppm NAA. Spraying with 5, 10, 15, and 20 ppm NAA at 30, 60, and 90-day intervals significantly enhanced plant height, branch and leaf numbers per plant, seed yield per plant, seeds per pod, 1,000-seed weight, and both gross and net returns (Dwivedi and Gurjar, 2023).

4.4 CONCLUSION

Trigonella foenum-graecum L., also called fenugreek, is a fragrant, therapeutic herb containing several significant compounds. Since a long *Fenugreek* is being utilized for crucial conventional, versatile medicinal plant in Chinese, Iranian, Indian, and Tibetan medical system. Its names are Methi in Hindi, Punjabi, Urdu, and Marathi, Moshoseitaro in Greek, Hulba in Arabic and Uluva in Malayalam. Owing to its economic and medicinal importance, the demand of this herbs has been increasing day by day and to meet the growing challenges its production is urgently and seriously needed. Though, there are several challenges to enhance the production such as varying climatic conditions, desertification, heavy metal in soil and deficiency in mineral nutrition, and so forth. The different scientific approaches employed have effectively offset losses in fenugreek production brought on by a number of abiotic stressors and play a crucial part in improving the productivity of the crop, thus boosting the synthesis of secondary metabolic products in both non-stress and abiotic stress conditions in *T. foenum graecum* L. The use of tissue culture technique may seem to be the most beneficial of these methods since they can boost up the synthesis of bioactive compounds from small amounts of biomass while overcoming periodic damage and stress that plants experience in their natural habitat. However, it is highly expensive, needs to be handled carefully, and be kept in aseptic conditions, making it out of the price range and comprehension of the average farmer. PGRs are essential for enabling yield enhancement in a model plant. In comparison to biotechnological methods, such PGRs are affordable and manageable. Due to the variety of stresses caused by degraded environmental conditions that fenugreek has to endure, including heavy metals, heat, drought, cold, flooding, and biotic strain. These PGRs efficiently mitigate and relieved the plant under stress by changing physiological mechanism and thereby triggering their defense mechanisms. PGR may be beneficial when applied individually to a variety of abiotic stresses. Apart from brassinosteroids, salicylic acid, and auxin (NAA), several other PGRs should be efficiently utilized against a variety of biotic and abiotic stress to enhance growth and development of the fenugreek thereby enabling enhanced output in terms of yield and secondary metabolite production.

KEYWORDS

- **auxin**
- **brassinosteroids**
- **growth regulators**
- **salicylic acid**
- **secondary metabolite production**
- **trigonelline**

REFERENCES

Abbas, S., Latif, H. H., & Elsherbiny, E. A. (2013). Effect of 24-epibrassinolide on the physiological and genetic changes in two varieties of pepper under salt stress conditions. *Pakistan Journal of Botany*, *45*, 1273–1284.

Abdelhameed, R. E., Abdel Latef, A. A. H., & Shehata, R. S. (2021). Physiological responses of salinized fenugreek (*Trigonella foenum-graecum* L.) plants to foliar application of salicylic acid. *Plants*, *10*(4), 657. https://doi.org/10.3390/plants10040657.

Abdelmoumen, H., Idrissi, M. M., & Al-Duraid, E. (2009). Germination, growth and nodulation of *Trigonella foenum-graecum* (Fenugreek) under salt stress. *African Journal of Biotechnology*, *8*, 2489–2496.

Al-Duraid, M. H., Al-Taey, K. A., & Al-Kikhani, A. H. J. (2019). Effect of phenylalanine and naphthaleneacetic acid on growth, yield, and antioxidant activity of fenugreek (*Trigonella foenum-graecum*). *IOP Conference Series: Earth and Environmental Science*, *388*(1), 012073. https://doi.org/10.1088/1755-1315/388/1/012073.

Arora, P., Bhardwaj, R., & Kumar Kanwar, M. K. (2010). 24-Epibrassinolide induced antioxidative defense system of *Brassica juncea* L. under Zn metal stress. *Physiology and Molecular Biology of Plants*, *16*(3), 285–293. https://doi.org/10.1007/s12298-010-0031-9.

Ashraf, M., Akram, N. A., Arteca, R. N., & Foolad, M. R. (2010). The physiological, biochemical and molecular roles of brassinosteroids and salicylic acid in plant processes and salt tolerance. *Critical Reviews in Plant Sciences*, *29*(3), 162–190. https://doi.org/10.1080/07352689.2010.483580.

Babar, S., Siddiqi, E. H., Hussain, I., Hayat Bhatti, K. H., & Rasheed, R. (2014). Mitigating the effects of salinity by foliar application of salicylic acid in fenugreek. *Physiological Journal*, *2014*, 1–6. https://doi.org/10.1155/2014/869058.

Balakrishnan, M., & Arunprasath, A. (2018). Impact of plant growth regulators and organic fertilizer on growth and biochemical composition of *Trigonella foenum-graecum* L. *International Journal of Biological Research*, *3*, 24–28.

Bashri, G., & Prasad, S. M. (2016). Exogenous IAA differentially affects growth, oxidative stress and antioxidants system in Cd-stressed *Trigonella foenum-graecum* L. seedlings: Toxicity alleviation by up-regulation of ascorbate-glutathione cycle. *Ecotoxicology and Environmental Safety*, *132*, 329–338. https://doi.org/10.1016/j.ecoenv.2016.06.015.

Cao, S., Xu, Q., Cao, Y., Qian, K., An, K., Zhu, Y., Binzeng, H., Zhao, H., & Kuai, B. (2005). Loss of function mutation in the DET2 gene leads to enhanced resistance to oxidative stress in Arabidopsis. *Physiologia Plantarum, 123*(1), 57–66. https://doi.org/10.1111/j.1399-3054.2004.00432.x.

Chauhan, R., Awasthi, S., Tripathi, P., Mishra, S., Dwivedi, S., Niranjan, A., Mallick, S., Tripathi, P., Pande, V., & Tripathi, R. D. (2017). Selenite modulates the level of phenolics and nutrient elements to alleviate the toxicity of arsenite in rice (*Oryza sativa* L.). *Ecotoxicology and Environmental Safety, 138*, 47–55. https://doi.org/10.1016/j.ecoenv.2016.11.015.

Dadrasan, M., Chaichi, M. R., Pourbabaee, A. A., Yazdani, D., & Keshavarz-Afshar, R. (2015). Deficit irrigation and biological fertilizer influence on yield and trigonelline production of fenugreek. *Industrial Crops and Products, 77*, 156–162. https://doi.org/10.1016/j.indcrop.2015.08.040.

Danesh Talab, S., Mehrafarin, A., Naghdi Badi, H., & Khalighi Sigaroodi, F. (2014). Changes in growth and trigonelline/mucilage production of fenugreek (*Trigonella foenum-graecum* L.) under plant growth regulators application. *Journal of Medicinal Plants, 13*(51), 15–25.

Davies, P. J. (2013). *Plant Hormones: Physiology, Biochemistry and Molecular Biology*. Springer Science and Business Media.

Dwivedi, B., & Gurjar, P. K. S. (2023). Effect of plant growth regulators and biofertilizers on productivity of fenugreek (*Trigonella foenum-graecum* L.). *The Pharma Innovation Journal, 12*, 2051–2055.

Espanany, A., Fallah, S., & Tadayyon, A. (2015). The effect of halopriming and salicylic acid on the germination of fenugreek (*Trigonella foenum-graecum*) under different cadmium concentrations. *Notulae Scientia Biologicae, 7*(3), 322–329. https://doi.org/10.15835/nsb.7.3.9563.

Fariduddin, Q., Hayat, S., & Ahmad, A. (2003). Salicylic acid influences net photosynthetic rate, carboxylation efficiency, nitrate reductase activity, and seed yield in *Brassica juncea*. *Photosynthetica, 41*(2), 281–284. https://doi.org/10.1023/B:PHOT.0000011962.05991.6c.

Fariduddin, Q., Yusuf, M., Ahmad, I., & Ahmad, A. (2014). Brassinosteroids and their role in the response of plants to abiotic stresses. *Biologia Plantarum, 58*(1), 9–17. https://doi.org/10.1007/s10535-013-0374-5.

Firoz, M., Wajid, M., & Bhat, M. (2019). Effect of salicylic acid sprays on the performance of fenugreek grown with graded levels of salinity. *The Saudi Journal of Life Sciences, 4*(10), 346–354.

Gao, S., Ouyang, C., Wang, S., Xu, Y., Tang, L., & Chen, F. (2008). Effects of salt stress on growth, antioxidant enzyme and phenylalanine ammonia-lyase activities in *Jatropha curcas* L. seedlings. *Plant, Soil and Environment, 54*(9), 374–381. https://doi.org/10.17221/410-PSE.

Gaspar, T., Kevers, C., Penel, C., Greppin, H., Reid, D. M., & Thorpe, T. A. (1996). Plant hormones and plant growth regulators in plant tissue culture. *In Vitro Cellular & Developmental Biology–Plant, 32*(4), 272–289. https://doi.org/10.1007/BF02822700.

Godara, A. S., Singh, R., Chouhan, G. S., & Nepalia, V. (2017). Yield and economics of fenugreek (*Trigonella foenum-graecum* L.) as influenced by fertility levels, biofertilizers and brassinosteroid. *Legume Research: An International Journal, 40*(OF), 165–169. https://doi.org/10.18805/lr.v0iOF.11192.

Gour, R., Naruka, I. S., Singh, P. P., Rathore, S. S., & Shaktawat, R. P. S. (2009). Effect of phosphorus and plant growth regulators on growth and yield of fenugreek (*Trigonella foenum-graecum* L.). *Journal of Spices and Aromatic Crops, 18,* 33–36.

Hamad, S. M., & Ali, A. H. (2021). Effect of treatment with salicylic acid on germination and some physiological traits of fenugreek (*Trigonella foenum-graecum* L.) growing under water stress conditions. *International Journal of Multidisciplinary Research Growth Evaluation, 2,* 158–163.

Hasanuzzaman, M., Nahar, K., & Bhuiyan, T. F. (2017). Salicylic acid: An all-rounder in regulating abiotic stress responses in plants. In *Phytohormones-Signal Mechanism Crosstalk in Plant Development and Stress Responses* (pp. 31–75). Springer.

Hayat, S., Ali, B., & Ahmad, A. (2007). Salicylic acid: Biosynthesis, metabolism and physiological role in plants. In *Salicylic Acid: A Plant Hormone* (pp. 1–14). Springer.

Heyland, D. K., Johnson, A. P., Reynolds, S. C., & Muscedere, J. (2011). Procalcitonin for reduced antibiotic exposure in the critical care setting: A systematic review and an economic evaluation. *Critical Care Medicine, 39*(7), 1792–1799. https://doi.org/10.1097/CCM.0b013e31821201a5.

Iqbal, S., Inam, A., Sahay, S., & Inam, A. (2015). Growth, physiological, yield and quality response in chili (*Capsicum annuum* L.) under wastewater irrigation and different levels of phosphorus. *American Journal of Experimental Agriculture, 5*(1), 70–81. https://doi.org/10.9734/AJEA/2015/9293.

Koca, H., Bor, M., Özdemir, F., & Türkan, İ. (2007). The effect of salt stress on lipid peroxidation, antioxidative enzymes and proline content of sesame cultivars. *Environmental and Experimental Botany, 60*(3), 344–351. https://doi.org/10.1016/j.envexpbot.2006.12.005.

Kumari, R., & Prasad, M. N. V. (2013). Medicinal plant active compounds produced by UV-B exposure. In *Sustainable Agriculture Reviews* (pp. 225–254). Springer. https://doi.org/10.1007/978-94-007-5961-9_8.

Kumari, R., & Prasad, M. N. V. (2014). Effect of UV-B pretreatment on essential oil components, health sensory secondary metabolites and antioxidant potential of *Coleus aromaticus*. *International Journal of Biological and Pharmaceutical Research, 5,* 675–688.

Maathuis, F. J. M., Ahmad, I., & Patishtan, J. (2014). Regulation of Na^+ fluxes in plants. *Frontiers in Plant Science, 5*, 467. https://doi.org/10.3389/fpls.2014.00467.

Mabrouk, B., Kâab, S. B., Rezgui, M., Majdoub, N., Teixeira da Silva, J. A., & Kâab, L. B. B. (2019). Salicylic acid alleviates arsenic and zinc toxicity in the process of reserve mobilization in germinating fenugreek (*Trigonella foenum-graecum* L.) seeds. *South African Journal of Botany, 124,* 235–243. https://doi.org/10.1016/j.sajb.2019.05.020.

Maher, S. S. A. A., & Al-Mohammad, H. S. (2018). Effect of spraying number and type of growth regulator on growth, leaves yield and antioxidant activity for fenugreek leaves. *Euphrates Journal of Agricultural Science, 1*(4), 88–96.

Mahesh, K., Balaraju, P., Ramakrishna, B., & Ram Rao, S. S. R. (2013). Effect of brassinosteroids on germination and seedling growth of radish (*Raphanus sativus* L.) under PEG-6000 induced water stress. *American Journal of Plant Sciences, 4*(12), 2305–2313. https://doi.org/10.4236/ajps.2013.412285.

Meena, S. S., Mehta, R. S., Bairwa, M., & Meena, R. D. (2014). Productivity and profitability of fenugreek (*Trigonella foenum-graecum*) as influenced by bio-fertilizers and plant growth regulators. *Legume Research, 37*(6), 646–650.

Métraux, J. P. (2002). Recent breakthroughs in the study of salicylic acid biosynthesis. *Trends in Plant Science, 7*(8), 332–334. https://doi.org/10.1016/s1360-1385(02)02313-0.

Moghaddam, S. S., Rahimi, A., Noorhosseini, S. A., Heydarzadeh, S., & Mirzapour, M. (2018). Effect of seed priming with salicylic acid on germinability and seedling vigour of fenugreek (*Trigonella foenum-graecum*). *YYU Journal of Agricultural Science, 28,* 192–199.

Moradi, N., & Moradi, K. (2013). Physiological and pharmaceutical effects of fenugreek (*Trigonella foenum-graecum* L.) as a multipurpose and valuable medicinal plant. *Global Journal of Medicinal Plant Research, 1*(2), 199–206.

Munns, R. (2005). Genes and salt tolerance: Bringing them together. *New Phytologist, 167*(3), 645–663. https://doi.org/10.1111/j.1469-8137.2005.01487.x.

Murtaza, I., Khurshid, A., Sidiqi, I., & Khan, F. A. (2021). Effect of 24-epibrassinolide pretreatments on mineral content of fenugreek (*Trigonella foenum-graecum* L.) sprouts and leaves. *Journal of Pharmaceutical Innovation, 10*(6), 1282–1285.

Nayak, A. K., Pal, D., & Das, S. (2013). Calcium pectinate-fenugreek seed mucilage mucoadhesive beads for controlled delivery of metformin HCL. *Carbohydrate Polymers, 96*(1), 349–357. https://doi.org/10.1016/j.carbpol.2013.03.088.

Negrão, S., Courtois, B., Ahmadi, N., Abreu, I., Saibo, N., & Oliveira, M. M. (2011). Recent updates on salinity stress in rice: From physiological to molecular responses. *Critical Reviews in Plant Sciences, 30*(4), 329–377. https://doi.org/10.1080/07352689.2011.587725.

Osman, E. H. M., El-Feky, S. S., Abo-Hamad, S. A., & Seliem, H. S. (2015). Improvement of harmful effects induced by temperature stress on *Trigonella foenum-graecum* L. by putrescine. *Egyptian Journal of Experimental Biology, 11,* 197–205.

Ouzir, M., El Bairi, K., & Amzazi, S. (2016). Toxicological properties of fenugreek (*Trigonella foenum-graecum*). *Food and Chemical Toxicology, 96,* 145–154. https://doi.org/10.1016/j.fct.2016.08.003.

Ozyigit, I. I., Yalcin, B., Turan, S., Saracoglu, I. A., Karadeniz, S., Yalcin, I. E., & Demir, G. (2018). Investigation of heavy metal level and mineral nutrient status in widely used medicinal plants' leaves in Turkey: Insights into health implications. *Biological Trace Element Research, 182*(2), 387–406. https://doi.org/10.1007/s12011-017-1070-7.

Panchal, K. J., Parmar, P., Dave, B., & Sudhir, A. (2013). Physiological, biochemical and molecular response of plants against heavy metals stress. *International Journal of Current Research, 5,* 80–89.

Petropoulos, G. A. (2002). *Fenugreek: The Genus Trigonella*. CRC Press.

Petropoulos, G. A. (2002a). *Fenugreek–The Genus Trigonella* (p. 255). Taylor & Francis.

Petropoulos, G. A. (2002b). *Fenugreek–The genus Trigonella* (p. 200). Taylor & Francis.

Pour, A. P., Farahbakhsh, H., Saffari, M., & Keramat, B. (2012). Effects of seed priming on germination and seedling growth under salinity stress in fenugreek. *International Journal of Agriculture and Crop Sciences, 4,* 779–786.

Radwan, S. S., & Kokate, C. K. (1980). Production of higher levels of trigonelline by cell cultures of *Trigonella foenum-graecum* than by the differentiated plant. *Planta, 147*(4), 340–344. https://doi.org/10.1007/BF00379843.

Rady, M. M., & Osman, A. S. (2012). Response of growth and antioxidant system of heavy metal contaminated tomato plants to 24-epibrassinolide. *African Journal of Agricultural Research, 7,* 3249–3254.

Rahnama, H., & Ebrahimzadeh, H. (2005). The effect of NaCl on antioxidant enzyme activities in potato seedlings. *Biologia Plantarum, 49*(1), 93–97. https://doi.org/10.1007/s10535-005-3097-4.

Ramakrishna, B., & Rao, S. S. R. (2013). Preliminary studies on the involvement of glutathione metabolism and redox status against zinc toxicity in radish seedlings by 28-homobrassinolide. *Environmental and Experimental Botany, 96,* 52–58. https://doi.org/10.1016/j.envexpbot.2013.08.003.

Rao, A. V. (2003). *Herbal Cure for Common Diseases*. Fusion Books.

Rashid, F., Hussain, S., & Ahmed, Z. (2018). Extraction, purification and characterization of galactomannan from fenugreek for industrial utilization. *Carbohydrate Polymers, 180,* 88–95. https://doi.org/10.1016/j.carbpol.2017.10.025.

Reddy, P. P., & Hore, J. K. (2020). Role of growth regulators on fenugreek (*Trigonella foenum-graecum* L.). *International Journal of Current Microbiology and Applied Sciences, 9*(7), 25–32. https://doi.org/10.20546/ijcmas.2020.907.004.

Rezaeian, S. (2011). Assessment of diosgenin production by *Trigonella foenum-graecum* L. in *in-vitro* condition. *American Journal of Plant Physiology, 6*(5), 261–268. https://doi.org/10.3923/ajpp.2011.261.268.

Rucińska-Sobkowiak, R. (2016). Water relations in plants subjected to heavy metal stresses. *Acta Physiologiae Plantarum, 38*(11), 257. https://doi.org/10.1007/s11738-016-2277-5.

Sadak, M. S. H. (2016). Mitigation of drought stress on fenugreek plant by foliar application of trehalose. *International Journal of ChemTech Research, 9*(2), 147–155.

Sadeghzadeh-Ahari, D., Hassandokht, M. R., Kashi, A. K., Amri, A., & Alizadeh, K. H. (2010). Genetic variability of some agronomic traits in the Iranian fenugreek landraces under drought stress and non-stress conditions. *African Journal of Plant Science, 4*(2), 12–20.

Sadura, I., & Janeczko, A. (2018). Physiological and molecular mechanisms of brassinosteroid induced tolerance to high and low temperature in plants. *Biologia Plantarum, 62*(4), 601–616. https://doi.org/10.1007/s10535-018-0805-4.

Sajjad, Y., Jaskani, M. J., Ashraf, M. Y., Qasim, M., & Ahmad, R. (2014). Response of morphological and physiological growth attributes to foliar application of plant growth regulators in gladiolus "White Prosperity." *Pakistan Journal of Agricultural Sciences, 51,* 123–129.

Schnelle, R. A., & Barrett, J. E. (2010). Paclobutrazol concentration and substrate moisture status impact efficacy of liner dips for size control of three bedding plants. *HortTechnology, 20*(4), 735–739. https://doi.org/10.21273/HORTTECH.20.4.735.

Sethy, S. K., & Ghosh, S. (2013). Effect of heavy metals on germination of seeds. *Journal of Natural Science, Biology and Medicine, 4*(2), 272–275. https://doi.org/10.4103/0976-9668.116964.

Sharma, I., Bhardwaj, R., & Pati, P. K. (2013). Stress modulation response of 24-epibrassinolide against imidacloprid in an elite indica rice variety Pusa Basmati-1. *Pesticide Biochemistry and Physiology, 105*(2), 144–153. https://doi.org/10.1016/j.pestbp.2013.01.004.

Sharma, I., Pati, P. K., & Bhardwaj, R. (2011). Effect of 28-homobrassinolide on antioxidant defense system in *Raphanus sativus* L. under chromium toxicity. *Ecotoxicology, 20*(4), 862–874. https://doi.org/10.1007/s10646-011-0650-0.

Sheikhi, S., Heidari, P., Ebrahimi, A., Amerian, M. R., & Rashidi, M. S. (2021). The effect of 24-epi-brassinosteroid on the expression of some genes involved in the diosgenin biosynthetic pathway of fenugreek (*Trigonella foenum-graecum* L.) under high temperature stress. *Iranian Journal of Field Crops Research, 52*, 11–24. https://doi.org/10.22059/IJFCS.2020.295770.654683.

Shivran, A. C., Jat, N. L., Singh, D., Sastry, E. V. D., & Rajput, S. S. (2016). Response of fenugreek (*Trigonella foenum-graecum* L.) to plant growth regulators and their time of application. *Journal of Spices and Aromatic Crops, 25,* 169–174.

Siddique, M. R. B., Hamid, A., & Islam, M. S. (2001). Drought stress effects on water relations of wheat. *Botanical Bulletin of Academia Sinica, 41,* 35–39.

Singh, M., Mahala, P., Yadav, T. V., & Kakraliya, B. L. (2018). Influence of brassinolide in mitigating the adverse effect of drought on yield attributes in fenugreek (*Trigonella foenum-graecum* Linn.) genotypes. *International Journal of Current Microbiology and Applied Sciences, 7*(6), 1475–1478. https://doi.org/10.20546/ijcmas.2018.706.175.

Swamy, K. N., Ramakrishna, B., Anuradha, S., Siddulu, N., & Rao, S. S. R. (2011). Cadmium toxicity is diminished by 24-epibrassinolide in seedlings of *Trigonella foenum-graecum* L. *Genetics and Plant Physiology, 1,* 163–175.

Swamy, K. N., Vardhini, B. V., Ramakrishna, B., Anuradha, S., Siddulu, N., & Rao, S. S. R. (2014). Role of 28-homobrassinolide on growth biochemical parameters of *Trigonella foenum-graecum* L. plants subjected to lead toxicity. *International Journal of Multidisciplinary Current Research, 2,* 2321–3124.

Tahjib-ul-Arif, M., Siddiqui, M. N., Sohag, A. A. M., Sakil, M. A., Rahman, M. M., Polash, M. A. S., Mostofa, M. G., & Tran, L. P. (2018). Salicylic acid-mediated enhancement of photosynthesis attributes and antioxidant capacity contributes to yield improvement of maize plants under salt stress. *Journal of Plant Growth Regulation, 37*(4), 1318–1330. https://doi.org/10.1007/s00344-018-9867-y.

Tchounwou, P. B., Yedjou, C. G., Patlolla, A. K., & Sutton, D. J. (2012). Heavy metal toxicity and the environment. *Molecular, Clinical and Environmental Toxicology, 101,* 133–164.

Whipker, B. W., Krug, B. A., & McCall, I. (2005). Preplant tuber soaks of 6-benzyladenine ineffective for chemical de-eyeing of caladium. *Plant Growth Regulation Society of America Quarterly, 33,* 16–20.

Wroblewska, K., & Debicz, R. (2013). Influence of time of benzyladenine application on rooting of cuttings and subsequent development of *Portulaca umbraticola* Kunth. *Acta Scientiarum Polonorum, 12(1),* 89–99.

Xia, X. J., Huang, Y. Y., Wang, L., Huang, L. F., Yu, Y. L., Zhou, Y. H., & Yu, J. Q. (2006). Pesticides-induced depression of photosynthesis was alleviated by 24-epibrassinolide pretreatment in *Cucumis sativus* L. *Pesticide Biochemistry and Physiology, 86*(1), 42–48. https://doi.org/10.1016/j.pestbp.2006.01.005.

Yadav, S. K., & Sehgal, S. (1997). Effect of home processing and storage on ascorbic acid and β-carotene content of bathua (*Chenopodium album*) and fenugreek (*Trigonella foenum-graecum*) leaves. *Plant Foods for Human Nutrition, 50*(3), 239–247. https://doi.org/10.1007/BF02436060.

Yao, D., Zhang, B., Zhu, J., Zhang, Q., Hu, Y., Wang, S., Wang, Y., Cao, H., & Xiao, J. (2020). Advances on application of fenugreek seeds as functional foods: Pharmacology, clinical application, products, patents and market. *Critical Reviews in Food Science and Nutrition, 60*(14), 2342–2352. https://doi.org/10.1080/10408398.2019.1635567.

Yuan, H. M., Liu, W. C., Jin, Y., & Lu, Y. T. (2013). Role of ROS and auxin in plant response to metal-mediated stress. *Plant Signaling & Behavior, 8*(7), e24671. https://doi.org/10.4161/psb.24671.

Zhang, H., Xu, J., Wang, M., Xia, X., Dai, R., & Zhao, Y. (2020). Steroidal saponins and sapogenins from fenugreek and their inhibitory activity against α-glucosidase. *Steroids, 161*, 108690. https://doi.org/10.1016/j.steroids.2020.108690.

Zhou, J., Chan, L., & Zhou, S. (2012). Trigonelline: A plant alkaloid with therapeutic potential for diabetes and central nervous system disease. *Current Medicinal Chemistry, 19*(21), 3523–3531. https://doi.org/10.2174/092986712801323171.

CHAPTER 5

Plant Growth Regulators and Their Effect on Fenugreek Growth

FILZA MUSTAFA

Department of Plant Biotechnology, Atta-Ur-Rahman School of Applied Biosciences, National University of Sciences and Technology H-12 Islamabad, Pakistan

ABSTRACT

Fenugreek also known as *Trigonella foenum graceum* L. is an important medicinal legume, a spice and a vegetable that belongs to family Fabaceae. It is a short living plant native to Indian subcontinent with ability of self-pollination. It is highly desirable plant due to its medicinal applications and a rich source of food for animals and humans. Generally, the fresh and dried seeds and leaves of fenugreek are used for food due to its richness in saponin, alkaloids, minerals, vitamin C, proteins, volatile oils, and different kind of polysaccharides. It is a very beneficial medicinal plant that has anticancer, antifertility, antidiabetic, antimicrobial, and antiparasitic activities. It is also used for the treating high blood pressure and reduces cholesterol level. Plant growth regulators are those organic compounds that not only used to improve growth parameters of plant body but also influence different physiological processes such as metabolism, photosynthesis, and morphogenesis.

They are available both in natural and synthetic forms. In case of plants, these PGRs also known as plant growth hormones. Commonly used plant growth regulators such as auxins, cytokinin, gibberellin, gibberellic acid, naphthaleneacetic acid, BAP, and so forth. Plants usually utilize a very small quantity of these beneficial substances to improve the nutrient uptake and enhance body efficacy to properly distribute these nutritional substances for maximum growth of different body parts. They also influence the

Fenugreek: Improvement in Yield Potential Through Plant Growth Regulators (PGRs) and Depolymerized Polysaccharides. Tariq Ahmad Dar, Bilal Ahmad, Moinuddin, and Khalid Rehman Hakeem (Eds.)

biosynthesis of phytochemicals even in cell suspension cultures. The growth of different body parts including formation of leaves, seeds germination, floral induction, flowering, stem, and root growth is affected by these PGRs. These hormones are also used to improve each other's efficiency as well for controlling growth of fenugreek.

These are helpful for fenugreek during the stress situations such as abiotic and abiotic factors and boost plant mechanisms to maintain growth of various body parts under stress elements. Different concentrations of these plant growth regulators and their time of application highly affect all growth aspects of fenugreek. These PGRs automatically reach the site of execution from the side of synthesis and control all activities of plant growth cycle. They are effective both for normal fenugreek plant growth as well as in case fenugreek is facing a growth deficiency due to external environmental stress inducers including light, temperature, drought, nutrient deficiency, and any kind of chemical or physical stress. Fenugreek responds towards these plant growth regulators and physical changes can properly be seen at every stage. Here in this chapter, we tried to explain the role of different plant growth regulators on various parts of fenugreek.

5.1 INTRODUCTION

The basic purpose of using plant growth regulators is their immediate role in plant improvement, also take less time for action and produced quantifiable advantages in targeted plants. They are commonly known as plant growth hormones responsible for regulating growth and developmental stages in plants (Prajapati et al., 2015). These plant growth regulators also known as signaling molecules that affect the plant growth and involved in different physiological activities such as cell enlargement and regulation of cell cycle. As functional message carriers, these PGRs totally change the complete plant life cycle and alter different processes such as flowering, leaf formation, propagation, maturation of fruit and radicalization. PGRs are used in various forms like powder, paste, liquid, and are used to stimulate different growth stages of plant (Alizadeh et al., 2013).

5.1.1 PLANT GROWTH REGULATORS

Plants need water, oxygen, sunlight, and certain minerals for proper growth and development. Not only these plants and their cell functions are probably regulated by certain chemical substances and hormones known as phytohormones or plant growth regulators. Plant growth regulators are either naturally present in plant body or supplemented from external sources such as foliar spray. They are critical to improve growth rate in fenugreek and helped the plant to initiate germination and growth under favorable growth circumstances that resemble the natural process of development. They are produced in certain areas of plant body and then transported to the site of actions specifically to any tissue or plant organ. Due to their organic nature, they are different from nutrients and vitamins. Currently plant growth regulators are known as the latest edition in agricultural chemicals due to their multiple benefits. A very minute amount of these PGRs is enough for speeding the growth stages of plants either by stimulating or modifying it. There are different types of plant growth regulators associated with specific functions. Some commonly being used PGRs are auxins, gibberellins, cytokinin, ethylene, abscisic acid, naphthaleneacetic acid, benzyl amino purine (BAP), gibberellic acid, and so forth (Prajapati et al., 2015).

The effect of plant growth regulators depends upon the quantity and method of their application. The foliar application of various plant growth regulators has a significant role in fenugreek growth status (Bairva et al., 2012). All growth attributes of fenugreek crop such as cell division, enlargement of cell, high rate of photosynthesis from excessive content of chlorophyll in leaves are associated with certain amounts of different plant growth regulators (Patel et al., 2021). A research study reported that PGRs has a great influence on callus induction when fenugreek explants especially leaf explants are grown in MS medium. The resulted callus will increase both in size and weight that corresponds the high rate of growth induction in callus. When fenugreek is exposed to any stress like salinity, water logging and drought, the growth is significantly decreased while the plants treated with PGRs are much healthy and stable under environmental changes (Talab et al., 2014).

PGRs are used in minute amounts to stimulate growth and development and a slight increase in their concentrations cause the plant injuries and promote the chances of plant death. They are used as cutting-edge technology to manage growth, development, and yield of plants by increasing shelf life of flower and seed in fenugreek for a longer period (Alam, 2018). Plant Growth Regulators are currently available in markets of different countries such a Europe, Japan, United States and are commercially being used in agricultural sectors. The foliar use of PGRs is associated with enhanced growth rate in fenugreek. They have the potential to improve intrinsic genetic potential of fenugreek that ultimately enhanced plant growth (Agha, 2022).

Seed germination is an initial stage of plant growth. To increase the rate of seed germination, plant growth regulators are quite beneficial when monitored concentrations are applied. The PGRs increase the rate of seed germination up to double and also protect them from external changes including day and night times cycles. Gibberellins, indole acetic acid, naphthaleneacetic acid, and so on, are the renowned PGRs used for seed germination across the world. Similarly, they are used to manipulate the growth and developmental circumstances of plants such as their height, number of leaves per plant, stem elongation, number of flowers, seed production and increase in dry matter (Prajapati et al., 2015).

All the growth stages are interlinked with the implement of PGRs during the growth time. Gibberellic acid has been reported to increase leaf area, shoot length, number of leaves while Gibberellins are good to introduce vegetative characters in plants. Along with the regulation of these features PGRs also regulate the sex expression in plants too. Gibberellic acid is associated with the formation of male and female flowers in different plants such as fenugreek. Moreover, it also controls the fruit diameter, its size and number of fruits per plant. Gibberellin has a profound effect on fruit production, sex expression and yield of fruits (Prajapati et al., 2015).

Another important role of Plant growth regulators is in plant regeneration. Naphthaleneacetic acid is beneficial for callus induction and regeneration of plants in MS media. Gibberellic acid is responsible for shoot formation in MS media while Auxins are used to induce root formation in MS media. However, a mixture of three different PGRs named IBA, BA, and GA3 induce both rooting and shooting in culture media simultaneously (Prajapati et al., 2015).

The induction of parthenocarpy in plants is regulated with plant growth regulators especially cytokinin, gibberellic acid and naphthaleneacetic acid. Furthermore, PGRs also responsible for the induction of fruit development, also increase fruit weight, length, and diameter (Prajapati et al., 2015). It was also determined that each plant growth regulator varies according to the type of culture and explants used (Agha, 2022). NAA controls the formation of branches in fenugreek plant and as NAA concentration increases, it increases the average number of branches (Jamil et al., 2021).

5.1.2 CHARACTERISTICS OF PLANT GROWTH REGULATORS

For plants to grow properly under different circumstances, they not only need oxygen, sunlight, water, nutrition but also need certain chemical components to maintain all stages of their growth and development. These plant growth regulators are known as phytohormones or plant growth hormones. All plant growth hormones should exhibit some characteristics such as they will be able to induce differentiation and elongation in various cells (Tavangar et al., 2021). They must be capable of initiating formation of leaves, stems, flowers, and fruits in plants. Some PGRs must regulate fruit ripening and wilting of leaves. Another important characteristic of PGRs is their role in seed dormancy (Tavangar et al., 2021; Figure 5.1).

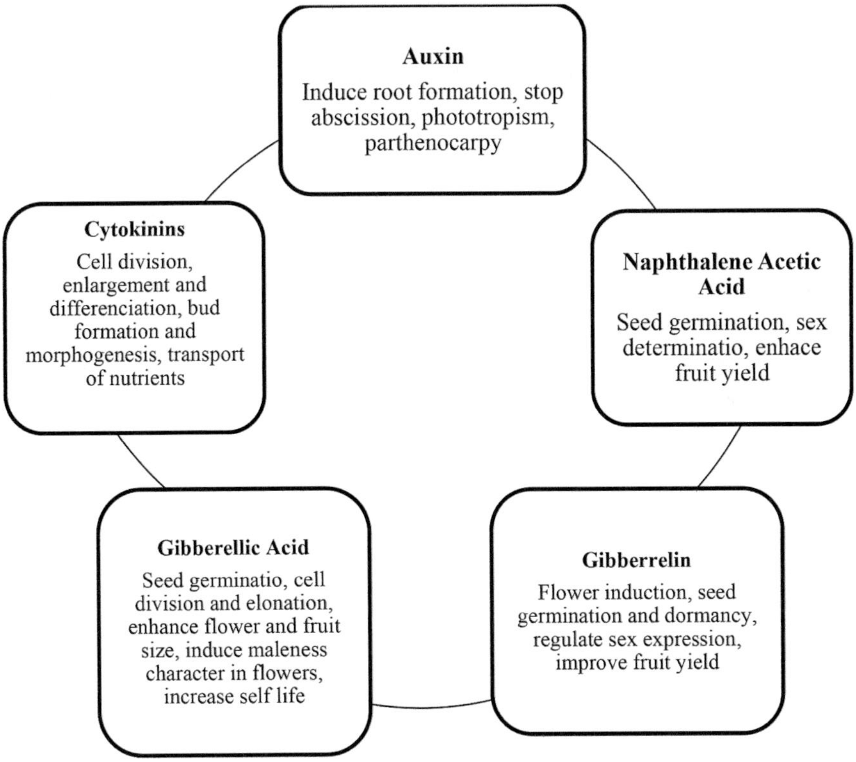

FIGURE 5.1 Role of different plant growth regulators in plant growth.

5.1.3 MAJOR TYPES OF PLANT GROWTH REGULATORS

There are two major types of plant growth regulators:

1. **Plant Growth Promoters:** These are used to initiate and promote cell division and enlargement, flower production, fruiting, and formation of seeds. Some commonly used plant growth regulators are auxins, cytokinin, and gibberellins (Student et al., 2011).

2. **Plant Growth Inhibitors:** These are those chemical substances that are involved in growth inhibition; also promote seed dormancy and abscission phases in plants to maintain the plant growth up to certain level. Abscisic Acid is a common example of plant growth inhibitors (Student et al., 2011).

5.2 FENUGREEK

Fenugreek *Trigonella foenum-graecum* L. commonly known as methi is a native crop of South Asian countries such as India, China, Pakistan, Bangladesh, and is also extended to different areas of world including Morocco, Ukraine, Turkey, Egypt, Greece, and so forth. It is an annual herbaceous plant with approximately 0.9 m height with light, green-colored leaves (Alam, 2018). The flowers are white to yellow in color to facilitate pollination by insects. It is a mid-summer plant that produces brownish yellow seed with high antioxidize activities (Alam, 2018). Fenugreek is a prime among all seed spices with different medicinal properties and act as aromatic, tonic, galactagogue, and carminative agent (Olfa et al., 2018). There are approximately 80 different species of fenugreek that are cultivated all over the world and it is used for different purposes such as in medicines, for making food, roasted grains for coffee, to control insect and even in perfume industry. It is highly rich in Vitamin A, B2, C, calcium, iron, and proteins that is used to reduce blood cholesterol level and help in managing blood sugar level. Normally its seeds are rich in aroma and are bitter in taste that is useful for adding flavor and nutritive elements in certain foods (Ahmad Dar et al., 2015; Almatroodi et al., 2021).

The plant is actively used for treating different diseases including rheumatism, neurasthenia, diabetes, atherosclerosis, gout, painful menstruation, anemia, gastric disorders, fevers, heartburn, inflammation, ulcers, swollen eyes, uterine problems, cancer, fertility, microbial infections and used as an appetizer (Al-Mohammad et al., 2019; Ghosh et al., 2015). Fenugreek seed are good source of fiber and minerals especially magnesium and iron. A high level of vitamin K is present in leaves. Seeds are mashed and mixed with yogurt to apply on scalp as a conditioner. These seeds are also used in making pickles (Aasim et al., 2010). Fenugreek is a wonderful plant because of its effectiveness in treating different health issues such as menopause, dysmenorrheal, Parkinson's tuberculosis mouth ulcer bronchitis kidney stones inflammation gastritis poly-cystic ovary syndrome, obesity, and so forth. It is also associated with other body systems such as improve and stimulate breast milk production improve weight loss boost sperm count and raise test to steer on reduce inflammation treat high blood pressure and cardiovascular diseases act as pain reliever. All these and many other health benefits of fenugreek are due to the presence of different alkaloids amino acids vitamins flavonoids saponins polyphenols carbohydrates lipids and amino acids present in its seeds leaves and other body parts of plant (Almatroodi et al., 2021).

It is cultivated mostly in arid and semi-arid areas that contain moderate concentration of salts. Soil salinity decreases rate of seed germination and halt fenugreek growth. Not only this almost all growth stages of fenugreek are disturbed by salinity, but seed germination and seedling are extremely affected by stress. The accumulation of salt in different cells of fenugreek alter the physiological and functional process linked with the growth as well (Olfa et al., 2018). The fenugreek group is highly impacted with water stress due to which not only the growth of different body parts but also the protein content in leaves section is disturbed. In those circumstances plant growth regulators are quite impressive to manage growth parameters of fenugreek and increase level of protein. The growth of fenugreek is affected by all the changes that take place in soil texture especially due to availability of water and temperature. When plant is exposed to these changes the growth of leaf and flowering are change and reduce both the qualitative and quantitative production of fenugreek (Battah et al., 2021; Moons, 2005).

5.2.1 GROWTH TIME PERIOD OF FENUGREEK

Growth is a major phenotypic expression that any plant shows with respect to suitable environmental conditions. Major body parts that undergo growth changes are leaf, stem, shoot, roots, fruits, and so forth (Olfa et al., 2018). The growing period of fenugreek is very crucial to obtain high yield. Usually for germination in soil 3–10 days are required. Once seed germination is initiated, the first leaf is produced after 6–10 days. In the next 5–8 days epicotyls region is not prominent. The growth is dependent to different conditions including temperatures, pH, and soil texture. When external conditions are cold, the growth of fenugreek is less. Due to which, the harvesting time is highly affected, and it causes less seed rate in plants (Student et al., 2011). The environmental factors also linked with the development of leaf, fruit, and seed. Once the seed is successfully germinated, it leads towards the growth of main plant that includes the growth and development of stems, pods, flowers, and seeds. The next stage is the maturation of these parts.

Especially in flower to increase the rate of pollination, in fruit to improve its ripening stage and in seed to complete its development for harvesting (Student et al., 2011).

5.3 MAJOR HORMONES AND THEIR ROLE IN PLANT GROWTH

5.3.1 AUXIN

Auxin is also known as first hormone to be discovered by Charles Darwin. It is available naturally and produce in plants specially in stem, root tips and then transported to other body parts. Auxins hormone is from the major class of plant growth regulators also known as first growth hormone discovered in plants. It is produced in various sections of plant including its stem, root tips and buds (Role et al., 2021). It is a major phytohormone responsible for controlling different stages of plant growth and development especially cell elongation, division, and differentiation (Ayaz et al., 2010). It is available either naturally or synthetically. Natural forms of Auxin are Indole acetic acid and Indole butyric acid while naphthaleneacetic acid and 2,4-dichloro-phenoxyacetic acid are synthetically available forms of auxin (Student et al., 2011).

Auxin is a major plant hormone that plays a significant role in plant growth such as cell division and differentiation, root formation and fruit development. The main purpose of auxin is to help plants during early stages of growth. It helps in the stimulation of plant cell elongation and responsible for the formation of apical meristem (Role et al., 2021). It is basically linked with apical dominancy in plants and responsible for separation of lateral buds. Auxins are responsible for root induction, apical dominance, callus formation, vascular differentiation, regulate flowering, photoperiodism, sex determination, differentiation of xylem in plants (Prajapati et al., 2015). Moreover, it is also determined for promoting breathing, prevent development of nodule, shortens internodes and inhibit lodging in plants (L et al., 2015). It has been proved as an essential regulator and controller for flower origination and development (Role et al., 2021).

An *in-vitro* study explained that the use of phytohormone especially auxin is responsible for inducing roots in fenugreek. When fenugreek explants are grown on culture media, auxin is a major hormone used to control the shoot length (Ishtiaq et al., 2021). Auxin is known as an indirect controller of flower development. It also acts as a vital part for originating flower primordial and impact root architecture. It is mostly produced in root apical meristem sections depending on Indole acetic acid biosynthesis. Many growth and developmental processes ranging from embryogenesis to senescence are controlled by auxin. Auxins are crucial for tackling environmental stress factors and required for the maintenance of growth under stress situation (Jamil et al., 2021). When fenugreek plants treated with auxins especially IAA for a longer time period, the development of roots is quick, and it improves the ability of nutrient uptake that aids the overall growth of the plants (Ayaz et al., 2010).

5.3.2 CYTOKININ

Cytokinin is another major class of plant growth regulators discovered by Skoog C. Miller and co-workers as an active agent to promote cell division. It is mainly produced in root meristem, young leaves, germinating seeds, developing fruits and sap streams (Xiong and Jiao, 2019). Cytokinin are plant growth regulators that controls pluripotency of shoot apical meristem, helpful in leaf formation, promotes cell expansion, and maintain all stages of cell proliferation (Xiong and Jiao, 2019).

Cytokinin is used to induce cotyledonary and hypocotyl growth stages in fenugreek explants. It has been demonstrated that when in culture media, fenugreek explants are treated with cytokinin, it highly affects the number of shoots per plant, the length and regeneration of shoot. When cotyledonary and hypocotyls segments of fenugreek explants are treated with cytokinin hormone, from callus it ultimately starts the initiation of shoot development and enhances the length of shoot area as well (Bairva et al., 2012). All growth stages of fenugreek are directly associated with the type and concentration of cytokinin (Jamwal, 2022). It is also responsible for cell division, branching, bud initiation and ripening of fruits (Bairva et al., 2012). A required content of cytokinin enhanced the shoot length in fenugreek (Aasim et al., 2010).

Cytokinin is generally useful for breaking dormancy of seeds and buds. Once dormancy is broken down, it initiates the growth of lateral buds, also boost apical dominance. It is quite efficient in increasing the flower freshness for a longer time and formation of adventitious shoots along with lateral shoots. To delay the ageing process of different crops especially fenugreek, cytokinin are used (Student et al., 2011). Cytokinin is interlinked with the formation of both leaves and chloroplast. Moreover, they are actively involved in nutrient mobilization for delaying leaf senescence (Moons, 2005).

5.3.3 GIBBERELLINS

Gibberellins hormone have been elucidated for simulating cell division, elongation, used for seed germination, induce flower ring during long days, prevent genetic dwarfism, dormancy, maleness (Prajapati et al., 2015). It is actively involved in increasing vegetative growth, stalk length, fruit cracking, and hastens maturity, improved fruit size and quality in different crops (2006). The use of gibberellin improves the process of flower formation and increase fruit development. It also promotes both grain quality and quantity in plants (Aasim et al., 2010). It has been reported that gibberellins are capable of inducing flowering in different varieties even in non-flowering varieties of plants (Prajapati et al., 2015).

They are terpenoids in nature and control all stages of growth, development, and physiological mechanisms of plants. Generally, gibberellins are used to induce germination and emergence of seeds, floral induction, flower, and fruit development and growth of leaf and stem. They are also important for improving root growth, floral inhibition, differentiation of bud, root hair abundance, dormancy of buds and seeds and regulate all the vegetative and reproductive processes. It is highly demanding plant growth regulator for cell division. Other advantages of Gibberellins are their role in boosting stalk length and fruit size in fenugreek along with enhancing both growth and flowering (Tufail et al., 2020).

In fenugreek gibberellins promote growth and development by producing different phenolic compounds to reduce the excessive production of auxins. The decrease in auxin production directly increases the gibberellins efficiency to improve plant height, elongate various body parts and help in vegetative growth. Not only this, but it also speeds up the process of cell division in sub-apical meristem to give a large size, to increase plant dry weight, produce a greater number of branches per plant, more seed quantity and promote the production of healthy seeds (Tavangar et al., 2021). Gibberellins are also utilized to break the dormancy and germination of seeds in fenugreek (Dewi and Pratiwi, 2016).

5.3.4 GIBBERELLIC ACID

Gibberellic Acid is a well-known group of phytohormones produced in plants that usually used for improving growth of various body parts. It is an important plant growth hormone that regulates crucial aspects of plant growth.

Growth including stem elongation, seed germination and protein synthesis in plant. It further regulates different mechanisms of higher plants that ultimately associated with growth and development of plant organs such as development of nodes and stimulation of dormancy, and so forth (Patel et al., 2021). It is commonly used in different agricultural practices and act as a growth stimulating agent (Valizadeh et al., 2018).

It is responsible for increasing plant height, seed production, root length, shoot length, fresh and dry weight of leaves, number of leaves and leaf area in fenugreek plant (Idrees, 2015). Gibberellic acid is associated with elongation of stems, seed germination and protein synthesis in fenugreek especially when it is used in the form of foliar spray especially on transplanted cuttings (Aasim et al., 2010). It is useful due to its various abilities of affecting plant growth and development by increasing the number of active constituents. Moreover, it is best known for its property of inducing an influx of calcium ions into the endoplasmic reticulum of guard cells that ultimately enhances the stomatal activities (Aasim et al., 2010).

In developing pods, Gibberellic Acid is used to improve the sink potential for enhancing the period of assimilate translocation especially to the reproductive parts to increase dry weight of pods. Once the dry weight of pods is increased from the normal rate, gibberellic acid also produces a greater number of seed per fruits when planting seed are soaked in gibberellic acid and potentially enhance the metabolic activity of different pathways that leads towards the production of secondary metabolites in leaves, stems, roots, and fruits. Metabolic constituents are higher in number in those plants treated with GA3 than those not treated with it (Aasim et al., 2010).

In fenugreek, gibberellic acid has shown improved vegetative growth and increased mucilage and trigonelline level especially when before planting seed was treated with it (Aasim et al., 2010). Another research study was performed to evaluate the effect of seed soaking in gibberellic acid and its linkage with increased growth of shoot, seed production, and so on, during water stress. The results concluded that when stressed fenugreek plants were treated with gibberellic acid, it improved the protein profile, enhanced shoot development, and developed healthier seeds (2006).

Gibberellic Acid is exceptionally crucial plant growth regulator for enhancing the rate of seed germination in fenugreek. All the morphological features of germinating seeds are improved by using GA3 (Kundlik and Pratishthan, 2020). When GA3 was applied to fenugreek, the extent of shooting has been dramatically increased. It has also shown a positive result both on fresh and dry weight of shoot and root. The percentage of seed germination was higher in plants treated with GA3 than untreated plants. The number of biochemical constituents, level of proteins and carbohydrate content was also high in GA3 treated seeds (Kundlik and Pratishthan, 2020).

Gibberellic acid has been proven to affect various physiological attributes of fenugreek especially growth of root and shoot sections. When GA 3 is applied to fenugreek, it activates various metabolic pathways to promote plant height, leaf length, to increase leaf area and to produce enlarged cells. It has also been reported that GA3 also improve the chlorophyll and carotenoids content in fenugreek (Idrees, 2015). It is directly associated with seed production in fenugreek and positive aspects are its capacity to induce growth and developmental progresses in fenugreek plant after seed production (Idrees, 2015).

The roots and stems produced in fenugreek by applying GA3 will be enriched with higher content of alkaloids compared to untreated plants (Idrees, 2015). A research study demonstrated positive role of different PGRs in seed germination and out of multiple PGRs, Gibberellic acid was found to be most beneficial to give highest germination percentage in plants (Agha, 2022).

5.3.5 NAPHTHALENEACETIC ACID

Naphthaleneacetic acid is a synthetic class of Auxin hormones that is responsible for cell division, elongation, differentiation, apical dominance, vascular tissue development, root initiation, fruit abscission, flowering, leaf senescence, and so forth. Furthermore, it is also beneficial to boost plant height, number of branches, leave formation and seed quantity in fenugreek. Different studies highlighted that the use of NAA in fenugreek is directly responsible for improving growth parameters and increase total leaf number, leaf area, fresh and dry weight, and chlorophyll content in plants (Ghosh et al., 2015).

Naphthaleneacetic acid is significantly involved for the initiation of shoot regeneration and induced callus formation in fenugreek. It efficiently increases somatic embryogenesis. The fenugreek explants treated with NAA are healthier than those produced without NAA inoculation. Moreover, NAA treatment is linked with shortening the time of tissue culture, reduce the chances of infection to improve the growth period of fenugreek (Naser and Mheidi, 2021).

NAA has been reported to be beneficial for fenugreek explants as it not only induces callus formation, but also increase the size of callus.

Furthermore, when fenugreek is grown in MS nutrient media, the addition of NAA in that media highly affects the color and texture of formed callus. Generally, when fenugreek is supplemented with NAA in MS media, the time of callus initiation in both hypocotyl and cotyledon sections is less than those grown without the addition of NAA (Bairva et al., 2012). The use of NAA has huge impact on all the growth stages of fenugreek. The most important aspects affected by NAA are development, growth, and yield of fruit in fenugreek (Talab et al., 2014).

Along with this it has also been reported that when fenugreek explants are grown on MS media enriched with NAA along with 2,4-dicholoro phenoxy acetic acid, kinetin, and coconut water, it induced callus formation from explants. Furthermore, it was also associated with shoot regeneration from organogenesis as well (Talab et al., 2014). The apical meristem and cotyledon leafy explants of fenugreek are used to produce shoots especially on MS media. It also depends on the type of explants used for callus formation and for regeneration of shoot area. Each section required a different quantity of naphthalene acetic acid hormone for callus induction with a significant increased rate from hypocotyls and cotyledon explants (Talab et al., 2014).

5.3.6 INDOLE ACETIC ACID

Indole Acetic Acid is an important plant growth regulator that actively controls both growth and developmental stages of plants, especially in fenugreek. It plays a pivotal role in cell elongation, in decreasing wall pressure, enhance number of osmotic elements of cells, and synthesize proteins. It is known as a key type of auxin hormone that regulates multiple processes of growth in fenugreek particularly division, enlargement, differentiation of cells and plant response toward light and gravity. IAA is linked with the root length, root fresh and dry weight both on seedling and vegetative stages (Tufail et al., 2021).

In some cases, Indole acetic acid gives significant results when it is applied exogenously along with gibberellic acid and improves foliage growth. IAA showed highest accumulation of calcium and potassium ions in roots and shoots of fenugreek. In shoots of fenugreek a high rate of calcium, sodium, and potassium accretion was noticed than in roots (Tufail et al., 2021). During seedling stages of fenugreek, introduction of IAA enhanced the number of leaves and branches in targeted plants. It is not only controlling the formation of flower but also involved in the identification of floral organs (Jamil et al., 2021).

5.4 CONCLUSION

The fenugreek is commercially a very useful crop due to its extremely high advantages and usage in different products as a food supplement, as spice, as a legume or a break crop for cereals. Plant growth regulators are crucial not only for fenugreek growth and development but also regulate its primary and secondary metabolic pathways too. When PGRs are externally applied, their effect starts from seed germination leading to all stages of cell cycle, flower development, fruit production and even during the production of seeds in fenugreek.

The fenugreek treated with PGRs efficiently withstands the environmental changes and keep the final production up to an increased level. One very necessary aspect is the time of PGRs utilization, their amount and method of implementation. Naturally produced PGRs give more quantifiable advantages than synthetically available forms due to their ability of tackling growth issues in fenugreek because of biotic and abiotic stress factors. Due to their environmentally safe nature and immediate impact, PGRs are highly appreciated for improving fenugreek cultivation. Hence overall, plant growth regulators are beneficial for controlling all the intercellular mechanisms of fenugreek including from plant initial growth stage till death, but their activities are associated with the need of targeted fenugreek plants.

KEYWORDS

- **auxin**
- **cytokinin**
- **fenugreek**
- **gibberellic acid**
- **gibberellin**
- **indole acetic acid**
- **naphthaleneacetic acid**
- **plant growth regulators**

REFERENCES

Aasim, M., Hussain, N., Umer, E. M., Zubair, M., & Bilal, S. (2010). *In vitro* shoot regeneration of fenugreek (*Trigonella foenum-graecum* L.) using different cytokinins. *African Journal of Biotechnology, 9*(42), 7174–7179. https://doi.org/10.5897/AJB10.956.

Agha, H. M. (2022). The influence of different concentrations of plant hormones *in vitro* on seed germination of fenugreek. *Journal of Experimental Biology and Agricultural Sciences, 12*(2), 104–113. https://doi.org/10.55493/5003.v12i2.4528.

Ahmad Dar, T., Uddin, M., Khan, M. M. A., Ali, A., & Hashmi, N. (2015). Cumulative effect of gibberellic acid and phosphorus on crop productivity, biochemical activities, and trigonelline production in *Trigonella foenum-graecum* L. *Cogent Food & Agriculture, 1*(1), 995950. https://doi.org/10.1080/23311932.2014.995950.

Alam, M. S. (2018). Effect of gibberellic acid on morphology, vegetative growth, and seed yield of fenugreek (*Trigonella foenum-graecum* L.) [Doctoral dissertation, Sher-e-Bangla Agricultural University]. *Sher-e-Bangla Agricultural University*.

Alizadeh, S. A., Pazoki, A., & Habibi, H. (2013). Effect of ascorbate and gibberellin on some morphological traits and relative water content in fenugreek (*Trigonella foenum-graecum* L.) under different levels of salinity stress. *Journal of Agricultural Sciences, 1*(11), 1436–1451.

Almatroodi, S. A., Almatroudi, A., Alsahli, M. A., Rahmani, A. H., Almatroodi, S. A., Alsahli, A., & Husain, A. (2021). Fenugreek (*Trigonella foenum-graecum*) and its active compounds: A review of its effects on human health through modulating biological activities. *Antioxidants, 13*(3), 813–821.

Al-Mohammad, M. H. S., Al-Taey, D. K., & Jasim, A. (2019). Effect of phenylalanine and naphthaleneacetic acid on growth, yield, and antioxidant activity of fenugreek (*Trigonella foenum-graecum*). *IOP Conference Series: Earth and Environmental Science, 388*(1), 012073. https://doi.org/10.1088/1755-1315/388/1/012073.

Aslam, M., Ahmad, E., Hussain, K., Ayaz, M., Ullah, I., Hussain, A., & Khan, D. I. (2010). Effect of plant growth regulator (NAA) and available soil moisture depletions on yield and yield components of chickpea.

Ayaz, M., Ullah, I., & Hussain, A. (2010). Effect of plant growth regulator (NAA) and available soil moisture depletions on yield and yield components of chickpea. *Journal of Agricultural Research, 26*(3).

Bairva, M., Meena, S. S., & Mehta, R. S. (2012). Effect of bio-fertilizers and plant growth regulators on growth and yield of fenugreek (*Trigonella foenum-graecum* L.). *Journal of Agricultural Sciences, 2*(January), 28–33.

Battah, M. G., Mostfa, M. A., Eladel, H. M., Soror, A. S., & Tantawy, M. M. (2021). Physiological response of fenugreek (*Trigonella foenum-graecum* L.) plant treated by farmyard manure and two selected seaweeds as biofertilizers. *Plant Archives, 21*(2), 115–124.

Crouch, I. J., & van Staden, J. (1993). Evidence for the presence of plant growth regulators in commercial seaweed products. *Plant Growth Regulation, 21*(1), 21–29. https://doi.org/10.1007/BF00207588.

Dewi, K., & Pratiwi, R. (2016). The effects of physical and hormonal treatments on dormancy breaking and the changes in seed coat ultrastructure of *Delonix regia*. *Bioscience Journal, 8*(1), 94–102. https://doi.org/10.13057/nusbiosci/n080117.

Ghosh, B., Chandra, I., & Chatterjee, S. (2015). Fenugreek (*Trigonella foenum-graecum* L.) and its necessity [Review paper]. *International Journal of Agricultural Sciences*, June.

Idrees, M. (2015). Phosphorus on crop productivity, biochemical cumulative effect of gibberellic acid and phosphorus on crop productivity, biochemical activities and trigonelline production in *Trigonella foenum-graecum* L. *Cogent Food & Agriculture*, *2*(1).

Ishtiaq, H., Bhardwaj, S., Ashraf, A., & Kapoor, D. (2021). Versatile role of auxin and its crosstalk with other plant hormones to regulate plant growth and development. *Plant Archives, 21*(1), 1621–1627. https://doi.org/10.51470/plantarchives.2021.v21.no1.221.

Jagtap, D. K., Gaikwad, R. B., Thakare, U. G., & Jakhi, P. S. (2013). Effect of vermiwash and gibberellic acid on seed germination in fenugreek (*Trigonella foenum-graecum* L.). *International Journal of Biotechnology and Biosciences*, *3*(4), 230–234.

Jamil, M., Saher, A., Javed, S., Farooq, Q., & Shakir, M. (2021). A review on potential role of auxins in plants, current applications, and future directions. *International Journal of Botany*.

Jamwal, K. (2022). Evaluation of effect of selected plant growth regulators on morphological traits and seed yield of *Fagopyrum esculentum* Moench of Himalayan region. *Journal of Pharmacognosy and Phytochemistry*, *18*(2), 111–121.

Kadhim, S. H., & Almukhtar, S. A. (2023). Effect of explants type and plant growth regulators on the stimulation and multiplication of vegetative branches of *Trigonella foenum-graecum* L. *in vitro*. *IOP Conference Series: Earth and Environmental Science*, *1158*(4), 042031. https://doi.org/10.1088/1755-1315/1158/4/042031.

Kundlik, J. D., & Pratishthan, V. (2020). Effect of vermiwash and gibberellic acid on seed germination in fenugreek (*Trigonella foenum-graecum* L.). *Journal of Biotechnology*, *No. 2*, 2–7.

L, T. F. G., Elnour, M. E. M., Ali, A. M. A., & Saeed, B. E. A. T. (2015). Effect of different concentrations of auxins and combination with kinetin on callus initiation of *Trigonella foenum-graecum*. *International Journal of Plant and Soil Science*, *3*(2), 117–122.

Mehrafarin, A., Labbafi, M., Qavami, N., Qaderi, A., & Badi, H. N. (2014). Responses of fenugreek (*Trigonella foenum-graecum* L.) to exogenous application of plant growth regulators (PGRs). *Journal of Food and Agricultural Sciences*, *12*(2), 142–148.

Moons, A. (2005). Regulatory and functional interactions of plant growth regulators and plant glutathione S-transferases (GSTs). *ScienceDirect, 72*, 155–202.

Naser, W. Y., & Mheidi, O. H. (2021). Fenugreek performance affected by foliar application of gibberellin and arginine acids. *IOP Conference Series: Earth and Environmental Science, 904*, 012059. https://doi.org/10.1088/1755-1315/904/1/012059.

Olfa, B., Maha, Z., Nada, B., & Zeineb, O. A. (2018). Effects of NaCl on plant growth and antioxidant activities in fenugreek (*Trigonella foenum-graecum* L.). *Bioscience Journal*, *34*(3), 683–696. https://doi.org/10.14393/BJ-v34n3a2018-38069.

Patel, S. R., Abhishek, D., Patel, H. N., Patil, S. S., & Patel, S. G. (2021). Development of somaclones through callus culture in fenugreek (*Trigonella foenum-graecum*) variety Gujarat Methi-2. *Legume Research, 44*(6), 1–9. https://doi.org/10.18805/lr-4467.

Prajapati, S., Jamkar, T., Singh, O. P., Raypuriya, N., Mandloi, R., Jain, P. K., Nehru, J., & Vishwa, K. (2015). Plant growth regulators in vegetable production: An overview. *International Journal of Plant Research, 15*(2), 619–626.

Role, V., Auxin, O. F., Crosstalk, I. T. S., Other, W., Hormones, P., & Growth, R. P. (2021). Versatile role of auxin and its crosstalk with other plant hormones. *Plant Archives*, *21*(1), 1621–1627.

Sadak, M. S. (2019). Impact of silver nanoparticles on plant growth, some biochemical aspects, and yield of fenugreek plant (*Trigonella foenum-graecum*). *Bulletin of the National Research Centre, 43*(1). https://doi.org/10.1186/s42269-019-0077-y.

Singh, U., Chamoli, M., Singh, K. P., Ram, L., Jangir, S., & Maheshwari, R. K. (2022). *Trigonella foenum-graecum* Linn. *International Journal of Plant & Soil Science, 4*(2), 19–27.

Student, M. A. P. D., D, R. S. P., & D, N. B. H. P. (2011). A review on biology, cultivation, and biotechnology of fenugreek (*Trigonella foenum-graecum* L.) as a valuable medicinal plant. *African Journal of Biotechnology, 10*(37), 6–24.

Talab, D., Mehrafarin, A., Badi, H. N., & Khalighi, F. (2014). Changes in growth and trigonelline/mucilage production of fenugreek (*Trigonella foenum-graecum* L.) under plant growth regulators application. *Journal of Agricultural Research, 13*(51), 15–25.

Tavangar, A., Karami, L., Hedayat, M., & Abdi, G. (2021). Effect of salinity and drought stress on morphological and biochemical properties of two Iranian fenugreek (*Trigonella foenum-graecum*) populations. *Notulae Botanicae Horti Agrobotanici Cluj-Napoca, 49*(2), 1–12. https://doi.org/10.15835/nbha49212038.

Tufail, M., Hussain, K., Iqbal, I., & Nawaz, K. (2020). Efficacy of IAA, GA_3, and riboflavin for morpho-biochemical and yield attributes of fenugreek (*Trigonella foenum-graecum* L.) in Pakistan. *Journal of Plant Nutrition, 43*(6), 780–787.

Tufail, M., Hussain, K., Saeed, Z., & Nawaz, K. (2021). Role of IAA, GA_3, and riboflavin for crop improvement in fenugreek (*Trigonella foenum-graecum* L.). *Iranian Journal of Plant Physiology, 11*(5), 3899–3910.

Valizadeh, M., Asia, S., & Asia, W. (2018). *In vitro* regeneration in medicinal plant fenugreek (*Trigonella foenum-graecum* L.). *International Journal of Plant Biology, 8*(2), 43–51.

Xiong, Y., & Jiao, Y. (2019). The diverse roles of cytokinins in regulating leaf development. *Plant, 8*(7), 243.

CHAPTER 6

Symbiotic Rhizobia, Nitrogen Fixation, and Fenugreek

MUHAMMAD DANISH TOOR,[1,2] LYUBKA KOLEVA,[3] MUHAMMAD MUGHEES UD DIN,[4] ABDURRAHMAN AY,[1] and SHER MUHAMMAD SHAHZAD[5]

[1]*Department of Soil Science and Plant Nutrition, Faculty of Agriculture, Ondokuz Mayis University, Samsun, Türkiye*

[2]*Department of Agrochemistry and Soil Science, Agricultural University, Plovdiv, Bulgaria*

[3]*Agricultural University, Department of Plant Physiology, Biochemistry, and Genetics, Plovdiv, Bulgaria*

[4]*Institute of Soil and Environmental Sciences, University of Agriculture, Faisalabad, Pakistan*

[5]*Department of Soil and Environmental Sciences, College of Agriculture, University of Sargodha, Pakistan*

ABSTRACT

Leguminous fenugreek (*Trigonella foenum-graecum* L.) has enormous economic and medicinal value. It is widely cultivated in South Asia, North Africa, and the Mediterranean region for its seeds, leaves, and roots used in cooking, herbal medicine, and cosmetics. Fenugreek grows well in soils deficient in nitrogen, making it a prime candidate for research into the potential advantages of rhizobia inoculation. The conversion of atmospheric nitrogen into forms that plants can use can significantly increase plant yield and growth. The symbiotic relationship between fenugreek and

Fenugreek: Improvement in Yield Potential Through Plant Growth Regulators (PGRs) and Depolymerized Polysaccharides. Tariq Ahmad Dar, Bilal Ahmad, Moinuddin, and Khalid Rehman Hakeem (Eds.)

rhizobia carries out this process. Rhizobia is a group of soil bacteria that can fix atmospheric nitrogen and transform it into a usable form for the plant. Rhizobia can thrive and perform nitrogen fixation in a specialized environment that legume plants offer in the form of root nodules. Significant effects on sustainable agriculture and food security result from the rise in soil fertility and plant productivity. Also, rhizobia inoculants can further improve fenugreek plant growth, yield, and nitrogen fixation, according to recent studies. The most suitable inoculant strain must be chosen, and the inoculation procedure must be optimized, for the inoculation to be successful.

For symbiosis to develop, rhizobia and leguminous plants must be compatible with one another. Flavonoid-like, specific signals from the plant draw rhizobia to the root system. Rhizobia produces nodulation factors in response to these signals, which encourage the development of root nodules. Rhizobia undergoes morphological and physiological changes once they are inside the nodules, turning into nitrogen-fixing bacteroids. The bacteroids then convert atmospheric nitrogen into ammonia. The plant can use the ammonia the bacteroids produce to grow and synthesize proteins. Establishing rhizobial inoculation in fenugreek depends heavily on the inoculant strains used. The capacity of an inoculant strain to outcompete local rhizobia in the soil and its compatibility with the plant determine its efficacy. Rhizobia strains that can improve nitrogen fixation and plant growth in fenugreek have been identified and characterized. The ability of these strains to boost plant biomass, seed yield, and nitrogen content has been demonstrated.

The effectiveness of rhizobial inoculation in fenugreek is also greatly influenced by the timing and technique of the inoculation. The sowing or transplanting period is the ideal time to inoculate fenugreek. The inoculation method can change depending on the type of inoculant and the planting technique. The three methods of vaccination used presently are seed coating, soil inoculation, and foliar spray. The cost, effectiveness, and efficiency of the method are a few variables influencing which inoculation method is most suitable. Environmental factors like soil pH, temperature, moisture, and salinity can also influence the effectiveness of rhizobial inoculation in fenugreek. Moisture and temperature significantly impact the growth and survival of rhizobia. Low temperatures and high moisture can increase the risk of disease, while high temperatures and low moisture can reduce the survival of rhizobia.

6.1 INTRODUCTION

Nitrogen is nutrient plants require to synthesize proteins, nucleic acids, and other important molecules. It is abundant in the atmosphere but in an unusable form for plants. Rhizobia are soil bacteria that have a symbiotic relationship with legume plants. The bacteria infect the plant's roots and cause nodules to form, which convert molecular nitrogen into a form the plant can use for growth, known as nitrogen fixation. This symbiotic relationship benefits both the plant and the bacteria, as the plant receives valuable nitrogen from the bacteria, and the bacteria receives energy and safety from the plant (Vessey, 2003). Rhizobia are highly specific to their host plants, with different rhizobia species infecting various legume plants. The rhizobial infection and nodule formation process is controlled by a complex signaling pathway involving the exchange of chemical signals and the activation of specific genes in both plants and rhizobia (Glick, 2012).

Fenugreek (*Trigonella foenum-graecum* L.) is a legume plant shown to benefit greatly from rhizobial inoculation. Studies have demonstrated that the use of specific rhizobial strains can significantly increase the growth and yield of fenugreek, as well as improve its nutrient content. In addition, using rhizobial inoculants can also reduce the use of synthetic fertilizers, which positively impact the environment (Sahu et al., 2019).

The nitrogen fixation process by rhizobia involves converting atmospheric nitrogen into ammonia, which is then assimilated into organic compounds by the plant. This process is carried out by a complex set of enzymes known as nitrogenase, which is highly sensitive to oxygen. As a result, the nodules formed by rhizobia have a unique structure that provides a low-oxygen environment for nitrogenase activity (Saleem et al., 2007). Overall, the symbiotic relationship between rhizobia and legume plants is a highly efficient and environmentally friendly means of nitrogen fixation. The use of specific rhizobial strains can greatly benefit nitrogen fixation. Further research into rhizobial infection and nitrogen fixation mechanisms could improve crop yields and more sustainable agricultural practices. Nitrogen-fixing crops can improve soil fertility and reduce soil erosion, making them a vital component of sustainable agriculture (Giller, 2001).

The ability of fenugreek to fix nitrogen makes it a valuable crop for farmers because it reduces the need for synthetic nitrogen fertilizers and improves soil health. Several studies have been conducted to determine the effectiveness of fenugreek as a nitrogen-fixing plant. According to one study, fenugreek plants fix nitrogen at a high rate, with up to 75% of the plant's nitrogen coming from atmospheric nitrogen (Kurdali et al., 2001). Another study discovered that fenugreek plants improved soil fertility by increasing soil nitrogen, phosphorus, and organic matter content (Abdul-Baki et al., 2001). Furthermore, even in the presence of nitrogen fertilizer, fenugreek had a high nitrogen fixation potential, implying that it could be a useful crop for reducing reliance on synthetic fertilizers (Hussain et al., 2018). Saini et al. (2019) investigate the nitrogen-fixing potential of rhizobacteria associated with fenugreek. The researchers discovered that several strains of bacteria could fix nitrogen and promote plant growth. Mandal and Datta (2020) studied the beneficial effects of fenugreek on soil fertility and crop growth in a maize-wheat cropping system. They found that fenugreek improved soil nitrogen content and increased maize yield.

6.1.1 RHIZOBIA-FENUGREEK SYMBIOSIS

Several studies on the Rhizobia-Fenugreek Symbiosis have been conducted, highlighting the importance of this relationship in improving the growth and yield of fenugreek plants. Inoculation with rhizobia, for example, improved fenugreek plants' growth, yield, and nutrient uptake under different soil conditions, according to Singh et al. (2020). The authors proposed that using rhizobia inoculants to boost fenugreek crop productivity could be a cost-effective and environmentally friendly approach.

Berraho et al. (2019) investigated the diversity and effectiveness of rhizobia strains isolated from fenugreek nodules in Morocco. They discovered that the rhizobia were highly effective in fixing nitrogen and enhancing fenugreek plant growth. They also discovered that the diversity of rhizobia strains varied depending on where the fenugreek plants were grown. Furthermore, Haddad et al. (2017) discovered that rhizobia inoculation increased fenugreek seeds' protein content and amino acid profile, which could have health benefits.

6.1.2 MECHANISMS OF RHIZOBIA-FENUGREEK SYMBIOSIS

The mechanisms of rhizobia-fenugreek symbiosis have been extensively studied, with several key mechanisms identified. The recognition and exchange of signals between the plant and the bacteria is one of the main mechanisms of rhizobia-fenugreek symbiosis. Flavonoids are released into the soil by the plant, which attracts rhizobia. Rhizobia respond by releasing nod factors, which are specific signal molecules that cause the formation of nodules on the plant's roots. This signal recognition and exchange is critical for establishing the two organisms' symbiotic relationship (Oldroyd, 2013).

The regulation of gene expression in both the plant and the bacteria is another important mechanism of rhizobia-fenugreek symbiosis. The development of nodules on plant roots necessitates the activation of specific genes in both the plant and the rhizobia. The plant, for example, activates genes involved in the production of leghemoglobin, which is required for oxygen transport and storage in nodules. Meanwhile, the rhizobia activate genes that produce nitrogenase, which is required to convert atmospheric nitrogen into a usable form for plants (Sprent, 2009). Additionally, the mechanisms of nutrient exchange between the plant and the rhizobia have been investigated. The plant provides carbohydrates to the rhizobia, which are necessary for the growth and survival of the bacteria within the nodules. In exchange, the rhizobia provide the plant with usable nitrogen, which is necessary for the plant's growth and development (Glick, 2014).

6.1.3 FACTORS AFFECTING RHIZOBIA-FENUGREEK SYMBIOSIS

Rhizobia-fenugreek symbiosis is influenced by several factors that can affect the plant's growth, nodulation, and nitrogen fixation. Understanding these factors is important to optimize the productivity of fenugreek crops and promote sustainable agriculture. In this section, we will discuss some of the key factors affecting rhizobia-fenugreek symbiosis (Table 6.1).

TABLE 6.1 Impact of Environmental Factors on Rhizobia–Fenugreek Symbiosis

Factor	Optimal Range	Effect on Rhizobia-Fenugreek Symbiosis	References
Soil pH	6.0–7.5	Enhances nodulation and nitrogen fixation	Kim et al. (2018)
Soil moisture	Moderate level	Maximizes productivity of fenugreek crops	Soliman et al. (2021)
Temperature	20°C–30°C	Promotes nodulation and nitrogen fixation	Singh et al. (2019)
Nutrient availability	Adequate phosphorus and potassium levels	Improves nodulation and nitrogen fixation	Saha et al. (2018)

Furthermore, Ahmed et al. (2022) investigated the effect of salinity stress on rhizobia-fenugreek symbiosis, focusing on nodulation, nitrogen fixation, and plant growth. The study discovered that salinity stress harmed the symbiotic relationship between rhizobia and fenugreek, resulting in decreased nodulation, impaired nitrogen fixation, and slowed plant growth. The findings emphasized the importance of managing salinity stress in saline-affected agricultural areas to ensure successful rhizobia-fenugreek symbiosis (Ahmed et al., 2022). Das et al. (2021) investigated the effect of heavy metal contamination on fenugreek nodulation, nitrogen fixation, and growth. The study found that heavy metal exposure harmed the symbiotic relationship between rhizobia and fenugreek, resulting in reduced nodulation, impaired nitrogen fixation, and compromised plant growth. The findings highlighted the potential risks of heavy metal contamination in the rhizobia-fenugreek symbiosis and emphasized the importance of remediation strategies to mitigate the negative effects (Das et al., 2021).

Sharma et al. (2020) investigated the effect of organic amendments on rhizobia-fenugreek symbiosis, focusing on nodulation, nitrogen fixation, and soil fertility. According to the findings, adding organic amendments influenced the symbiotic relationship between rhizobia and fenugreek. It increased nodulation, improved nitrogen fixation, and improved soil fertility. The findings highlighted the importance of organic amendments in improving soil health and optimizing rhizobia-fenugreek symbiosis (Sharma et al., 2020). Verma et al. (2020) investigated the role of mycorrhizal symbiosis in enhancing fenugreek nodulation and nitrogen fixation. The study found that mycorrhizal association influenced the rhizobia-fenugreek symbiosis, leading to increased nodulation and nitrogen fixation. The findings highlighted the potential of mycorrhizal symbiosis as a useful strategy for improving fenugreek crop performance (Verma et al., 2020).

6.1.4 BENEFITS OF RHIZOBIA-FENUGREEK SYMBIOSIS

The rhizobia-fenugreek symbiosis is a beneficial relationship between the leguminous plant fenugreek (*Trigonella foenum-graecum* L.) and rhizobia bacteria. This symbiotic relationship improves fenugreek plants' growth, yield, and nutritional quality and has several implications for sustainable agriculture. Increased nitrogen availability for the crop is one of the primary advantages of Rhizobia-Fenugreek symbiosis. This increases the plant's nitrogen content, improving the productivity of fenugreek crops and the soil used for the next crop. Several studies have found that using rhizobia inoculants can improve fenugreek plant growth and yield under various soil conditions (Berraho et al., 2019; Singh et al., 2020).

Another advantage of Rhizobia-Fenugreek symbiosis is Rhizobia's ability to improve the nutritional quality of fenugreek plants. Fenugreek is prized for its high protein and mineral content, and rhizobia inoculants have been shown to improve fenugreek seeds' protein and amino acid profile (Haddad et al., 2017). This could have health benefits because fenugreek is traditionally used for its medicinal properties. Furthermore, Rhizobia-Fenugreek symbiosis can contribute to sustainable agriculture by reducing the need for chemical fertilizers. Using rhizobia inoculants can reduce the amount of nitrogen fertilizer required, which can have economic and environmental benefits such as reducing soil erosion and salination (Mishra et al., 2021).

6.2 NITROGEN FIXATION IN FENUGREEK

Nitrogen fixation is a critical process in which atmospheric nitrogen is converted into organic compounds that plants can use. Fenugreek (*Trigonella foenum-graecum* L.) is an important leguminous crop with a high nitrogen fixation capacity. Several studies have been conducted to understand the nitrogen fixation process in fenugreek better. Prasad and Rao (1994) discovered that fenugreek plants convert atmospheric nitrogen at 28 to 40 kg N per hectare per year. Singh et al. (2003) discovered that fenugreek has the potential to fix atmospheric nitrogen at a rate of up to 120 kg N per hectare per year in another study (Table 6.2).

TABLE 6.2 Factors Influencing Nitrogen Fixation in Rhizobia–Fenugreek Symbiosis

Factors	Study	Findings
Abiotic factors	Gupta et al. (2013)	Investigated the influence of abiotic factors on nitrogen fixation in rhizobia-fenugreek symbiosis.
Soil pH	Reddy et al. (2004)	Higher nitrogen fixation rates were observed in soils with pH 7.5 compared to pH 6.5.
Moisture, pH, and temperature	Bekele et al. (2012)	Studied the effects of moisture, pH, and temperature on the growth and survival of rhizobia strains.
Salinity	Pathak et al. (2016)	Examined the impact of salinity on growth, nodulation, and enzyme activity of *Rhizobium* strains.
Temperature and soil moisture	Singh et al. (2013)	Investigated the effects of temperature and soil moisture on rhizobial populations.
Plant beneficial microorganisms	Parnell et al. (2016)	Provided an industrial perspective on the role of plant-beneficial microorganisms, including rhizobia.
Temperature, pH, and salt	Gupta et al. (2012)	Explored the influence of temperature, pH, and salt on growth, nodulation, and nitrogen fixation.

6.2.1 OVERVIEW OF THE PROCESS OF NITROGEN FIXATION

Nitrogen fixation is an important biological process that converts atmospheric nitrogen into a form that plants and other organisms can use. Nitrogen-fixing bacteria carry out this process by breaking the strong triple bond of atmospheric nitrogen and converting it to ammonia. Nitrogen fixation is critical for many organisms' growth and survival because it provides the nitrogen required to synthesize amino acids, nucleotides, and other essential biomolecules (Dixon and Kahn, 2004). It is a complicated biochemical pathway involving numerous enzymes and cofactors. The enzyme nitrogenase performs the first step in the process: converting atmospheric nitrogen to ammonia. Nitrogenase is a complex enzyme that requires the presence of several cofactors, including ATP and electron carriers (Oldroyd and Dixon, 2014). It is composed of two subunits, Fe-protein and MoFe-protein. Because nitrogen fixation is an energy-intensive process, nitrogen-fixing bacteria have evolved several strategies to optimize the process, including symbiotic associations with plants and the production of specialized structures known as heterocysts (Rubio and Ludden, 2008).

The nitrogen-fixing process has been extensively researched in free-living and symbiotic nitrogen-fixing bacteria. Recent research has focused on elucidating the genetic and biochemical mechanisms underlying nitrogen fixation and identifying new nitrogen-fixing bacterial strains for use in agriculture and biotechnology (Mus et al., 2021). Nitrogen fixation is an important biological process with major implications for agriculture, biotechnology, and environmental sustainability. Understanding the mechanisms underlying nitrogen fixation is critical for developing new strategies for increasing nitrogen utilization efficiency in agriculture and reducing the environmental impact of nitrogen fertilizer use.

6.2.2 FACTORS AFFECTING NITROGEN FIXATION IN FENUGREEK

Nitrogen fixation, in general, is a complex process influenced by various factors. We investigate the factors that influence nitrogen fixation in fenugreek and summarize the existing literature in this review. The availability of soil nutrients is one of the most important factors influencing nitrogen fixation in fenugreek. Nitrogen is a necessary nutrient for plant growth and development, and its availability in the soil significantly impacts nitrogen fixation. Several studies have found that nitrogen fertilizers applied to the soil can reduce nitrogen fixation in fenugreek (Ali et al., 2018; Sarwar et al., 2013).

Nitrogen fixation is also affected by the type and source of fenugreek inoculum used. Rhizobium strains are the most common inoculants used in legume crops, including fenugreek. Using effective *Rhizobium* strains has been shown in studies to significantly increase nitrogen fixation in fenugreek (Gupta et al., 2017; Singh et al., 2015). Furthermore, fenugreek's genetic makeup influences nitrogen fixation. Several studies have shown that selecting nitrogen-fixing fenugreek cultivars with high yields can significantly increase nitrogen fixation (Ghosh et al., 2014; Singh et al., 2015). All factors must be managed properly to maximize nitrogen fixation in fenugreek and achieve sustainable crop production.

6.2.3 SOIL CONDITIONS AFFECTING RHIZOBIA–FENUGREEK SYMBIOSIS

Rhizobia are soil bacteria that form mutualistic relationships with leguminous plants, providing nitrogen in exchange for carbohydrates. However, soil conditions can influence this symbiosis, which can either promote or inhibit rhizobia growth and survival in the rhizosphere. This literature review aims to provide an overview of the soil conditions that affect rhizobia-fenugreek symbiosis, as well as the mechanisms involved. Soil pH is one of the most important factors influencing rhizobia-fenugreek symbiosis. A slightly acidic to neutral soil pH of 6.5 to 7.0 is ideal for rhizobia and fenugreek growth and survival (Daryanto et al., 2021). Conversely, excess acidic or alkaline soil conditions can harm symbiosis by inhibiting rhizobia growth and survival or altering their metabolic activity. According to Kumar et al. (2019), soil pH below 5.5 significantly reduces the number of nodules formed in fenugreek roots, while soil pH above 8.0 forms small and ineffective nodules.

The texture and structure of the soil are also important in the rhizobia-fenugreek symbiosis. A well-aerated soil with a high organic matter content and good water-holding capacity benefits rhizobia growth and survival, and the development of fenugreek roots (Daryanto et al., 2021). Compacted soils with poor water infiltration and drainage, on the other hand, can inhibit root development and nodule formation, resulting in lower nitrogen fixation and crop productivity (Kumar et al., 2019). Another important factor influencing rhizobia-fenugreek symbiosis is soil fertility. Nitrogen is the most important nutrient for rhizobia growth and the formation of nodules in fenugreek roots.

Soils with low nitrogen availability may limit rhizobia growth and survival, resulting in reduced nodule formation and nitrogen fixation (Mugwe et al., 2019). Excess nitrogen fertilization, conversely, can harm symbiosis by reducing the need for nitrogen fixation and altering the carbon: nitrogen balance in the rhizosphere (Daryanto et al., 2021).

Salinity, heavy metal contamination, pathogens, and other microorganisms are other soil factors that may influence rhizobia-fenugreek symbiosis. High salinity can inhibit rhizobia and fenugreek plant growth and survival, reducing nitrogen fixation and crop productivity (Mugwe et al., 2019). Heavy metal contamination can also be toxic to rhizobia and fenugreek, reducing nitrogen fixation and inhibiting growth and survival (Daryanto et al., 2021). Pathogens and other microorganisms may compete with rhizobia for resources in the rhizosphere, reducing their population and ability to nodulate and fix nitrogen (Mugwe et al., 2019).

6.2.4 RHIZOBIA STRAIN SPECIFICITY IN NITROGEN FIXATION

Vincent and Stowers (1954) conducted one of the earliest studies on rhizobia strain specificity, discovering that different strains of *Rhizobium trifolii* D. had varying degrees of effectiveness in fixing nitrogen for different strains of clover. Numerous studies have since been conducted to investigate the factors contributing to this specificity. Host plant genetics is one of the most important factors influencing rhizobia strain specificity. For example, Sprent et al. (1983) discovered that different cowpea genotypes had varying degrees of specificity in their nodulation response to different rhizobia strains. Similarly, Huygens et al. (2004) discovered different genotypes of the tropical legume *Stylosanthes guianensis* Aubl., which were more or less effective at nitrogen fixation with different rhizobia strains.

Rhizobia genetics also contributes to strain specificity. The genome sequence of *Rhizobium leguminosarum bv. viciae* 3841, for example, revealed that this strain has many genes involved in the production of exopolysaccharides, which are important in establishing symbiosis with host plants (Young et al., 2006). Other research has discovered that certain rhizobia strains have specific genetic adaptations that allow them to colonize specific host plants more effectively (Grönemeyer et al., 2014).

Researchers have developed various methods for selecting and breeding rhizobia strains that are more effective at fixing nitrogen for specific legume hosts and investigating the factors that contribute to rhizobia strain specificity.

For example, Kiers et al. (2003) used a model system to demonstrate how host plant selection can drive the evolution of rhizobia strains that are more effective at fixing nitrogen for that host. On the other hand, Sanginga et al. (2000) used a combination of laboratory and field experiments to select rhizobia strains that were more effective at fixing nitrogen for soybean under specific environmental conditions.

6.2.5 EFFECTS OF PLANT GENOTYPE ON NITROGEN FIXATION

Plant genotype is important in nitrogen fixation. Several studies have shown that the ability of different plant genotypes to form symbiotic relationships with nitrogen-fixing bacteria varies. For example, Nandasena et al. (2007) discovered that some legume genotypes, such as soybeans and common beans, had a higher ability to fix nitrogen than others. This variation was attributed to differences in the plant's ability to excrete flavonoids, the compounds that attract nitrogen-fixing bacteria to the plant roots.

Furthermore, research has shown that the effectiveness of nitrogen fixation can be affected by the strain of bacteria used. For example, Rodríguez-Navarro et al. (2014) investigated the effects of different *Rhizobium* strains on the nitrogen-fixing ability of common bean genotypes. The results revealed that different *Rhizobium* strains had varying nitrogen fixation abilities, with some strains being more efficient than others. Plant genotypes can also influence nitrogen fixation efficiency by influencing the morphology and physiology of the plant roots. For example, a study by Li et al. (2017) found that different soybean genotypes had different root features, such as root length and diameter, which influenced the amount of nitrogen fixed. The study found that soybean genotypes with longer and thicker roots had a higher ability to fix nitrogen than those with shorter and thinner roots.

Furthermore, plant genotype can influence the rhizosphere microbial community, which includes nitrogen-fixing bacteria. Saia et al. (2015) investigated the effects of different wheat genotypes on the rhizosphere microbial community. The study discovered that the microbial community in the rhizosphere differed significantly between wheat genotypes, implying that plant genotype can influence the composition of the microbial community and, thus, nitrogen fixation. The effects of plant genotype on nitrogen fixation are complex and multifaceted in general. Nitrogen fixation efficiency can depend on various factors, including the type of plant genotype, the strain of bacteria, root morphology and physiology, and the composition of the microbial community in the rhizosphere. Understanding these mechanisms is essential for developing sustainable agricultural practices that rely on nitrogen fixation as a source of plant nutrition.

6.3 APPLICATIONS OF RHIZOBIA–FENUGREEK SYMBIOSIS

1. **Improving Soil Fertility:** Rhizobia-fenugreek symbiosis can improve soil fertility by increasing nitrogen availability to plants. Nitrogen is required for plant growth, and leguminous plants can fix atmospheric nitrogen via their symbiotic relationship with rhizobia. This process benefits the host plant and increases nitrogen availability in the soil, which can benefit other crops grown in the same area. Several studies have shown that inoculating soil with rhizobia can improve soil fertility. Ramakrishnan et al. (2019), for example, discovered that inoculating fenugreek seeds with rhizobia increased crop yield by 33.3%. Similarly, Singh et al. (2016) discovered that rhizobia inoculation increased fenugreek yield by 45.6% compared to the control group.
2. **Increasing Crop Yields:** Rhizobia-fenugreek symbiosis has also been shown to increase fenugreek crop yields. Nitrogen fixation can increase nitrogen availability in the soil, resulting in improved plant growth and development and higher crop yields. Several studies have shown that rhizobia inoculation improves fenugreek crop yields. For example, El-Awadi et al. (2019) discovered that rhizobia inoculation increased fenugreek yield by 25.8% compared to the control group. Similarly, Sarwar et al. (2014) discovered that rhizobia inoculation increased fenugreek yield by 37% compared to the control group.
3. **Producing Bioactive Compounds:** Fenugreek has medicinal properties and is used in traditional medicine to treat various ailments. The plant contains several bioactive compounds responsible for its medicinal properties, including alkaloids, flavonoids, and saponins. The symbiotic relationship between rhizobia and fenugreek has been shown to increase the production of these bioactive compounds, making fenugreek a valuable resource for the pharmaceutical industry. Several studies have shown that rhizobia inoculation increases the production of bioactive compounds in fenugreek. For example, El-Mawla et al. (2019) discovered that rhizobia inoculation

increased flavonoid production in fenugreek by 32% compared to the control group. Similarly, a study by Salama et al. (2016) found that rhizobia inoculation increased the production of saponins in fenugreek by 43% compared to the control group.

4. **Stress Tolerance:** Rhizobia inoculation has been shown to improve fenugreek plant tolerance to various environmental stresses, including drought, salinity, and heavy metal toxicity (El-Awadi et al., 2019; Ramakrishnan et al., 2019). This is thought to be due to rhizobia's ability to produce plant growth-promoting substances and increase nutrient availability in the soil.

5. **Soil Improvement:** Fenugreek is a good cover crop that can improve soil's physical and chemical properties. It can also improve soil biological properties by increasing microbial activity and nitrogen accumulation when inoculated with rhizobia (Sarwar et al., 2014).

6. **Food and Medicinal Applications:** Fenugreek is frequently employed in traditional medicine for various health benefits, including its ability to reduce blood sugar and cholesterol levels. Inoculating fenugreek seeds with Rhizobia have increased the concentration of these beneficial compounds, making it a more valuable crop for food and medicinal applications (Salama et al., 2016).

7. **Sustainable Agriculture:** Rhizobia inoculation is a low-cost, long-term method of increasing crop yields while reducing the need for chemical fertilizers. It also benefits the environment by decreasing greenhouse gas emissions and improving soil health (El-Mawla et al., 2019).

8. **Crop Rotation:** Fenugreek is a legume used in crop rotation to increase soil fertility and reduce soil-borne diseases. It can fix more atmospheric nitrogen when inoculated with rhizobia, which can be used by subsequent crops in the rotation (Singh et al., 2016).

6.3.1 BENEFITS OF USING RHIZOBIA TO ENHANCE NITROGEN FIXATION IN FENUGREEK ALONG WITH CHALLENGES AND LIMITATIONS

Fenugreek has been found to benefit greatly from the presence of rhizobia. This section will discuss the benefits of using rhizobia to enhance nitrogen fixation in fenugreek. Also, challenges and limitations of using rhizobia-fenugreek symbiosis in agriculture. Studies have shown that inoculating fenugreek seeds with rhizobia can increase nitrogen fixation by up to 50% (Ali et al., 2015). One of the primary advantages of using rhizobia to improve nitrogen fixation in fenugreek is that it can reduce the need for synthetic nitrogen fertilizers, which can harm the soil on multiple levels in the long run. Synthetic nitrogen fertilizers are widely used in agriculture to increase crop yields but can have negative environmental consequences. They can easily leach into groundwater, pollute surface water, and contribute to greenhouse gas emissions under certain conditions. Farmers can reduce their reliance on synthetic nitrogen fertilizers by using rhizobia to boost nitrogen fixation in fenugreek, which can benefit both the environment and their bottom line.

Using rhizobia to improve nitrogen fixation in fenugreek can increase crop yields while reducing the need for synthetic nitrogen fertilizers. Inoculating fenugreek seeds with rhizobia has been shown in studies to increase plant growth, seed yield, and seed protein content (Ali et al., 2015). This is due to the increased nitrogen fixation provided by the rhizobia, which allows the plant to synthesize more amino acids and other nitrogen-containing compounds, both of which are required for plant growth and development.

Using rhizobia to improve nitrogen fixation in fenugreek can also improve soil health. When legume plants use rhizobia to fix atmospheric nitrogen, they also release organic compounds into the soil that can improve soil structure and fertility. This can benefit the soil's overall health and the growth and development of other plants in the crop rotation system (Tables 6.3 and 6.4).

TABLE 6.3 Benefits of Rhizobia Inoculation in Agriculture

Sl. No.	Effects	References
1.	Increased drought tolerance	Sgolastra et al. (2019)
2.	Reduced greenhouse gas emissions	Herridge et al. (2008)
3.	Improved soil structure	Gopalakrishnan et al. (2015)
4.	Cost-effective	Sarathambal et al. (2015)
5.	Enhanced plant growth and yield	Shukla et al. (2021)
6.	Increased nitrogen use efficiency	Kumar et al. (2021)
7.	Improved soil health	Rai et al. (2020)
8.	Reduced environmental pollution	Chen et al. (2020)
9.	Improved nutritional quality	Rani et al. (2021)

TABLE 6.4 Challenges and Limitations of Using Rhizobia–Fenugreek Symbiosis in Agriculture

Factors	Description	References
1. Compatibility	Different rhizobia strains interact differently with fenugreek, affecting their ability to colonize the roots and fix nitrogen. The right strain must be chosen for a successful symbiosis.	Bora and Singh (2017)
2. Environmental factors	The rhizobia-fenugreek symbiosis is influenced by soil pH, temperature, moisture, and nutrient availability. Soil pH should be neutral to slightly alkaline for nitrogen fixation. High temperatures and moisture stress can reduce the effectiveness of rhizobia. Nitrogen fixation can be hampered by other nutrients such as phosphorus.	Khan et al. (2018)
3. Competition with other microorganisms	Rhizobia competes with soil fungi and bacteria, preventing them from colonizing fenugreek roots and reducing nitrogen fixation. This competition reduces plant growth and yield.	Meghvansi et al. (2016)
4. Variability in symbiosis effectiveness	The efficacy of rhizobia-fenugreek symbiosis varies according to rhizobia strain and fenugreek cultivar. Some strains are more effective at nitrogen fixation than others, and some cultivars form better symbiotic relationships than others. The proper strains and cultivars must be chosen for optimal symbiosis.	Mishra and Singh (2017)
5. Seed treatment	Coating fenugreek seeds with suitable rhizobia strains before sowing is crucial for successful symbiosis. The process can be challenging in large-scale agriculture, requiring specialized equipment and expertise. Compatibility of the coating material with the seed and its impact on germination and growth must be considered.	Soltani, Rahmani, and Bahramnejad (2019)
6. Sustainability	It is critical to ensure the long-term viability of the rhizobia-fenugreek symbiosis. Repeated use of the same strain or cultivar can result in the development of resistant strains, as well as pests and diseases that have a negative impact on growth and yield.	Vyas and Gulati (2017)
7. Inoculation method	The method used determines the efficacy of rhizobia inoculation. Methods such as seed coating, soaking, and root dipping produce varying results and may not be appropriate for all legumes.	Soltani et al. (2019)

TABLE 6.4 *(Continued)*

Factors	Description	References
8. Soil conditions	A suitable soil environment, including adequate moisture, temperature, and pH levels, is required for successful rhizobia-fenugreek symbiosis. Unfavorable conditions, such as drought, high temperatures, or soil acidity, can impede the symbiosis's establishment and survival.	Bora and Singh (2017)
9. Nutrient availability	The availability of nutrients, particularly phosphorus, is critical for the rhizobia-fenugreek symbiosis. Low phosphorus levels in the soil can limit the symbiosis's effectiveness.	Meghvansi et al. (2016)
10. Genotypic variation	The efficacy of the rhizobia-fenugreek symbiosis can vary depending on the genotypic characteristics of both the rhizobia and the legume. This variability has an impact on the establishment and sustainability of the symbiosis in various environments.	Soltani et al. (2019)
11. Limited rhizobia strains	In comparison to other legumes, the availability of effective rhizobia strains for fenugreek is limited, potentially affecting the success of the symbiosis.	Khan et al. (2018)

6.4 CONCLUSION

In conclusion, fenugreek is an adaptive and highly valued crop with several uses in various industries. The symbiotic relationship of fenugreek and rhizobia is critical to plant growth and yield. The ability of Rhizobia to fix atmospheric nitrogen and transform it into usable for plants form significantly increases the productivity of fenugreek. The success of rhizobia inoculation in fenugreek, on the other hand, depends on selecting the best inoculant strain and optimizing the inoculation process. Furthermore, environmental factors such as soil pH, temperature, moisture, and salinity can influence the efficacy of rhizobia inoculation. Despite these challenges, rhizobia inoculants can greatly enhance soil fertility and plant productivity, ultimately contributing to sustainable agriculture and food security. Identifying and characterizing rhizobia strains that improve nitrogen fixation and plant growth in fenugreek is a crucial area of research that can pave the way for further advancements in the field. The potential benefits of rhizobia inoculation in fenugreek make it a promising area for continued exploration and development.

KEYWORDS

- **fenugreek**
- **inoculant strains**
- **nitrogen fixation**
- **nodulation factors**
- **rhizobia**
- **sustainable agriculture**
- **symbiosis**

REFERENCES

Abdul-Baki, A. A., Al-Qadi, F., & Teasdale, J. R. (2001). Sustainable production of tomatoes using organic and conventional soil management in a soil-solar greenhouse in the Mediterranean environment. *Journal of Sustainable Agriculture*, *17*(4), 5–22.

Ahmed, S., Hasanuzzaman, M., & Bhuyan, M. H. M. B. (2022). Impact of salinity stress on rhizobia-fenugreek symbiosis: Nodulation, nitrogen fixation, and plant growth. *Plants*, *11*(2), 338.

Ali, M., Hassan, F., & Mirza, M. S. (2015). Impact of *Rhizobia* inoculation on growth, yield, and protein content of fenugreek (*Trigonella foenum-graecum* L.) under different levels of nitrogen. *Journal of Plant Nutrition*, *38*(2), 269–280. https://doi.org/10.1080/01904167.2014.930793.

Ali, N., Ahmad, N., & Bano, A. (2018). Impact of different doses of nitrogen on yield and biological nitrogen fixation of fenugreek. *Journal of Soil Science and Plant Nutrition*, *18*(2), 308–318.

Bekele, A., Afzal, A., Ahmad, I., Mirza, M. S., & Hameed, S. (2012). Effects of moisture, pH, and temperature on the growth and survival of rhizobia strains nodulating *Trigonella foenum-graecum* L. *The Scientific World Journal*, *2012*, 561794.

Berraho, E. L., Sanguin, H., Boubaker, A., Filali-Maltouf, A., & de Lajudie, P. (2019). Diversity and effectiveness of rhizobia isolated from fenugreek (*Trigonella foenum-graecum* L.) nodules in Morocco. *Rhizosphere*, *12*, 100168.

Bora, T., & Singh, V. (2017). *Rhizobia*: An overview. *International Journal of Current Microbiology and Applied Sciences*, *6*(11), 4077–4085.

Chen, X., Li, L., Li, Z., & Li, Y. (2020). The effect of *Rhizobium* inoculation on growth, nitrogen fixation, and nitrogen leaching in a fenugreek-maize cropping system. *Frontiers in Environmental Science*, *8*, 100.

Daryanto, S., Wang, L., & Jacinthe, P. A. (2021). Soil conditions affecting rhizobia-fenugreek symbiosis: A review. *Agronomy*, *11*(3), 503. https://doi.org/10.3390/agronomy11030503.

Das, P., Chakrabarti, K., & Chaudhuri, D. (2021). Influence of heavy metal contamination on nodulation, nitrogen fixation, and growth of fenugreek (*Trigonella foenum-graecum* L.). *Journal of Environmental Science and Health, Part B*, *56*(9), 947–956.

Dixon, R., & Kahn, D. (2004). Genetic regulation of biological nitrogen fixation. *Nature Reviews Microbiology*, *2*(8), 621–631.

El-Awadi, M. E., Hassan, N. A., & Aly, H. F. (2019). Enhancing growth, yield, and quality of fenugreek (*Trigonella foenum-graecum* L.) plants by rhizobial inoculation. *Journal of Plant Nutrition*, *42*(11, 12), 1472–1483.

El-Mawla, A. M., El-Morsy, E. M., El-Sayed, M. A., & El-Tahan, M. (2019). Response of *Trigonella foenum-graecum* L. to inoculation with *Rhizobium*, humic acid, and compost. *Journal of Applied Sciences Research*, *5*(1), 39–45.

Ghosh, P. K., Maiti, T. K., Basu, P. S., & Banerjee, H. (2014). Influence of rhizobial inoculation and nitrogen fertilization on growth, nodulation, and yield of fenugreek (*Trigonella foenum-graecum* L.). *Annals of Biological Research*, *5*(3), 87–97.

Giller, K. E. (2001). *Nitrogen Fixation in Tropical Cropping Systems* (2nd ed.). CABI Publishing.

Glick, B. R. (2012). Plant growth-promoting bacteria: Mechanisms and applications. *Scientifica, 2012*, 963401.

Glick, B. R. (2014). Bacteria with ACC deaminase can promote plant growth and help to feed the world. *Microbiological Research*, *169*(1), 30–39.

Gopalakrishnan, S., Sathya, A., Vijayabharathi, R., Varshney, R. K., Gowda, C. L. L., & Krishnamurthy, L. (2015). Plant growth-promoting traits of biocontrol potential bacteria isolated from rice rhizosphere. *SpringerPlus*, *4*(1), 1–13. https://doi.org/10.1186/s40064-015-1170-1.

Grönemeyer, J. L., Hurek, T., Bünger, W., & Reinhold-Hurek, B. (2014). Rhizobia with specific symbiotic genes produce nodules on *Arabidopsis thaliana*. *PLoS ONE*, *9*(9), e108657. https://doi.org/10.1371/journal.pone.0108657.

Gupta, A., Gopal, M., & Tilak, K. V. B. R. (2012). Influence of temperature, pH, and salt on growth, nodulation, and N_2 fixation by a *Rhizobium* strain nodulating *Trigonella foenum-graecum*. *Acta Physiologiae Plantarum*, *34*(5), 1775–1781.

Gupta, A., Singh, K. N., & Mishra, A. (2017). Role of *Rhizobium* inoculation in improving yield and nitrogen fixation of fenugreek (*Trigonella foenum-graecum* L.) in acid soil. *International Journal of Current Microbiology and Applied Sciences*, *6*(1), 216–224.

Gupta, G., Parihar, S. S., Ahirwar, N. K., Snehi, S. K., & Singh, V. (2013). Nitrogen-fixing legumes: A boon to soil fertility. *Journal of Indian Society of Soil Science*, *61*(1), 1–12.

Haddad, N., Karam, N., & Dib, J. (2017). Effect of rhizobial inoculation on the protein and amino acid profiles of fenugreek seeds. *Journal of Agricultural and Food Chemistry*, *65*(4), 717–723.

Herridge, D. F., Peoples, M. B., & Boddey, R. M. (2008). Global inputs of biological nitrogen fixation in agricultural systems. *Plant and Soil*, *311*(1, 2), 1–18. https://doi.org/10.1007/s11104-008-9668-3.

Hussain, M., Qamar, Z., Abbas, T., Mubeen, M., & Majeed, A. (2018). Nitrogen fixation potential of fenugreek (*Trigonella foenum-graecum* L.) under different doses of nitrogen fertilizer. *Journal of Plant Nutrition*, *41*(15), 2033–2043.

Huygens, D., Tytgat, T., Vanholme, B., De Meyer, T., Van Montagu, M., Coosemans, J., & Holsters, M. (2004). Structure and evolution of a novel class of the endosymbiont-derived Nod factor receptors. *Molecular Plant-Microbe Interactions*, *17*(2), 226–233.

Khan, N., Bano, A., & Rahman, M. A. (2018). Fenugreek (*Trigonella foenum-graecum* L.): An overview. *Journal of Pharmaceutical Research International*, *22*(3), 1–15.

Kiers, E. T., Rousseau, R. A., West, S. A., & Denison, R. F. (2003). Host sanctions and the legume-rhizobium mutualism. *Nature*, *425*(6953), 78–81.

Kim, J. H., Lim, S. W., Jang, H. G., Kim, Y. H., & Chung, B. N. (2018). Optimization of soil pH and inoculum density for the growth and nodulation of fenugreek (*Trigonella foenum-graecum* L.) with rhizobia. *Korean Journal of Soil Science and Fertilizer*, *51*(6), 657–663.

Kumar, A., Kumar, A., Kumar, A., Singh, A., Singh, V., & Singh, R. (2021). Effect of *Rhizobium* inoculation and nitrogen levels on growth, yield, and quality of fenugreek (*Trigonella foenum-graecum* L.) in calcareous soil. *International Journal of Agriculture, Environment and Biotechnology*, *14*(2), 381–386.

Kumar, V., Sharma, S., & Jindal, V. (2019). Effect of soil pH and different nitrogen sources on nodulation and yield of fenugreek (*Trigonella foenum-graecum* L.). *International Journal of Chemical Studies*, *7*(5), 983–987. https://doi.org/10.22271/chemi.2019.v7.i5a.8424.

Kurdali, F., Al-Shamma'a, M., Al-Ain, F., & Smith, G. D. (2001). Nitrogen fixation by chickpea, lentil, and fenugreek in different soils of Syria. *Biology and Fertility of Soils*, *33*(6), 442–447.

Li, Y., Liu, X., Zhuang, L., & Zhu, H. (2017). Genetic variation in root morphology and nitrogen-fixing efficacy in soybean (*Glycine max* (L.) Merr.). *Plant and Soil*, *419*(1, 2), 251–263.

Mandal, S., & Datta, J. K. (2020). Beneficial effects of *Trigonella foenum-graecum* L. on soil fertility and growth of maize in a maize-wheat cropping system. *Communications in Soil Science and Plant Analysis*, *51*(1), 1–10.

Meghvansi, M. K., Singh, S. K., & Kumbhar, M. B. (2016). Fenugreek (*Trigonella foenum-graecum* L.): A review on its nutraceutical properties and utilization in various food products. *Journal of Food Science and Technology*, *53*(2), 583–594.

Mishra, A. K., & Singh, A. (2017). Recent advances in microbial strategies for nitrogen fixation in agriculture. In *Microbial Strategies for Crop Improvement* (pp. 15–35). Springer, Singapore.

Mishra, A. K., Tripathi, A. K., & Singh, P. (2021). Rhizobium-legume symbiosis: An approach towards sustainable agriculture. In *Advances in Agrotechnology* (pp. 167–192). Springer.

Mugwe, J., Mugendi, D., & Mucheru-Muna, M. (2019). Soil properties affecting rhizobia-legume symbiosis in smallholder farms of Central and Western Kenya. In M. A. Khan, S. Z. Husaini, & M. S. Zaman (Eds.), *Advances in Rhizobium Research* (pp. 187–205). Springer. https://doi.org/10.1007/978-981-13-8491-2_10.

Mus, F., Alleman, A. B., Pence, N., Seefeldt, L. C., & Peters, J. W. (2021). Exploring the limits of biological nitrogen fixation. *Nature Reviews Chemistry*, *5*(10), 691–710.

Nandasena, K. G., O'Hara, G. W., Tiwari, R. P., Sezmiş, E., & Howieson, J. G. (2007). High levels of genetic diversity among ineffective rhizobia nodulating diverse Australian native legumes include unusual *Bradyrhizobium* species. *Applied and Environmental Microbiology*, *73*(23), 7568–7578.

Oldroyd, G. E. (2013). Speak, friend, and enter: Signaling systems that promote beneficial symbiotic associations in plants. *Nature Reviews Microbiology*, *11*(4), 252–263.

Oldroyd, G. E., & Dixon, R. (2014). Biotechnological solutions to the nitrogen problem. *Current Opinion in Biotechnology*, *26*, 19–24.

Parnell, J. J., Berka, R., Young, H. A., Sturino, J. M., Kang, Y., Barnhart, D. M., & Madsen, E. L. (2016). From the lab to the farm: An industrial perspective of plant beneficial microorganisms. *Frontiers in Plant Science*, *7*, 1110.

Pathak, P. V., Mishra, A., Rajput, M., Verma, P., & Maheshwari, D. K. (2016). Influence of salinity on growth, nodulation, and enzyme activity of salt-tolerant and salt-sensitive strains of *Rhizobium meliloti* D. isolated from fenugreek (*Trigonella foenum-graecum* L.). *Archives of Agronomy and Soil Science*, *62*(6), 853–867.

Prasad, R., & Rao, J. V. (1994). Nodulation and nitrogen fixation in fenugreek (*Trigonella foenum-graecum* L.). *Indian Journal of Plant Physiology*, *37*(1), 51–54.

Rai, R., Pandey, P., & Singh, R. (2020). Impact of rhizobia inoculation on microbial biomass, enzyme activities, and carbon and nitrogen cycling in a fenugreek-rice cropping system. *Journal of Environmental Biology*, *41*(5), 1084–1088.

Ramakrishnan, K., Renganathan, M., & Ganesan, M. (2019). Effect of *Rhizobium* inoculation on growth and yield of fenugreek (*Trigonella foenum-graecum* L.) under varying levels of nitrogen. *Journal of Pure and Applied Microbiology*, *13*(3), 1595–1600.

Rani, S., Pandey, S., & Bharti, S. (2021). Effect of *Rhizobium* inoculation on the protein and iron content of fenugreek (*Trigonella foenum-graecum* L.) seeds. *Indian Journal of Agricultural Sciences*, *91*(8), 23–26.

Reddy, M. S., Ali, M., Srinivasan, S., & Kumar, P. S. (2004). Nitrogen fixation and yield attributes of fenugreek (*Trigonella foenum-graecum* L.) as influenced by phosphorus, molybdenum, and lime. *Indian Journal of Agronomy*, *49*(1), 29–32.

Rodríguez-Navarro, D. N., Temprano, F. J., Pérez-Sanz, A., de Felipe, M. R., Fernández-Pascual, M., & Lucas, M. M. (2014). Effects of *Rhizobium* strain differences and plant genotype on nitrogen fixation in common bean (*Phaseolus vulgaris* L.). *Biology and Fertility of Soils*, *50*(3), 413–421.

Rubio, L. M., & Ludden, P. W. (2008). Biosynthesis of the iron-molybdenum cofactor of nitrogenase. *Annual Review of Microbiology*, *62*, 93–111.

Saha, S., Mukherjee, S., Saha, B., & Mukherjee, S. K. (2018). Phosphorus and potassium application to fenugreek (*Trigonella foenum-graecum* L.) for improving nodulation and nitrogen fixation. *International Journal of Agricultural Technology*, *14*(2), 215–227.

Sahu, S. K., Sharma, M., & Sharma, A. K. (2019). Legume-rhizobia symbiosis: An overview. *Legume Research: An International Journal*, *42*(1), 1–14.

Saia, S., Rappa, V., Ruisi, P., Abenavoli, M. R., Sunseri, F., & Giambalvo, D. (2015). Rhizosphere soil microbial index of biotic soil quality and plant disease suppression perspective. *Applied Soil Ecology*, *86*, 121–130.

Saini, S., Singh, R., Sharma, R., & Sharma, M. (2019). Characterization of nitrogen-fixing rhizobacteria associated with fenugreek (*Trigonella foenum-graecum* L.) from the Indian Thar Desert. *Journal of Basic Microbiology*, *59*(4), 385–395.

Salama, M. S., Abdalla, M. A., El-Din, M. A. G., & Haggag, W. M. (2016). Response of fenugreek plant to bio-fertilization by *Rhizobium* and/or *Bacillus* species. *Journal of Applied Sciences Research*, *12*(10), 69–80.

Saleem, M., Arshad, M., Hussain, S., & Bhatti, A. S. (2007). Perspective of plant growth-promoting rhizobacteria (PGPR) containing ACC deaminase in stress agriculture. *Journal of Industrial Microbiology and Biotechnology*, *34*(10), 635–648.

Sanginga, N., Vanlauwe, B., & Danso, S. K. A. (2000). Nitrogen fixation in Africa: History, challenges, and prospects. Springer.

Sarathambal, C., Shanthi, V., & Saravanan, S. (2015). Effect of *Rhizobium* and phosphorus on yield and quality of fenugreek (*Trigonella foenum-graecum* L.). *International Journal of Agriculture, Environment and Biotechnology*, *8*(2), 379–382. https://doi.org/10.5958/2230-732X.2015.00049.8.

Sarwar, M., Kiran, S., Iqbal, M., & Shafiq, M. (2014). Response of fenugreek to *Rhizobium* inoculation under different nitrogen levels. *International Journal of Agriculture and Biology*, *16*(1), 163–168.

Sarwar, M., Saleem, M., & Khalid, A. (2013). Effect of different levels of nitrogen on nodulation, nitrogen fixation, and yield of fenugreek. *International Journal of Agriculture and Biology*, *15*(6), 1222–1226.

Sgolastra, F., Mancuso, S., & Li, Z. (2019). *Rhizobia* inoculation improves water stress tolerance of legume crops. *Agronomy*, *9*(5), 255. https://doi.org/10.3390/agronomy9050255.

Sharma, P., Gaur, R., Sharma, S., & Kumar, A. (2020). Effect of organic amendments on rhizobia-fenugreek symbiosis: Nodulation, nitrogen fixation, and soil fertility. *Communications in Soil Science and Plant Analysis*, *51*(17), 2187–2201.

Shukla, A., Singh, S. K., & Singh, R. (2021). Impact of rhizobia and inorganic fertilizers on growth, yield, and nutrient uptake of fenugreek (*Trigonella foenum-graecum* L.). *Indian Journal of Agricultural Sciences*, *91*(2), 199–204.

Singh, D., Sharma, S., Upadhyay, A. K., & Singh, J. S. (2013). Effects of temperature and soil moisture on rhizobial populations of *Trigonella foenum-graecum* in a North Indian agricultural field. *Journal of Basic Microbiology*, *53*(3), 238–246.

Singh, M., Singh, R. K., Singh, S., & Upadhyay, R. S. (2019). Effect of temperature on growth and nitrogen fixation of fenugreek (*Trigonella foenum-graecum* L.) with rhizobia. *International Journal of Chemical Studies*, *7*(5), 2556-2560.

Singh, M., Singh, R. K., Singh, S., & Upadhyay, R. S. (2020). Impact of Rhizobium inoculation on growth, yield, and nutrient uptake of fenugreek (*Trigonella foenum-graecum* L.) under different soil conditions. *Communications in Soil Science and Plant Analysis*, *51*(6), 746–758.

Singh, R., Kumar, A., & Kumar, V. (2015). Selection of high-yielding and nitrogen-fixing cultivars of fenugreek (*Trigonella foenum-graecum* L.). *Annals of Agricultural Research*, *36*(4), 471–475.

Singh, S., Singh, B., Singh, R. S., Singh, S. P., & Sharma, S. (2016). Effect of *Rhizobium* inoculation on growth and yield of fenugreek (*Trigonella foenum-graecum* L.). *International Journal of Agricultural Sciences*, *8*(2), 310–313.

Singh, V., Singh, B., Kumar, A., Singh, R. P., & Chaudhary, R. P. (2003). Studies on the yield attributes, yield, and N fixation of fenugreek (*Trigonella foenum-graecum* L.) varieties as influenced by different levels of nitrogen and sulfur. *Indian Journal of Agricultural Sciences*, *73*(9), 506–508.

Soliman, S. S., Othman, E. S., & Al-Sherif, E. A. (2021). Effects of soil moisture on growth, nodulation, and nitrogen fixation of fenugreek (*Trigonella foenum-graecum* L.) inoculated with rhizobia. *Journal of Plant Nutrition*, *44*(6), 1068–1082. https://doi.org/10.1080/01904167.2021.1870788.

Soltani, B. M., Rahmani, H. A., & Bahramnejad, B. (2019). Biological nitrogen fixation in legumes: Perspectives and challenges. *Archives of Agronomy and Soil Science*, *65*(6), 742–755.

Sprent, J. I. (2009). *Legume nodulation: A global perspective*. John Wiley and Sons.

Sprent, J. I., Minchin, F. R., & Summerfield, R. J. (1983). Legume nodulation and nitrogen fixation. In *Handbook of Plant and Crop Physiology* (pp. 313–351). CRC Press.

Verma, S., Patra, A. K., Verma, S. S., & Meena, V. S. (2020). Mycorrhizal symbiosis enhances nodulation and nitrogen fixation in fenugreek (*Trigonella foenum-graecum* L.). *Journal of Applied Horticulture*, *22*(3), 244–248.

Vessey, J. K. (2003). Plant growth-promoting rhizobacteria as biofertilizers. *Plant and Soil*, *255*(2), 571–586.

Vincent, J. M., & Stowers, M. D. (1954). Clover strain specificity in *Rhizobium trifolii*. *Journal of General Microbiology*, *11*(1), 25–34.

Vyas, P., & Gulati, A. (2017). *Rhizobia* as PGPR: Potential benefits and challenges. In *Plant Growth Promoting Rhizobacteria for Sustainable Stress Management* (pp. 33–44). Springer, Singapore.

Young, J. P., Crossman, L. C., Johnston, A. W., Thomson, N. R., Ghazoui, Z. F., Hull, K. H., Wexler, M., Curson, A. R., Todd, J. D., Poole, P. S., & Mauchline, T. H. (2006). The genome of *Rhizobium leguminosarum* F. has recognizable core and accessory components. *Genome Biology*, *7*(4), R34.

CHAPTER 7

Fenugreek-Enabled Green Synthesis of Nanoparticles and Its Impact on Growth Factors

ARINDAM PAL,[1] PRIYA,[1] DEEPIKA JEENGAR,[2] PARIKSHIT UCHAI,[1] VYSHNAVI SUNIL,[3] and VINOD KUMAR DUBEY[4]

[1]*Department of Entomology, Post-Graduate College of Agriculture, Dr. Rajendra Prasad Central Agricultural University, Pusa, Samastipur, Bihar, India*

[2]*Department of Entomology, College of Agriculture, Govind Ballabh Pant University of Agriculture and Technology, Pantnagar, Uttarakhand, India*

[3]*Department of Entomology, College of Agriculture, Kerala Agricultural University, Vellayani, Trivandrum, Kerala, India*

[4]*Department of Genetics & Plant Breeding, College of Agriculture, Acharya Narendra Deva University of Agriculture and Technology, Ayodhya, Uttar Pradesh, India*

ABSTRACT

The medicinal herb fenugreek (*Trigonella foenum-graecum* Linn.), known for its valuable nutritional properties, is one of the most economically significant herbs in the world. It belongs to the Fabaceae family. Both the green leaves and seeds of fenugreek have several uses, including as an anti-diabetic, anti-carcinogenic, hypocholesterolemic, hypoglycemic agent, antioxidant, antibacterial agent, gastric stimulant, anti-anorexic, and for the green synthesis of nanoparticles.

Fenugreek: Improvement in Yield Potential Through Plant Growth Regulators (PGRs) and Depolymerized Polysaccharides. Tariq Ahmad Dar, Bilal Ahmad, Moinuddin, and Khalid Rehman Hakeem (Eds.)

Nanoparticles (NPs) are classified as microscopic ultrafine particles with at least one dimension smaller than 100 nm. The physical and chemical properties of nanoparticles, including their catalytic activity, optical properties, electronic properties, antibacterial properties, and magnetic properties, have attracted scientists' attention to develop more effective and cost-efficient synthesis techniques.

Traditional wet chemistry methods of nanoparticle synthesis face challenges such as contamination, hazardous by-products, toxic solvents, and difficulties in large-scale production. This highlights the necessity of green synthesis. Fenugreek leaves and seeds are used to reduce a variety of nanoparticles in a green synthesis process, including silver nanoparticles, palladium nanoparticles, cobalt, and iron oxide nanoparticles, graphene-based biomaterials, selenium nanoparticles, gold nanoparticles, zinc oxide nanoparticles, copper nanoparticles, and magnesium oxide nanoparticles.

When silver nanoparticles (AgNPs) were incorporated into fenugreek, the plants grew faster, acquired more photosynthetic pigments, contained more indole acetic acid (IAA), had higher yields, and produced more diosgenin, a bioactive molecule. Silver (Ag) nanoparticles also improved germination properties and increased tolerance to salinity. Silicon (Si) nanoparticles impact several characteristics of plants, including Si absorption, translocation, accumulation, lignification of cell walls, and the production of stress enzymes. Conversely, aluminum and zinc oxide nanoparticles negatively impact seed germination, morphological characteristics, and genetic variation, especially at higher concentrations.

In this chapter, we provide comprehensive information about the green synthesis of various nanoparticles and their effects on the growth parameters of the fenugreek plant.

7.1 INTRODUCTION

Fenugreek is a commonly known herb around the world containing several medicinal and therapeutic properties. Its seed contains saponin (fenugrin B), gitogenin, tigogenin, and diosgenin with several alkaloids (Deshmukh et al., 2019). In addition to fenugreek seed, other major bioactive compounds which include polyphenolic compounds such as rhaponticin and isovietin (Wani and Kumar, 2018). In recent times fenugreek has gotten newfound attention for its capability of green synthesis of several nanoparticles. Nanoparticles can be created using various components of the fenugreek plant, such as its leaves and seeds. Currently, nanotechnology stands as one of the most strategically mapped-out domains in material science due to its novel or enhanced attributes (like particle size, distribution, morphology, etc.) in contrast to larger-scale counterparts (Kaviya et al., 2019). Different approaches or methods including physical, chemical, and biological can be used to synthesize nanoparticles. The two main types of nanoparticles are organic carbon-containing particles and inorganic nanoparticles. Nanoparticles such as the magnetic ones, the noble metal nanoparticles (such as gold and silver), and the semiconductor nanoparticles (such as titanium oxide and zinc oxide) all belong to this group (Varghese et al., 2019). Chemical method of nanoparticle synthesis takes only short time but involve the use of highly lethal chemicals in addition to several disadvantages. So, it's important to switch to green synthesis of nanoparticles.

7.2 ROLES OF FENUGREEK IN GREEN SYNTHESIS OF NANOPARTICLES

Fenugreek has showed a great potential towards green synthesis of several nanoparticles, here we discussed about some successful examples are shown in (Figure 7.1).

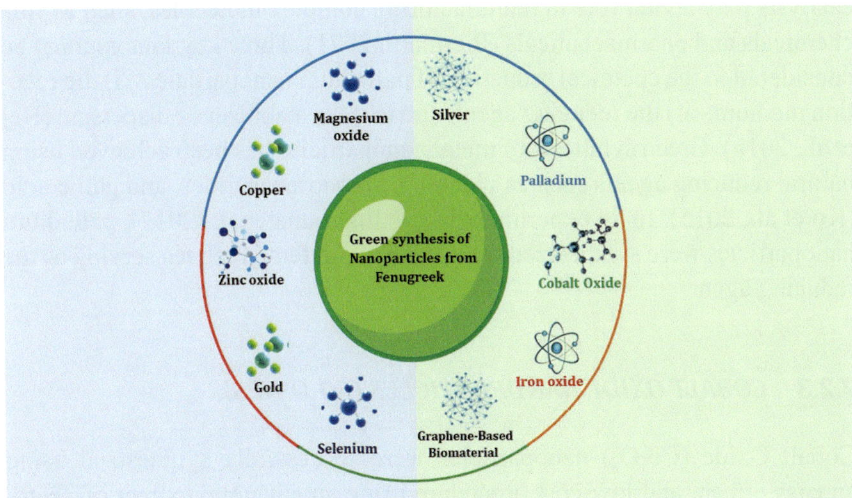

FIGURE 7.1 Nanoparticles synthesized from fenugreek.

7.2.1 SILVER NANOPARTICLES (AGNPS)

It is widely known that silver nanoparticles (AgNPs) can inhibit cancer cell growth and defend against microbial pathogens that cause disease (Varghese et al., 2019). Much research has shown that, silver nanoparticles can be synthesized from fenugreek plant parts (leaves and seeds). Silver nanoparticles synthesized from *fenugreek* leaves showed higher inhibition against *Pseudomonas aeruginosa* compared to other bacteria which is an environment friendly and inexpensive method. Plant-mediated AgNP might interact with intracellular organelles and adhere to the upper surface of cell and disrupt the cell membrane which might be one of its possible modes of action (Ghoshal and Singh, 2022). In a similar experiment silver nanoparticles were formed by the action of silver nitrate and fenugreek seed extract (Angelina et al., 2013; Singh et al., 2011). Also synthesized silver nanoparticles through aqueous $AgNO_3$ and fenugreek leaf extract.

7.2.2 PALLADIUM NANOPARTICLES (PDNPS)

In recent years, significant attention has been directed toward designing efficient, low-cost, and environmentally sustainable palladium nanoparticles (PdNPs) via green synthesis methods in the chemical industry, as Pd catalysts play a vital role in manufacturing complex molecules, such as fine chemicals and pharmaceuticals (Piermatti, 2021). Three key factors must be considered in the chemical reduction of palladium nanoparticles: (i) the reaction medium; (ii) the reducing agent; and (iii) the stabilizer or dispersant (Bej et al., 2016). Green synthesis of metal nanoparticles has been achieved using natural reducing agents such as vitamin C, monosaccharides, and gallic acid (Ko et al., 2015). In an experiment by Mallikarjuna et al. (2017), palladium nanoparticles were synthesized from $PdCl_2$, with fenugreek tea serving as the reducing agent.

7.2.3 COBALT OXIDE NANOPARTICLES (CO_3O_4NPS)

Cobalt Oxide (Co_3O_4) nanoparticles were successfully synthesized using an easy, green, and low-cost procedure using an ethanolic extract of *Trigonella foenum graceum* Linn. The organic bioactive compounds present in fenugreek leaf extract acted as both reducing and stabilizing agents during the synthesis of metal nanoparticles using cobalt chloride hexahydrate as a precursor. FT-IR, FTIR spectroscopy, SEM, AFM, and UV-Visible spectroscopy were used to characterize the synthesized cobalt nanoparticles (Akhlaghi et al., 2020).

7.2.4 *IRON OXIDE NANOPARTICLES (FE_2O_3NPS)*

Iron oxide nanoparticles have garnered considerable attention owing to their distinctive attributes, encompassing super paramagnetism, an advantageous surface-to-volume ratio, expanded surface area, and facile separability. Several methods have been used to synthesize iron oxide nanoparticles, including mechanochemical as well as chemical methods (Sun et al., 2014). But as a sustainable green alternative of synthesizing iron oxide nanoparticle, fenugreek can be used. Ultrasound assisted iron oxide nanoparticles were synthesized by using fenugreek seed extract containing enhanced antibacterial and antioxidant properties (Deshmukh et al., 2019).

7.2.5 *GRAPHENE-BASED BIOMATERIAL*

Fenugreek can be a very effective solution for synthesis of graphene-based biomaterials. Recently Singh et al. (2016) has demonstrated synthesis of palladium nanoparticles (PdNPs) on reduced graphene oxide (GO), where fenugreek seed extract was used as reducing agent.

7.2.6 SELENIUM NANOPARTICLES (SENPS)

The biocompatibility, bioavailability, and minimal toxicity of selenium nanoparticles (SeNPs) have captured the interest of numerous researchers. With the help of fenugreek seed extract, a simple biological procedure was used to produce selenium nanoparticles (SeNPs). Under ambient conditions, SeNPs can be produced in a size range of 50–150 nm with this method. SeNPs inhibit cell growth of human breast-cancer cells (MCF-7) in a dose dependent manner (Ramamurthy et al., 2013).

7.2.7 ZINC OXIDE NANOPARTICLES (ZNONPS)

One of the most important inorganic metal oxide nanoparticles is zinc oxide (ZnO), which is an insoluble white powder. Zinc oxide nanoparticles are naturally known for their high microbe resistance. Zinc oxide nanoparticles are widely used for biological labelling, biological sensing, drug delivery, gene delivery, and nanomedicine. In comparison to other oxides, fenugreek gives a sustainable and ecofriendly route for synthesizing ZnONPs. ZnONPs were synthesized using the fenugreek seed extract and then was characterized by scanning electron microscope (SEM), energy dispersive X-ray (EDX), X-ray diffraction (XRD), UV-V spectroscopy and transmission electron microscope (TEM). The synthesized ZnONPs displayed both cytotoxic effect, and antitumor activity (Radwan et al., 2021).

7.2.8 GOLD NANOPARTICLES (AUNPS)

AuNPs is one of the most valuable metallic nanoparticles. AuNPs synthesized using green plants are eco-friendly, inexpensive, have good antimicrobial activities against microorganisms. By controlling the synthesis parameters, gold nanoparticles of different sizes can be obtained between 15 and 25 nm. Gold nanoparticle is synthesized by using the aqueous extract of fenugreek seeds (as reducing and protecting agent). In this pathway, chloroauric acid is reduced by the extract of fenugreek seeds (Aromal and Philip, 2012). This method is simple, efficient, economical, and nontoxic.

7.2.9 MAGNESIUM OXIDE NANOPARTICLES (MgONPs)

Green synthesis of MgONPs is regarded as a potential and environmentally responsible method. Fenugreek leaf extract can be used for successful synthesis of MgONPs. Synthesis of MgNPs involves reaction of magnesium nitrate (as a precursor), sodium hydroxide and fenugreek leaf extract. Then the synthesized MgONPs are characterized by different techniques like UV-Vis spectrophotometers, FT-IR, GC-MS, X-ray diffraction (XRD), and SEM with EDX (Vergheese and Vishal, 2018).

7.2.10 COPPER NANOPARTICLES (CuNPs)

Manufacturers find CuNPs exceptionally appealing for a multitude of reasons. These include their high thermal conductivity, substantial surface area to volume ratio, reduced production expenses, potent antimicrobial properties, catalytic capabilities, and captivating visual attributes. Fermented fenugreek powder was used to prepare copper nanoparticles under the influence of *Pleurotus ostreatus* (which reduced and preserved the particles). Using inexpensive polysaccharides, CuNPs were prepared through gamma irradiation of aqueous fermented fenugreek powder under the influence of *Pleurotus ostreatus* which was used as reducing and preserving agent. By gamma irradiating aqueous fermentation fenugreek powder with inexpensive polysaccharides, CuNPs can be prepared under the influence of *Pleurotus ostreatus*, a reducing agent and preservative (El-Batal et al., 2018).

7.3 IMPACT OF NANOPARTICLES ON GROWTH PARAMETERS OF FENUGREEK

7.3.1 EFFECT OF AGNPS ON FENUGREEK

In Agriculture, AgNPs have been studied and used extensively. Because of the novel properties present in silver nanoparticles, it has been used as an antimicrobial (antibacterial, antifungal, anti-viral) for controlling plant diseases, insecticide, seed protecting agent, packaging material in post-harvest management for increasing shelf life, and, most importantly, plant growth enhancer (Kale et al., 2021). The cause of induced plant growth by AgNPs may be due to the inhibition of ethylene signaling (Rezvani et al., 2012). Here we discussed the influence of AgNPs on several growth factors of fenugreek.

7.3.1.1 EFFECT ON PLANT GROWTH INDICATORS

Silver nanoparticles (AgNPs) enhance several plant growth indicators. In an experiment where varying concentrations of AgNPs were applied as a foliar spray on fenugreek plants, shoot length (cm), number of leaves per plant, and dry weight of shoots per plant increased significantly at 60 days after sowing (DAS). The 40 mg/L AgNP treatment proved to be the most effective across all criteria (Sadak, 2019). A similar study by Jasim et al. (2017) reported comparable positive effects on growth indicators.

7.3.1.2 INCREASE IN DIOSGENIN PRODUCTION

The bioactive molecule diosgenin is a steroid sapogenin present in fenugreek that acts as a natural antioxidant and also includes therapeutic properties.

AgNPs have also shown potential for increasing diosgenin production in fenugreek plants. AgNP treated plants expressed enhanced concentration of diosgenin (214.06 ± 17.07 μg/mL) compared to the control (164.49 ± 7.67 lg/mL). In the HPLC chromatogram, the apex corresponding to the standard diosgenin, and the sample treated with Ag-NPs was detected at a retention time of 4.8 (Jasim et al., 2017).

7.3.1.3 EFFECT ON PHOTOSYNTHETIC PIGMENTS AND INDOLE-3-ACETIC ACID (IAA)

Applying varying concentrations of silver nanoparticles (AgNPs) to fenugreek plants resulted in an increase in the content of photosynthetic pigments, including chlorophyll a, chlorophyll b, carotenoids, and total pigments, along with an increase in indole-3-acetic acid (IAA). In both cases, Sadak (2019) identified the optimal dosage as 40 mg/L.

7.3.1.4 EFFECT ON GERMINATION OF FENUGREEK

In order for plants to grow and yield, germination is considered one of the most fundamental factors. The application of AgNPs results in a notable rise in the percentage of seed germination, mean germination time, seedling vigor index, and seed germination index for fenugreek seedlings. Comparing different concentrations applied during treatment, 10 μg/mL was the best and showed highest results (Hojjat and Hojjat, 2015). Seedling germination is a dynamic process that is highly susceptible to abiotic stresses. In arid regions, soil salinity is a major stress factor that severely affects germination. AgNps have shown promising effects on the germination of fenugreek under saline conditions. Treatment of 30 μg/mL AgNPs was found to be the most effective against salinity and also enhanced the germination capabilities of fenugreek (Hojjat and Kamyab, 2017).

7.3.1.5 EFFECT ON YIELD PARAMETERS

The use of silver nanoparticles (AgNPs) enhances both the quality and quantity of yield in fenugreek. Foliar application of AgNPs positively influences yield components such as the number of pods per plant, number of seeds per pod, seed weight per plant, and seed index. Additionally, an increase in total organic compounds—including carbon, protein, phenolic compounds, flavonoids, and tannins—was observed in fenugreek plants. Sadak (2019) found that a concentration of 40 mg/L AgNPs produced the most promising results (*see* Figure 7.2).

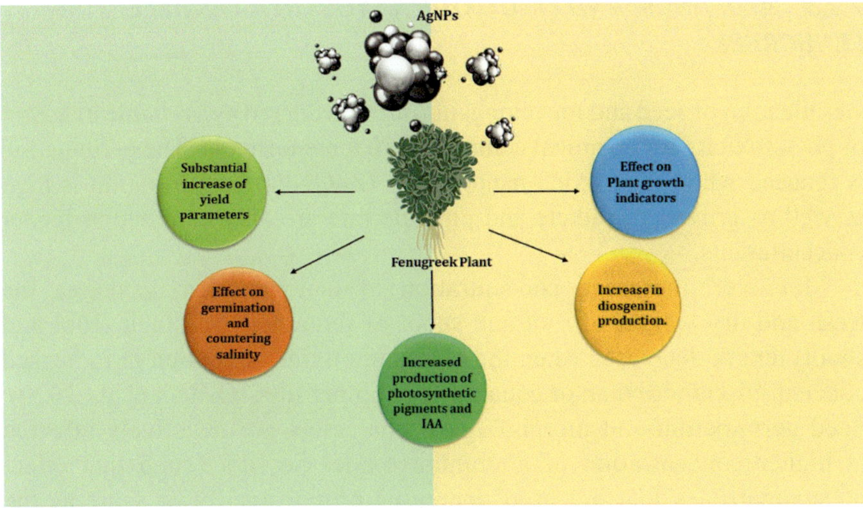

FIGURE 7.2 Effect of AgNPs on fenugreek.

7.3.2 EFFECT OF SILICON NANOPARTICLES (SINPS) ON FENUGREEK

Fertilizing crops with silicon (Si) can improve crop yield, as it can be applied in soluble form as silicates (Si_3O_5). Phytoliths, which are amorphous SiO_2 particles similar to nano-scale SiO_2 particles, are a major source of plant-available Si (Imtiaz et al., 2022). There are several benefits of silicon for plants, including protecting and stabilizing their structure. It enhances plant growth, resilience to salt and drought, and addresses nutrient imbalances (Zhu and Gong, 2014). Fenugreek treatment with silicon nanoparticles and silicates has been studied by Nazaralian et al. (2017). It was found that both the silicate and the SiNP treatments took up Si, and adding sodium 13 silicate increased Si translocation to the shoot but adding SiNP did not increase it. In the presence of sodium silicate, PST was more highly up regulated than when it was present in the presence of SiNP. There was no difference in the effects of sodium silicate and SiNP treatments on the other parameters. A significant increase in xylem wall lignification was noted in both treatments, as well as a significant increase in PAL activity and protein concentration in seedlings, but not in antioxidative enzyme activity. Leaves with Si addition showed a strong decrease in Tfgd 1 expression. Due to the release of silicate and absorption of silicate by SiNP, there are similar effects between SiNP and silicate.

7.3.3 ALUMINUM AND ZINC OXIDE NANOPARTICLES AFFECTING FENUGREEK

Germination of seed and root elongation are considered as a reliable indicator of phytotoxicity for chemical compounds that are unstable. The germination is reduced when Zinc oxide nanoparticle (ZnONPs) concentration is high as well as growth of radicle and plumule rate are also inhibited at higher concentrations.

Moreover, increased concentration of nanoparticles decreases the fresh and dry weight of root and shoot of fenugreek and their (root and shoot) length decreases when the aluminum oxide is applied in increased concentration more than or equal to 100 mg per liter (Osman et al., 2020). Seed germination and morphological characters are negatively affected at higher concentration of aluminum oxide. So, it's found that effect of nanoparticles that are used, increases proportionally or same as the concentration rises. But when ZnOPs are applied at lower concentration it works efficiently and increases fresh weight and dry weight of Fenugreek seedling and it is found that when fenugreek seeds are treated with the concentration of 30 mg per liter ZnOPs the flavones and phenol content tends to increase in the seedling (Elsherif et al., 2023) and it influences the biosynthesis of carbohydrates in leaves and changes protein content of plants that depends upon zinc oxide concentration (Rao et al., 2007; Zhao et al., 2014). ZnO_2 notably influences the total amount of protein and carbohydrate of fenugreek seedlings by increasing their total amount with the increase in zinc oxide concentration till the 30 mg per liter (Elsherif et al., 2023).

7.4 GREEN BIOSYNTHESIS OF NANOPARTICLE OVER TRADITIONAL METHODS

Green biosynthesis uses biological methods to synthesize nanoparticles like, Plants, microorganisms, viruses, and their byproducts, like proteins and lipids, are all examples. Green biosynthesis method offers a promising alternative route for synthesizing stable nanoparticles that are biocompatible. This method is not only sustainable but also eliminates several problems that are associated with traditional chemical methods like, environmental hazards, high cost, long processing time, and so forth. The major merits of green biosynthesis of nanoparticles over traditional methods are discussed here:

7.4.1 MERITS OF GREEN SYNTHESIS

- A convenient way to synthesize agents;
- As possible alternatives to chemical and physical methods that are both environmentally and economically friendly;
- As the monodispersity and size and shape of nanoparticles play a major role in evaluating their synthesis, efficient control over their morphology and monodispersity should be explored;
- The preparation of well-characterized nanoparticles can be achieved by screening organisms that are able to produce large quantities of NPs in a controlled reaction environment;
- As a result, it can be possible to synthesize at rates that are faster than those achieved through chemical or physical methods;
- The conventional method of synthesizing NPs typically requires the use of harmful reductants like hydrazine and sodium borohydride, additionally, the procedure often entails multiple steps, including heat treatments;
- The synthesis procedure produces hazardous by-products, including heat treatments, which lead to hazardous waste;
- The most energy efficient way to synthesize products is to have the reaction occur at ambient temperature and pressure under neutral pH;
- Biological systems are then the most appropriate factory for reacting such natural chemical conditions;
- It has been discovered that some microbes can be used to manufacture nanomaterials, specifically silver metal nanoparticles, in an eco-friendly manner;
- It is found in plants that the most effective compounds are used for synthesis, which increases the rate of synthesis is depicted in (Figure 7.3).

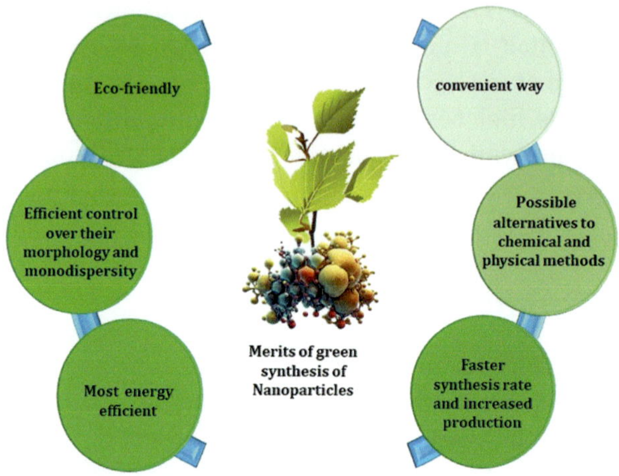

FIGURE 7.3 Merits of green synthesis.

7.5 CONCLUSION

Over the past few years, scientists have made great efforts to develop green nanoparticles. As a method of synthesis for nanoparticles, fenugreek shows great promise, since it is eco-friendlier and easier to use than traditional methods. This method preserves nearly uniform particle size and allows for relatively small particle sizes to be achieved. In addition to being cheap, green synthesis can create nanoparticles at room temperature as well. Considering that the fenugreek is easily available and widely distributed, it can also be handled safely. In the future, we will have a great deal to do with this methodology of synthesis.

KEYWORDS

- **bio-reducing agent**
- **extraction**
- **fenugreek**
- **gold nanoparticles**
- **green nanoparticles**
- **monodispersity**
- **nanoparticles**
- **silver nanoparticles**

REFERENCES

Akhlaghi, N., Najafpour-Darzi, G., & Younesi, H. (2020). Facile and green synthesis of cobalt oxide nanoparticles using ethanolic extract of *Trigonella foenum-graecum* (fenugreek) leaves. *Advances in Powder Technology*, *31*(8), 3562–3569.

Angelina, E. D. R., Bavyaa, R., & Rajagopal, R. (2013). Green synthesis and characterization of silver nanoparticles using fenugreek seed extract. *International Journal of Scientific Research*, *3*(7), 1–3.

Aromal, S. A., & Philip, D. (2012). Green synthesis of gold nanoparticles using *Trigonella foenum-graecum* and its size-dependent catalytic activity. *Spectrochimica Acta Part A: Molecular and Biomolecular Spectroscopy*, *97*, 1–5.

Bej, A., Ghosh, K., Sarkar, A., & Knight, D. W. (2016). Palladium nanoparticles in the catalysis of coupling reactions. *RSC Advances*, *6*(14), 11446–11453.

Deshmukh, A. R., Gupta, A., & Kim, B. S. (2019). Ultrasound-assisted green synthesis of silver and iron oxide nanoparticles using fenugreek seed extract and their enhanced antibacterial and antioxidant activities. *BioMed Research International*, *2019*, 1–14.

El-Batal, A. I., Al-Hazmi, N. E., Mosallam, F. M., & El-Sayyad, G. S. (2018). Biogenic synthesis of copper nanoparticles by natural polysaccharides and *Pleurotus ostreatus*-fermented fenugreek using gamma rays with antioxidant and antimicrobial potential towards some wound pathogens. *Microbial Pathogenesis*, *118*, 159–169.

Elsherif, D. E., Abd-ElShafy, E., & Khalifa, A. M. (2023). Impacts of ZnO as a nano-fertilizer on fenugreek: Some biochemical parameters and SCoT analysis. *Journal of Genetic Engineering and Biotechnology*, *21*(1), 1–10.

Ghoshal, G., & Singh, M. (2022). Characterization of silver nanoparticles synthesized using fenugreek leaf extract and its antibacterial activity. *Materials Science and Engineering Technology*, *5*, 22–29.

Hojjat, S. S., & Hojjat, H. (2015). Effect of nano-silver on seed germination and seedling growth in fenugreek seeds. *ETP International Journal of Food Engineering*, *1*(2), 106–110.

Hojjat, S. S., & Kamyab, M. (2017). The effect of silver nanoparticles on fenugreek seed germination under salinity levels. *Russian Agricultural Sciences*, *43*, 61–65.

Imtiaz, H., Shiraz, M., Mir, A. R., & Hayat, S. (2022). Green synthesis of nanoparticles and their effect on plant growth and development: A review. *Plant Archives*, *22*(1), 343–350.

Jasim, B., Thomas, R., Mathew, J., & Radhakrishnan, E. K. (2017). Plant growth and diosgenin enhancement effect of silver nanoparticles in fenugreek (*Trigonella foenum-graecum* L.). *Saudi Pharmaceutical Journal*, *25*(3), 443–447.

Kale, S. K., Parishwad, G. V., & Patil, A. S. H. A. S. (2021). Emerging agriculture applications of silver nanoparticles. *ES Food & Agroforestry*, *3*, 17–22.

Kaviya, S., Santhanalakshmi, J., & Viswanathan, B. (2019). Green synthesis of silver nanoparticles using *Polyalthia longifolia* leaf extract along with D-sorbitol: Study of antibacterial activity. *Journal of Nanotechnology*, *2019*, 1–6.

Ko, Y. L., Krishnamurthy, S., & Yun, Y. S. (2015). Facile synthesis of monodisperse Pt and Pd nanoparticles using antioxidants. *Journal of Nanosci. Nanotechnol.*, *15*, 412–417.

Mallikarjuna, K., Bathula, C., Buruga, K., Shrestha, N. K., Noh, Y. Y., & Kim, H. (2017). Green synthesis of palladium nanoparticles using fenugreek tea and their catalytic applications in organic reactions. *Materials Letters*, *205*, 138–141.

Nazaralian, S., Majd, A., Irian, S., Najafi, F., Ghahremaninejad, F., Landberg, T., & Greger, M. (2017). Comparison of silicon nanoparticles and silicate treatments in fenugreek. *Plant Physiology and Biochemistry*, *115*, 25–33.

Osman, H. E., Al-Jabri, M., El-Ghareeb, D. K., & Al-Maroai, Y. A. (2020). Impact of aluminum and zinc oxides on morphological characters, germination, metals accumulation and DNA in fenugreek (*Trigonella foenum-graecum*). *Journal of the Saudi Society of Agricultural Sciences*, *19*(8), 510–520.

Piermatti, O. (2021). Green synthesis of Pd nanoparticles for sustainable and environmentally benign processes. *Catalysts*, *11*(11), 1258.

Radwan, A. M., Aboelfetoh, E. F., Kimura, T., Mohamed, T. M., & El-Keiy, M. M. (2021). Fenugreek-mediated synthesis of zinc oxide nanoparticles and evaluation of its *in vitro* and *in vivo* antitumor potency. *Biomedical Research and Therapy*, *8*(8), 4483–4496.

Ramamurthy, C., Sampath, K. S., Arunkumar, P., Kumar, M. S., Sujatha, V., Premkumar, K., & Thirunavukkarasu, C. (2013). Green synthesis and characterization of selenium nanoparticles and its augmented cytotoxicity with doxorubicin on cancer cells. *Bioprocess and Biosystems Engineering*, *36*(8), 1131–1139.

Rao, R. D., Kodandaramaiah, J., Reddy, M. P., Katiyar, R. S., & Rahmathulla, V. K. (2007). Effect of VAM fungi and bacterial biofertilizers on mulberry leaf quality and silkworm cocoon characters under semiarid conditions. *Caspian Journal of Environmental Sciences*, *5*, 111–117.

Rezvani, N., Sorooshzadeh, A., & Farhadi, N. (2012). Effect of nano-silver on growth of saffron in flooding stress. *International Journal of Agricultural Biosystems Engineering*, *6*(1), 11–16.

Sadak, M. S. (2019). Impact of silver nanoparticles on plant growth, some biochemical aspects, and yield of fenugreek plant (*Trigonella foenum-graecum*). *Bulletin of the National Research Centre*, *43*(1), 1–6.

Singh, C., Ali, M. A., & Sumana, G. (2016). Green synthesis of graphene-based biomaterial using fenugreek seeds for lipid detection. *ACS Sustainable Chemistry & Engineering*, *4*(3), 871–880.

Singh, R. P., Magesh, S., & Rakkiyappan, C. (2011). Formation of fenugreek (*Trigonella foenum-graecum*) extract-mediated Ag nanoparticles: Mechanism and applications. *International Journal of Bioengineering and Science Technology*, *2*(3), 64–73.

Sun, S. N., Wei, C., Zhu, Z. Z., Hou, Y. L., Venkatraman, S. S., & Xu, Z. C. (2014). Magnetic iron oxide nanoparticles: Synthesis and surface coating techniques for biomedical applications. *Chinese Physics B*, *23*(3), 037503.

Varghese, R., Almalki, M. A., Ilavenil, S., Rebecca, J., & Choi, K. C. (2019). Silver nanoparticles synthesized using the seed extract of *Trigonella foenum-graecum* L. and their antimicrobial mechanism and anticancer properties. *Saudi Journal of Biological Sciences*, *26*(1), 148–154.

Vergheese, M., & Vishal, S. K. (2018). Green synthesis of magnesium oxide nanoparticles using *Trigonella foenum-graecum* leaf extract and its antibacterial activity. *Journal of Pharmaceutical and Phytochemistry*, *7*(3), 1193–1200.

Wani, S. A., & Kumar, P. (2018). Fenugreek: A review on its nutraceutical properties and utilization in various food products. *Journal of the Saudi Society of Agricultural Sciences*, *17*(2), 97–106.

Zhao, L., Peralta-Videa, J. R., Rico, C. M., Hernandez-Viezcas, J. A., Sun, Y., Niu, G., & Gardea-Torresdey, J. L. (2014). CeO_2 and ZnO nanoparticles change the nutritional qualities of cucumber (*Cucumis sativus*). *Journal of Agricultural and Food Chemistry*, *62*(13), 2752–2759.

Zhu, Y., & Gong, H. (2014). Beneficial effects of silicon on salt and drought tolerance in plants. *Agronomy for Sustainable Development*, *34*, 455–472.

CHAPTER 8

Biotic Stress Alleviation in Fenugreek

AB WAHEED WANI,[1] HARJINDER KAUR,[1] THEMMEICHON CHAMROY,[1] SANJEEV KUMAR,[2] ANIS A. MIRZA,[1] RUPESH KAUSHIK,[1] SHAKIR ALI,[3] ZARINA,[4] and ANMOL[1]

[1]*Department of Horticulture, School of Agriculture, Lovely Professional University, Phagwara, Punjab, India*

[2]*Department of Fruit Science, Dr. Yashwant Singh Parmar University of Horticulture and Forestry, Nauni, Solan, Himachal Pradesh, India*

[3]*Division of Fruit Science, Sher-e Kashmir University of Agricultural Science and Technology, Kashmir, India*

[4]*Division of Vegetable Science, Sher-e Kashmir University of Agricultural Science and Technology, Jammu, India*

ABSTRACT

The term "biotic stress" refers to the harm caused to plants by living organisms, such as infections and pests. Implementing strategies to mitigate the damage caused by these organisms is essential for alleviating biotic stress in fenugreek. Several techniques can be employed to reduce biotic stress in fenugreek.

One such technique is biological control, which is an important part of integrated pest management (IPM) strategies. This method uses beneficial organisms to reduce insect populations, promoting ecological balance and reducing the dependence on synthetic pesticides. Another effective strategy for alleviating biotic stress in fenugreek is crop rotation. Rotating the cultivation of fenugreek with non-host crops can help reduce the population of specific diseases and pests by depriving them of their preferred host.

Genetic resistance plays a crucial role in reducing biotic stress. Breeding programs are being used to develop fenugreek varieties with enhanced

Fenugreek: Improvement in Yield Potential Through Plant Growth Regulators (PGRs) and Depolymerized Polysaccharides. Tariq Ahmad Dar, Bilal Ahmad, Moinuddin, and Khalid Rehman Hakeem (Eds.)

tolerance to diseases and pests. This involves identifying and incorporating resistance genes into cultivated varieties. Additionally, cultural practices help in reducing biotic stress. Proper spacing, irrigation, and nutrient management improve plant vigor, making plants more resistant to pests and diseases.

Regular monitoring and early detection of biotic stress symptoms enable prompt action, minimizing damage. Organic farming techniques can also be adopted by farmers to reduce reliance on synthetic chemicals. These methods have fewer negative environmental impacts and promote the natural balance of beneficial organisms.

In conclusion, a combination of techniques, including integrated pest management, crop rotation, genetic resistance, cultural practices, and organic farming, can be employed to reduce biotic stress in fenugreek. By implementing these strategies, farmers can mitigate the negative effects of diseases and pests, leading to healthier and more productive fenugreek crops.

8.1 INTRODUCTION

Due to global warming and potential climate anomalies, crops often face combinations of abiotic and biotic stress, negatively impacting their growth and productivity (Atkinson et al., 2013; Mahalingam, 2015; Suzuki et al., 2014). Plant stress is described by Lichtenthaler (1996) as "any unfavorable conditions or substances that affect or block a plant's metabolism, growth, or development." Plant stressors can be categorized as either "internal" or "external" based on their origin. The terms "stress factors" and "stressors" are typically used to describe external and internal strains, respectively (Kranner et al., 2010).

The term "biotic stress" refers to damage caused by various living organisms, such as weeds, insects, parasitic nematodes, bacteria, viruses, fungi, and other native or cultivated plants. Plant pathogens such as viruses, bacteria, fungi, and nematodes are responsible for most plant diseases. After infecting different plant organs, these pathogens cause symptoms such as spots, wilts, and cankers. Nematodes target every part of the plant, consuming plant cell material and causing nutrient deficiencies, which can result in stunting or wilting. They also facilitate soil-borne diseases by making it easier for pathogens to enter the root system. However, nematodes rarely kill the host. Plants are also harmed when mites and insects lay eggs or feed on them. Piercing-sucking insects, for instance, can act as virus vectors, transmitting viruses to plants through their stylets (Schumann and D'Arcy, 2010).

Trigonella foenum-graecum L., commonly known as fenugreek, is an annual, drought-tolerant, multipurpose crop (used as an herb, spice, vegetable, forage, and legume) with significant potential for adoption in suitable agroclimatic zones worldwide. The plant is widely cultivated in countries including Argentina, the Indian subcontinent, Egypt, and Algeria, and it originates from Southwest Asia. The somatic chromosome numbers of fenugreek are 2n = 14, 16, 30, and 46, and these chromosomes have been used to study its genome. The genome size of fenugreek is approximately 685 Mbp (Vaidya et al., 2013; Young et al., 2003). Over 260 species of *Trigonella* have been identified globally (Acharya et al., 2006), but fenugreek farming is still primarily concentrated in India (Verma et al., 2020).

Fenugreek is largely grown in Indian states such as Gujarat, Madhya Pradesh, Uttar Pradesh, and Rajasthan. Rajasthan is the largest production center, with over 80% of its area devoted to fenugreek cultivation. Green fenugreek leaves contain carotene and ascorbate, and fenugreek seeds are widely used in pharmaceutical, nutraceutical, and medicinal applications (Thomas et al., 2011). Fenugreek seeds are also rich in fiber, iron, calcium, and zinc (Hooda and Jood, 2003). Additionally, fenugreek is beneficial in managing conditions such as diabetes, hyperglycemia, and hypercholesterolemia.

8.2 PLANTS' REACTIONS TO BIOTIC STRESSORS

Numerous microorganisms, including bacteria, fungi, and insect herbivores, pose a risk to plants (Ramegowda and Senthil-Kumar, 2015). For their protection, plants have evolved a variety of defense mechanisms, some of which are activated in response to pathogen invasion (Durner et al., 1997). Plants' extracellular receptors, located on cell membranes, sense microbial attacks and detect pathogen-associated molecular patterns before harmful microorganisms can spread throughout the plant (Boller and He, 2009; Chisholm et al., 2006).

As pathogens develop mechanisms to overcome primary defenses, plants acquire a more specific immune response known as effector-triggered immunity (ETI; Bilgin et al., 2010). In ETI, resistance (R) genes are typically expressed, enabling the plant to recognize pathogenic molecules that enter host cells. This interaction between effector molecules from pathogens and plant receptors initiates changes within the plant (Miedes et al., 2014). A subsequent protective mechanism involves a form of programmed cell death, or apoptosis, intended to prevent infections from spreading by restricting nutrients and water to the infected area. This response requires an oxidative burst in the form of reactive oxygen species (ROS) (Sato et al., 2010).

8.3 RESPONSES OF PLANTS TO VIRAL INFECTIONS

As pathogens, viruses harm plants and reduce production (Anderson et al., 2004). Because viruses have a small genome and the capacity to spread disease, they must come into contact with plants. Plant proteins and viral proteins interact to complete the viral life cycle. Molecular processes and tolerance mechanisms are still present in viral illnesses; however, many remain unknown (Kang et al., 2005). To explore the interactions between plants and viruses, proteomics uses subcellular fractionation (Haynes and Roberts, 2007). For example, *Nicotiana benthamiana* is impacted by the pepper mottle virus, where proteomics research has characterized changes in several proteins involved in the Calvin cycle and respiratory chain (Pineda, Sajnani, and Barón, 2010). To combat viral genetic resistance, it is necessary to control vectors, use virus-free plant materials, and reduce transmission methods (Durrant and Dong, 2004; Kang et al., 2005; Wu et al., 2010).

8.4 PLANTS RESPONSES TO FUNGAL INFECTION

Typically, plants communicate resistance to biotrophs by utilizing salicylic acid (SA) to signal tolerance against pathogenic attack. On the other hand, jasmonic acid (JA) and ethylene (ET) signaling play a crucial role in resistance to necrotrophs (Caarls et al., 2015). As a result, during biotrophic fungal attacks, gene resistance and SA-dependent protective strategies are both employed (Thomma et al., 2001). Although JA-dependent gene responses do not begin immediately, they have been shown to be highly effective when activated (Glazebrook, 2005). Under biotic stress, nitrogen's role in synthesizing photosynthetic proteins is diminished, with the majority being used to produce compounds essential for a sensitive pea cultivar's biotic defense mechanism against fungi (Amey et al., 2008). Continuous changes in protein concentration during rust infection were discovered through a proteomics investigation in the bean leaf (*Phaseolus vulgaris*; Rodríguez-Moreno et al., 2008). It has been found that R gene expression, which is involved in producing defensive proteins in plants, enhances the plant's defense against fungal pathogens (Collins et al., 2003). This gene expression plays a significant role in the plant's natural defense mechanism and contributes actively to strengthening resistance mechanisms (Lee et al., 2009).

8.5 RESPONSES OF PLANTS TO HERBIVORY

Insects and plants' similar evolutionary histories demonstrated varied strategies for creating mode of tolerance for one another (Verhage et al., 2010). Because of this connection, plants have evolved strong defense mechanisms that can detect impulses coming from tissues damaged by insect attack (Hare, 2011). Both directly and indirectly, plants defend themselves from pests by altering their reproductive behavior (direct defense) and by luring other creatures who are pest adversaries or predators (indirect defense; Dudareva et al., 2006). Using indirect defenses entails the pests' predators' attraction to them for food and lodging (Arimura et al., 2009). Plants have created resistant genotypes in the race against herbivores, nonetheless (Sharma et al., 2009). The complex and active defense mechanism used by plants in response to herbivore wounds includes physical barriers and harmful substances (Karban, 2011). According to Miranda et al. (2007), plant response is an important strategy for the control of pests and insects in crops. These are crucial if they are made to respond to pests' infection and are meant to lessen frequent pressures (Miranda et al., 2007). Directly occurring induced responses are very beneficial to crops since they reduce subsequent herbivore attacks and improve the overall adaptability of valued crops (Agrawal, 2011). The emergence of insect-plant communication has corrected our understanding of how plants evolved defense mechanisms against herbivory (Tables 8.1–8.3; Karban, 2011).

TABLE 8.1 Genes for Host Defense Against Viral Infections

Sl. No.	Host	Pathogen	R-Gene	Place in the Genome	Avr-Gene	References
1.	*Solanum peruvianum*	The TSWV (tomato spotted wilt virus)	*Sw-5*	Long arm of chr 9	–	Roselló et al. (2001)
2.	*Nicotiana glutinosa* L.	TMV, or tobacco mosaic virus	Metallothionein-like gene	–	–	Chen et al. (2007)
3.	*Capsicum chinense*	The TMV (tobacco mosaic virus)	*L3*	–	Coat protein	Back et al. (2002)
4.	Potato	X-potato virus	*Rx1*	Top arm of chr 12	Coat protein	Perveen and Qaiser (2007)
5.	Capsicum	Turning-pot mosaic virus	*L1, L2, L3*	–	Coat protein	Gururani et al. (2012)

TABLE 8.2 Genes That Confer Resistance to Fungi

Sl. No.	Host	Pathogen	R-Gene	Place in the Genome	Avr-Gene	References
1.	*Solanum peruvianum*	*Cladosporium fulvum*	*Cf-2*	Short arm of Chr 6	Avr-2	Dixon et al. (1998)
2.	*Pimpinellifolium*	*Fusarium oxysporum*	*I-2*	–	Avr1	Song et al. (2017)
3.	*Nicotiana benthamiana*	*Colletotrichum destructivum*	*NbGSTU1, NbGSTU2, NbGSTU3, NbGSTF1,*	–	–	Song et al. (2017)

TABLE 8.2 *(Continued)*

Sl. No.	Host	Pathogen	R-Gene	Place in the Genome	Avr-Gene	References
4.	*Solanum torvum*	*Verticillium dahliae*	*StoVe1*	–	–	Song et al. (2017)
5.	*Capsicum chacoense*	*Xanthomonas campestris*	*Bs2*	–	Avr-Bs2	Liu et al. (2012)

TABLE 8.3 Pests' Increasing Host Horizon in Vegetable Crops

Sl. No.	Insect Pest	Old Host Crops	New Host	References
1.	*Liriomyza trifolii*	*Lycopersicum esculentum*	*Solanum melongena*, *Phaseolus vulgaris*, leafy vegetables, cucurbits	Durairaj et al. (2016)
2.	*Helicoverpa armigera*	*Lycopersicum esculentum*, *Brassica oleracea* var. capitate	*Pisum sativum*, *Capsicum frutescens*, *Solanum melongena*, *Abelmoschus esculentus*	Durairaj et al. (2016)
3.	*Pieris brasicae*	*Brassica oleracea* var. capitata *Brassica oleracea* var. botrytis, and *Brassica nigra*	*Brassica oleracea*, *Raphanus sativus*	David and Asamoah (2011)
4.	Gall midge	*Solanum melongena*	*Capsicum frutescens*, *Solanum melongena*, *Capsicum annum*	David and Asamoah (2011)
5.	*Tetranychus* sp.	–	*Solanum melongena*, *Abelmoschus esculentus*, *Lablab purpureus*	David and Asamoah (2011)

8.6 HOW DO PLANTS PROTECT THEIR SELVES FROM BIOTIC STRESS?

In order to fight against biotic stresses, plants often have a highly evolved system of immunity. First, plants have physical defenses like waxy substances, thick cuticles, and distinctive trichomes that keep pathogens and insects from attaching to them. In order to protect themselves from pathogens and herbivores, plants also create chemical substances (Zaynab et al., 2018). Plants can identify infections thanks to two systems that start defensive responses. The first route is a receptor for pattern recognition that can identify pathogen-associated molecular patterns (PAMPs; Monaghan and Zipfel, 2012). Plant resistance (R) proteins, the second route of the immune system, recognize the specific effectors of pests or pathogens (Avr proteins) and trigger the plant defense response via an effector-triggered immunity (ETI) pathway. The hypersensitive reactions (HR), which involve programmed cell death in the damaged cells and their surrounds, can be brought on by this kind of mechanism, according to Spoel and Dong (2012). Among the signaling pathways generated by PTI and ETI, a number of plant hormones, including ET, JA, and SA, stand out. The three hormones SA, JA, and ET signal two strong defense mechanisms against plant pathogens in general. After primary infection with a necrotizing pathogen, systemic acquired resistance (SAR), the first kind, is triggered and is followed by increased levels of SA and associated pathogenesis proteins (Grant and Lamb, 2006). The second route of resistance of plant is known as induced systemic resistance (ISR), which is signaled by particular species of non-pathogenic bacteria that colonize the roots. JA and ET are needed for this type of resistances signaling. Some pests and insects make plants to release volatile substances to attract their enemies and warn their neighboring plants of approaching hazards by detecting retained herbivore-associated stimulants of the invaded insect (Santamaria et al., 2013).

8.6.1 DIRECT DEFENSES

Enzyme inhibitors, alkaloids, and phenolic compounds all individually reduced tomato growth when ingested, but they also interacted with one another and the digestion of bugs (Carrière et al., 2015). The following modifications take place in reaction to insects.

8.6.2 MORPHOLOGICAL STRUCTURES

The first line of protection to herbivory is the development of plant frameworks, such as the synthesis of a waxy cuticle and the development of trichomes, thorns, and so on (Sharma et al., 2009). Physical defenses are made up of physical and internal characteristics that assist crops physiologically by preventing pests from becoming infected directly (Agrawal and Konno, 2009), as well as minute changes in the cell wall's makeup that result in variations in thickness (He et al., 2011). Physical traits like thorns, trichomes, and stiffened leaves play a big part in plant defense system to pest and insect attack (Chamarthi et al., 2014).

8.6.3 TRICHOMES

Trichomes are essential for defending plants against a variety of pests (Hanley et al., 2007). Trichomes also hinder herbivory and alter other arthropods' infections on plant surfaces, which prevents them from coming into touch with the epidermis of leaf (Agrawal and Konno, 2009). Terpenoids, flavonoids, and other secondary metabolites that can repel, injure, and catch insects and other pests are secreted by glandular trichomes, resulting in a variety of plant defenses (Sharma et al., 2009). The extension of a pest infection may result in more cell division, whereas thickness remained constant (Agrawal and Konno, 2009).

8.6.4 PLANT METABOLITES

8.6.4.1 PRIMARY METABOLITES

For cell development, primary metabolites are necessary. They assist fundamental metabolic functions including respiration and photosynthesis and are produced during the many stages of cell growth (Apel and Hirt, 2004). Amino acids, tricarboxylic acids, carbohydrates, structural components, and energy sources are among the most well-known metabolites in the majority of organisms (Zientz et al., 2004). Proteins, nucleic acids, and polysaccharides are also regarded as primary metabolites in addition to the aforementioned components (Tettelin et al., 2001).

8.6.4.2 SECONDARY METABOLITES

Secondary metabolites are created from primary metabolites. According to Bourgaud et al. (2001), the generation of secondary metabolites is not constant. These substances are the byproducts of primary metabolites. They typically do not participate in the fundamental growth and an organism's developmental processes are essential for survival (Walker et al., 2003); their absence puts the organism's defense at risk (Kumar et al., 2014). Various organisms, including plants, algae, fungi, bacteria, and mammals, naturally create chemicals known as secondary metabolites (Gribble, 2003). They only last a short time, and plants can continue to grow without them, demonstrating that they are not necessary for plant growth. Another essential component of metabolism is secondary metabolism. The cellular machinery, vital enzymes, and energy required for the plants' long-term survival are all produced by the fundamental metabolism (Roze et al., 2011).

8.6.5 SECONDARY METABOLITES AND PLANT DEFENSE

The production of secondary metabolites reduces the palatability of plant tissues and is not fundamentally associated with normal plant growth and development (Boege and Marquis, 2005). According to Howe and Jander (2008), they are also referred to as defensive metabolites since they accumulate in inactive forms before becoming active in response to the pest invasion. Originally known as phytoanticipins, they are now called phytoalexins. The -glucosidase activates phytoanticipins, which regulate the formation of many chemical substances, during herbivory (Morant et al., 2008). The development of anti-insect chemicals depends on phytoalexins, which include alkaloids, terpenoids, and other metabolites (Walling, 2008). They enhance the capacity of the plants and shield them from various challenges. Plants are protected from insect pests by lignin, which is phenolic in nature (War et al., 2012). Pest infestations cause its production, and its rapid deposition has an impact on pathogen fertility. The accumulation of these substances in amino acids reduces the insects' ability to digest proteins, which has an impact on pest proliferation. Phenols also inhibit the formation of defense enzymes by lowering the levels of singlet oxygen and hydroxide radicals involved in a number of reactions (Sharma et al., 2012). Simple phenolics called salicylates act as insect herbivores, Salix leaves act as an antifeedant; SA is a highly important preventative agent (Hardie et al., 2016).

8.6.5.1 PRIMARY GROUPS

1. **Terpenes:** One of the main categories of chemicals that share a biosynthetic origin from glycolytic metabolites is the terpenes. The majority of plant terpenes are secondary metabolites that have toxic effects on insects and other herbivores. The following categories apply to terpenes.
2. **Monoterpenes (C10):** Monoterpenes are common chemical substances released by plants that have a significant role in mediating insect toxicity. For instance, different Chrysanthemum species and their foliage and flowers produce monoterpene esters that exhibit potent responses of pesticide to pests like moths and are a common component in effective pesticides because of their low environmental stability and low mammalian toxicity (Taft et al., 2015). Monoterpenes, particularly myrcene and limonene, are secreted by resin-secreting cells found in the stem of conifers. All monoterpenes are toxic to a variety of pests (Sarria et al., 2014).
3. **Sesquiterpenes (C15):** These are known for serving as antiherbivore agents in plant defense. They have a five-membered lactone

ring and effectively deter various herbivorous insects from feeding. Sesquiterpene abscisic acid plays regulating functions in the maintenance of seed and bud dormancy, and plants respond to water stress by altering membrane characteristics. It also works as a transcription activator. Additionally, it raises the amount of calcium in the cytosol (Cheval et al., 2013).

4. **Diterpenes (C20):** The leguminous plants make the diterpene abietic acid. It is present in resins that the tree trunk's resin canals exude; when pests pierce these canals, the release of resin may severely impede eating. Diterpene, produced by Euphorbiaceae plants, irritates the epithelium and has toxic effects inside the body of pest, animals, and insects. Additionally, gibberellin is a diterpene that has a variety of functions during different stages of plant development, including leaf growth, and fruiting (Ouzounidou and Ilias, 2005).
5. **Triterpenes (C30):** Triterpenes, several of these triterpenes have a crucial role in the structure of plant cell membranes by regulating protein channels and maintaining the ability of the membrane to absorb small particles by shrinking lipid globules. Triterpenes called phytoecdysones offer some self-protective properties against insects by changing how they develop, combined with some undesirable adverse impacts (Fatope et al., 1990).
6. **Polyterpenes (C5):** These are a variety of high-molecular-weight terpenes, such as tetra and polyterpenes, which are produced by plants. The primary tetraterpenes, or pigments, are carotenoids; another one is rubber, which is found in laticifers and provides defense by acting as a mechanism for healing infections and as a resistance to pests (Silvestre and Gandini, 2008).
7. **Phenolic Compounds:** According to Wuyts et al. (2006), plants synthesize several secondary products that have a phenol group, an OH functional group connected to a benzene ring, and are an important part of their defense mechanism against herbivorous pests. Through the shikimic acid system, vascular plants create simple phenolic compounds in a variety of ways to aid in various plant defense activities against insect herbivores and fungus (Ghasemzadeh et al., 2011). According to Brooker et al. (2008), they are a particularly potent category of chemicals that have antipathogenic effects on bacteria and fungus. We discover that

these cyclic molecules in plants provide natural pesticide defense and alter a variety of antifungal action. *In vitro*, the halogenated chemical coumarin effectively inhibits the growth of fungi (Serghini et al., 2001).

8. **Furano-Coumarins**: These compounds are characteristic of Umbelliferae plants and are noteworthy due to their phytotoxicity. When activated by UV light, furano-coumarins can be harmful as they inhibit DNA transcription and repair, ultimately leading to cell death (Sajjadi et al., 2013). One particularly important furano-coumarin is psoralen, which plants commonly use as a defense mechanism against pathogenic fungal growth.

9. **Lignin:** Coniferyl, sinapyl, and coumaryl are the phenylpropanoid groups for a big branching polymer in lignin. These three distinct alcohols could be used as a starting point for the production of lignin. Lignin is resistant to herbivorous feeding and penetration due to its physical toughness and chemical strength, respectively (Walters, 2011).

10. **Flavonoids:** About 5000 different kinds of flavonoids have been identified in plants, according to Yao et al. (2004). Plants are protected by flavonoids against a variety of stressors, such as insect infestations. They help plants resist herbivory, control reactive oxygen species (ROS), and reduce ROS generation by chelating metals. They are separated into several groups, including proanthocyanidins, flavonols, chalcones, flavanones, and dihydroflavonols. Numerous flavones have been investigated as potential insect feeding inhibitors. For Spodoptera exempta, flavonoids such flavones act as feeding inhibitors (da Silva et al., 2017).

11. **Isoflavonoids:** These are flavonone derivatives that are mostly found in plants. They are essential for plant development and are produced by legumes. They also serve a critical function in promoting the establishment of root nodules by symbiotic bacteria (Sreevidya et al., 2006). Plants actively combat ROS by producing these flavonoids (Posmyk et al., 2009).

12. **Tannins**: Tannins are a class of phenolic polymers with anti-herbivory properties commonly found in plants. These natural toxins act as feeding deterrents for pests and significantly inhibit the growth and sustainability of many pests (Barbehenn et al., 2008). Mammals

often avoid plants with high levels of tannic acid. Chlorogenic and protocatechuic acids play various roles in disease resistance among different plants. Many disease-resistant plant varieties produce chlorogenic acid, which is believed to oxidize into quinones that combat fungal infections (Haldar et al., 2015).

Tannins inhibit insect growth by binding with proteins, reducing the efficiency of nutrient absorption, and causing lesions in the midgut (Barbehenn and Constabel, 2011). They are odorless compounds that irritate insects during feeding (Eneh et al., 2016). Tannin consumption by herbivores lowers the nutritional value of plants, as tannins make plant proteins more challenging to digest (Teferedegne, 2000).

13. **Plant Defensive Proteins:** Based on microarray and proteomic approaches, it has been demonstrated that a wide variety of PRPs is necessary for plants to be protected from insect attack. Numerous signaling pathways are activated as a result of the various feeding strategies of arthropods, resulting in the synthesis of JA, SA, and ET and influencing the arthropod-inducible proteins (Arimura et al., 2009).

8.6.5.2 SECONDARY METABOLITES OF SULFUR

This group consists of phytoalexins, glycosphingolipids, thionins, allinins, and defensins, and so on, are used for the purpose of protecting them from microbial diseases (Halkier and Gershenzon, 2006):

1. **GSH:** An organic substance is glutathione (GSH). S is found in a variety of amino acids, including cysteine and methionine. It is also present in the soluble portion of GSH, which works as a transportable reservoir of S during growth and development settling. During times of stress, it also functions as an antioxidant in the cell by bolstering the plant's defense mechanism. Particular cells, such as trichomes, demonstrated an enhanced synthesis of GSH synthesis-related enzymes and a few other chemicals necessary for plants to defend themselves against heavy metal stress. The primary cellular regions of plants have been discovered to contain a concentration of GSH between 3 and 10 mM (Nocito et al., 2002).
2. **Phytoalexins:** Plants create phytoalexins during biotic stress to help inhibit the spread of infections by clumping together around the

infection site; this appears to be a common strategy used by many plants to defend themselves against insect pests (Jeandet et al., 2013). Many variations are connected to a quick, appropriate response that kills plants. This response is known as a hypersensitive response. Many families of plants produce phytoalexins of an organic nature, such as isoflavonoids in the Leguminosae and sesquiterpenoids in the Solanaceae, but only the Crucifereae produces the S metabolites that set it apart from the other GSL chemicals. Cruciferous crops are crucial because they contain a lot of phytoalexins, is a feature of this family (Monde et al., 2003). They also have a role in pest defense, for an example of this is the production of derivatives of furano-acetylenic following a pest attack on *Vicia fava* (Soylu et al., 2002).

3. **Defensins, Thionins, and Lectins:** Under biotic stress, proteins rich in defensins, thionins, and lectins are generated; the majority of these proteins protect plants from fungal infection. In addition, infections partially stimulate the genes for defensins (Gu et al., 2002).

8.7 VARIOUS BREEDING TECHNIQUES TO IMPROVE BIOTIC STRESS RESISTANCE

8.7.1 BREEDING FOR RESISTANCE

One of the numerous illnesses which harm the *Trigonella foenum-graecum*, which is a typical issue since it lowers production, quality as well as productivity of crop, is the powdery mildew. To find the underlying causes, it is desirable to utilize screening techniques primarily for generations in the case of breeding resistance. In the early years of fenugreek cultivation, diseases were less of a problem, but with breeding for resistance is currently necessary due to the spread of many diseases and pests, the introduction of crops in new locations, and other serious challenges. The screening process must be simple, practical, and efficient to make varieties resistant to various biotic stresses. In addition to the illness, unfavorable climatic conditions like salinity, high temperature, low temperature, and various other factors automatically reduce the production of seeds. The pressures have a significant impact on physiological, biochemical, and morphological processes. Fenugreek is vulnerable to salt issues in the growing region, but effective breeding techniques can provide resistant cultivars that are acceptable for the region. Fenugreek crops can also be produced with high quality through resistance breeding, which eliminates the need to use pesticides, which frequently leave behind undesired residues. By using the same resistant breeding approach as that used to create salt tolerant cultivars, you can stop the emergence of powdery mildew. In a study it was found that overall, the cultivars of fenugreek that are resistant to the powdery mildew disease include RMT-1, RMT-143, and RMT-305. As a result, RMT-305 has been identified as a sustainable powdery mildew disease resistant variety (Soylu et al., 2002).

8.7.2 INCREASING GERMPLASM PRESERVATION AND COLLECTION

Scientists have created attractive, productive cultivars of *Trigonella foenum-graecum* by using the proper selection methods from stocks of native or foreign material. The careful gathering and maintenance of native germplasm, alien material, particularly from primary sites of origin, and subsidiary centers of origin diversity must also be introduced. According to reports, mass methods of selection have been used frequently to generate fenugreek, a hardy self-pollinating crop. Selection from a single plant or a pure line has been utilized frequently. Compared to other techniques like pest-resistant genes, different technologies of screening of germplasm and various sources used in resistance of disease are significant as well as impactful (Malhotra and Vashishtha, 2007). By collecting and analyzing germplasm, the seed spice (fenugreek) advances genetic potential attributing and ultimately increase variability which result in the creation of improved varieties. The majority of traditional crops are preserved on farms, which preserve crop genetic variety and traditional knowledge while growing them frequently on small plots of land and in backyard gardens (Padulosi et al., 2013). At various centers of the All India Coordinated Research Projects, the 989 fenugreek seed accessions are gathered, assessed, maintained, and conserved (Lal, 2018). On the other hand, the adoption of enhanced varieties could result in the irretrievable loss of valuable genes if gene collection and conservation are not quickly secured. The quantity of diversity and germplasm obtained might be evaluated for commercially significant traits connected to the crop development program (Lal, 2018).

8.7.3 MUTATION BREEDING

In recent years, the use of mutation breeding as a successful strategy for crop improvement across several crops has grown in popularity. The report

(Thomas et al., 2006) demonstrated successful crop enhancement through fenugreek mutant breeding. Through mutation breeding, it is possible to discriminate between normal plant types, which may be governed by a single dominant gene, and large mutant plant types, which may be governed by a recessive allele. The treatment of seeds of *Trigonella foenum-graecum* with chemical mutagens, such as ethyl methane sulfonate (EMS), can damage the chromosomes, leading to individual mutations as well as the addition or deletion of chromosomes from cells (Petropoulos, 2002). Mutagenesis was used in agriculture as a novel alternative to traditional breeding techniques for crop plant improvement. The fenugreek crop has a pretty wide range of adaptation in many locales, making it more significant economically, agronomically, and environmentally. Mutation breeding can be used to fix a large number of alleles at once for a specific characteristic of interest and to provide new genetic diversity to an existing gene pool for fenugreek traits (Zandi et al., 2017). A crucial tool for agricultural development, mutagenesis is exempt from the legal constraints imposed on genetically modified plants. According to agricultural improvement discoveries, the FAO officially released and registered 3,248 mutant varieties (Sureshkumar et al., 2018). In a study it was found that the most effective method for applying mutagens is 0.4% of plant growth is stimulated by EMS dosage (Kavina et al., 2020). Chemical mutagenesis has been demonstrated to improve *Trigonella foenum-graecum* crop quality like seed yield (Basu et al., 2008). According to Raina et al. (2016), produced varieties by mutation improve biodiversity and provide breeding material for traditional plant breeding, directly contributing to conservation and the use of plant genetic resources.

8.7.4 TECHNIQUE OF TISSUE CULTURE

A multidisciplinary cooperation approach is promised since the disease resistance and quality improvement have a unique place in seed spices like *Trigonella foenum-graecum* crop to fulfil the goal of creating higher-quality, disease-resistant, and high-yielding cultivars. Effective control efforts have so far prevented powdery mildew in fenugreek. Variation can be produced using the tissue culture technique (Malhotra and Vashishtha, 2007). *In vitro* mutagenesis and mass population screening for tolerance to biotic and abiotic stressors can increase the frequency of somaclonal diversity, but regeneration is also challenging, and repeatability is low (Malhotra and Vashishtha, 2007). The creation of uniform and fake disease epiphytotics is necessary for the identification of the pathogen causing the disorders. The rapid resistance screening program, on which research must be started to make resistance breeding effective, is aided by the use of tissue culture. However, because it eliminates the need for the application of pesticides, which frequently leaves unwanted residue, resistance breeding is also directly related to produce quality (Mulat, 2017).

8.7.5 BREEDING WITH MARKER-ASSISTED SELECTION

The performance of conventional breeding is largely dependent on the superior genotypes' selection phenotypically. Therefore, phenotypic selection has a built-in flaw. Quite little despite advances in sequencing technology, efforts have been made to quantify the genetic diversity among fenugreek genotypes in recent years. Aasim et al. (2018) conducted several studies on the genetic diversity of fenugreek using a variety of conventional genetic markers, including the use of 14 ISSR and 22 RAPD markers. Eight landraces were recently explored using a combination of six SRAP primers (Amiriyan et al., 2019), 90 genotypes were investigated using 13 SSR and 49 RAPD markers, and 17 accessions were estimated (Dangi et al., 2004). At the moment, it is common practice to select plants using DNA markers that have genomic areas implicated in the expression of desired features. Through marker-assisted selection, plant genetic traits might be given more weight and precise gene transfer, which would hasten the recovery of recurring parent genomes like that of fenugreek (Young et al., 2003).

8.7.6 USES OF METAL AND METAL OXIDE NANOPARTICLES

In addition to promoting plant growth and resilience under various stress conditions, metal-based nanoparticles (NPs) are promising agents for managing plant diseases, pests, and insects. These NPs improve plant biomass and crop yield through mechanisms such as crop protection (nano-pesticides), stress tolerance, soil enhancement, and growth promotion (Kah et al., 2019; Khan et al., 2021; Tripathi et al., 2022). Metal NPs employ multiple antibacterial mechanisms that directly impact microorganisms. By releasing ions and penetrating cells, they damage DNA, disrupt cell membranes, and inhibit electron transport. They also produce reactive oxygen species (ROS), which can damage DNA, alter enzyme function, and denature proteins (Dizaj et al., 2014). While metal-based NPs have positive effects on agriculture, some studies have investigated their toxicological impacts on human health and ecosystems. Key physicochemical properties, such as size, stability, and shape, are critical in determining their toxicological effects. To maximize benefits while minimizing risks, dose-response studies have identified optimal concentrations that provide positive impacts without adverse effects (Missaoui et al., 2018).

This section focuses on the applications of silver (Ag) and copper (Cu) NPs (Chen et al., 2020; Dizaj et al., 2014; Farhana et al., 2022; Khan et al., 2021; Mikhailova, 2020). For example, Ag NPs enhanced the biochemical profile of *Silybum marianum*, promoting seed germination and increasing metabolite levels (Salman Khan et al., 2016). Another study found that Ag-priming of cabbage seeds improved yield, seedling growth, and germination speed. AgNP seed priming also increased plant nutritional value, boosting Fe levels and key amino acid content in cabbage leaves (Zhou et al., 2022). Cu NPs act as effective nanopesticides, enhancing plant nutrition and growth. According to Dizaj et al. (2014), Cu NPs exhibit antibacterial activity by penetrating bacterial cell membranes and damaging critical enzymes, effectively controlling microorganisms and phytopathogens such as *Fusarium*, *Alternaria alternata*, *Curvularia lunata*, *Phoma destructive*, *Fusarium fujikuroi*, *Aspergillus niger*, and *Drechslera sorghicola* (Kanhed et al., 2014; Shende et al., 2016). In tomatoes, Cu NPs applied topically improved growth and chlorophyll content, while reducing *Fusarium wilt* incidence and severity by 68% and 66.5%, respectively (López-Vargas et al., 2018).

In addition to Ag and Cu, metal-based titanium dioxide (TiO_2) NPs show potential for reducing biotic stress in crops. By generating ROS and causing lipid peroxidation, TiO_2 NPs induce oxidative stress, which increases membrane fluidity and compromises pathogen cell integrity (Dizaj et al., 2014). Zinc oxide (ZnO) NPs are another noteworthy metal-based NP. ZnO NPs can disrupt cell membranes, release zinc ions, generate ROS on their surfaces, and penetrate pathogen cells (Gu et al., 2002; Luksiene et al., 2020). In one approach, soil inoculation with ZnO NPs derived from *Matricaria chamomilla* reduced *Ralstonia solanacearum* populations responsible for bacterial wilt in tomatoes, while enhancing plant growth. Zn^{2+} ions released by ZnO NPs likely caused bacterial cell deformation, including membrane and cell wall rupture and leakage of cellular contents (Khan et al., 2021).

It is essential to recognize that NP effects are dose-dependent, and improper NP use may lead to phytotoxicity. NPs can directly disrupt plant homeostasis by binding to DNA bases or indirectly by generating ROS, which alters antioxidant enzyme activity. Such disruptions can shift the redox balance in plant cells, causing oxidative damage to biomolecules and altering cell signaling pathways (Arif et al., 2018). Silicon has been reported to benefit plant health in various ways (Gillman et al., 2003; Van Bockhaven et al., 2013). Silicon inhibits disease spread by reinforcing the epidermal layers of leaves and fruits and increasing low molecular weight compounds (Gillman et al., 2003; Samuels et al., 1991). Silicon not only provides physical defense against biotic and abiotic stressors but also strengthens cell walls (Pilon-Smits et al., 2009; Sangster et al., 2001). As a preventive measure, silicon delays or reduces disease onset and, as a biological inducer, activates plant defense mechanisms (Van Bockhaven et al., 2013). Silicon forms a network with cellulose in cell walls, acting as a barrier to pathogens, potentially resisting pest attacks by limiting lesion spread and spore formation for secondary infection (Seebold et al., 2001). Silicon crystals in plant tissues can also inhibit insect feeding by damaging insect mandibles, thus serving as a physical barrier. Additionally, silicon helps plants conserve water (Meyer and Keeping, 2001).

8.8 DEFENSE ELICITORS

Pest secretions are frequently reported to regulate plants affected by herbivore infections (Pauwels et al., 2009). According to Hettenhausen et al. (2014), the primary components of insect oral discharges are fatty acid-amino acid conjugates (FACs). The FACs have also been reported to boost JA production in tobacco plants, which results in the expression of genes that respond to herbivores (Falk et al., 2014).

8.8.1 JASMONIC ACID

Jasmonic acid is one of the important plant hormones in the defense towards carnivore parasites and regulates the commencement of intermediate as well as direct defensive strategies (Shivaji et al., 2010). Plants are protected from pests and insects by a variety of phytohormones. Because JA is derived from linoleic acid, when insects damage plant parts, defense signal transduction pathways are activated. The formation of trichome and the synthesis of JA, which contain antioxidant enzymes, are two examples of plant defense responses (Pauwels et al., 2009). JAs control a number of genes that help plants defend themselves from biotic stress (Shivaji et al., 2010). According to Ribot et al. (2008), the generation of Jasmonates causes plants to produce the protective chemical indole glucosinolate, which alters the intracellular calcium levels and redox status of cells (Walter et al., 2007). JA also increases plant resistance by generating chemicals that protect them from insects (Barbehenn et al., 2008).

8.8.2 SALICYLIC ACID

One of the key phytohormones that supports sustaining the plant defense and serves as a precursor in SA synthesis is benzoic acid (War et al., 2011). It is a key phytohormone which affects plant growth and development as well as a number of defensive responses. SA offers increased protection against pests and insects that pierce and suck the plant as compared to the insects which chew (Maffei, 2010; Zhao et al., 2009).

8.8.3 ETHYLENE

One of the crucial growth regulators that shields plants from various insect attacks is ethylene (van Loon et al., 2006). The infections and herbivores are defended against directly and indirectly. According to reports, HIPV manufacturing plays a part in its indirect role. In the expression of plant defense mechanisms against plant pathogens and herbivorous insects, it signaling is connected to the JA. By using ET- and JA-related signaling, PI production in response to herbivorous insects is seen in tomato (Kunkel and Brooks, 2002). According to Horiuchi et al. (2003), the ET precursor 1-amino-cyclopropane-1-carboxylic acid increases the formation of unstable chemicals from leaves treated with JA.

8.8.4 IONS OF CALCIUM (Ca^{2+})

Different signal transduction pathways are initiated in plants in response to herbivore attacks in order to defend them from this biotic stress. Many plants produce CDPKs when under stress, and they are useful for helping plants deal with both biotic and abiotic challenges (Kanchiswamy et al., 2010). According to reports, in tobacco under biotic stress, CDPKs activate MAPKs (Satti et al., 2021).

8.9 CONCLUSIONS AND FUTURE PERSPECTIVES

In conclusion, biotic stressors' effects on horticulture crops pose a serious threat to economies and food security around the world. The use of biostimulant compounds, microbes, phytochemicals, secondary metabolites, and antimicrobial peptides, however, has been demonstrated in recent studies as viable techniques to increase crop tolerance and productivity. Researchers are also making use of cutting-edge technologies like epigenomics, genomics, proteomics, and metabolomics to better understand the mechanisms underlying plant stress responses and create fresh ideas for enhancing crop durability and productivity. The development of resilient and sustainable horticulture crop production systems that can adjust to changing human activities and the global climate is dependent on ongoing research in this area. Researchers can aid in the development of more robust and successful breeding programs by looking for genotypes with high levels of dry matter and biochemical activity, choosing desired features associated to resistance, and utilizing the most recent tools and techniques. In the end, these initiatives could considerably increase crop plants' ability to withstand stress and help provide high-quality, nourishing food that is sustainable for the expanding global population.

KEYWORDS

- **biochemical activity**
- **breeding**
- **dry matter**
- **epigenomics**
- **genotypes**
- **global population**
- **metabolomics**
- **organic farming**
- **stress**

REFERENCES

Aasim, M., Baloch, F. S., Nadeem, M. A., Bakhsh, A., Sameeullah, M., & Day, S. (2018). Fenugreek (*Trigonella foenum-graecum* L.): An underutilized edible plant of the modern world. In *Global Perspectives on Underutilized Crops* (pp. 381–408).

Acharya, S., Srichamroen, A., Basu, S., Ooraikul, B., & Basu, T. (2006). Improvement in the nutraceutical properties of fenugreek (*Trigonella foenum-graecum* L.). *Songklanakarin Journal of Science and Technology*, *28*(1), 1–9.

Agrawal, A. A. (2011). Current trends in the evolutionary ecology of plant defense. *Functional Ecology*, *25*(2), 420–432.

Agrawal, A. A., & Konno, K. (2009). Latex: A model for understanding mechanisms, ecology, and evolution of plant defense against herbivory. *Annual Review of Ecology, Evolution, and Systematics*, *40*, 311–331.

Amey, R. C., Schleicher, T., Slinn, J., Lewis, M., Macdonald, H., Neill, S. J., & Spencer-Phillips, P. T. N. (2008). Proteomic analysis of a compatible interaction between *Pisum sativum* (pea) and the downy mildew pathogen *Peronospora viciae*. In *The Downy Mildews: Genetics, Molecular Biology and Control* (pp. 41–55).

Amiriyan, M., Shojaeiyan, A., Yadollahi, A., Maleki, M., & Bahari, Z. (2019). Genetic diversity analysis and population structure of some Iranian fenugreek (*Trigonella foenum-graecum* L.) landraces using SRAP markers. *Molecular Biology Research Communications*, *8*(4), 181.

Anderson, P. K., Cunningham, A. A., Patel, N. G., Morales, F. J., Epstein, P. R., & Daszak, P. (2004). Emerging infectious diseases of plants: Pathogen pollution, climate change and agrotechnology drivers. *Trends in Ecology & Evolution*, *19*(10), 535–544.

Apel, K., & Hirt, H. (2004). Reactive oxygen species: Metabolism, oxidative stress, and signal transduction. *Annual Review of Plant Biology*, *55*, 373–399.

Arif, N., Yadav, V., Singh, S., Tripathi, D. K., Dubey, N. K., Chauhan, D. K., & Giorgetti, L. (2018). Interaction of copper oxide nanoparticles with plants: Uptake, accumulation, and toxicity. In *Nanomaterials in Plants, Algae, and Microorganisms* (pp. 297–310). Academic Press.

Arimura, G. I., Matsui, K., & Takabayashi, J. (2009). Chemical and molecular ecology of herbivore-induced plant volatiles: Proximate factors and their ultimate functions. *Plant and Cell Physiology*, *50*(5), 911–923.

Atkinson, N. J., Lilley, C. J., & Urwin, P. E. (2013). Identification of genes involved in the response of *Arabidopsis* to simultaneous biotic and abiotic stresses. *Plant Physiology*, *162*(4), 2028–2041.

Back, M. A., Haydock, P. P. J., & Jenkinson, P. (2002). Disease complexes involving plant parasitic nematodes and soilborne pathogens. *Plant Pathology*, *51*(6), 683–697.

Barbehenn, R. V., & Constabel, C. P. (2011). Tannins in plant–herbivore interactions. *Phytochemistry*, *72*(13), 1551–1565.

Barbehenn, R. V., Jaros, A., Yip, L., Tran, L., Kanellis, A. K., & Constabel, C. P. (2008). Evaluating ascorbate oxidase as a plant defense against leaf-chewing insects using transgenic poplar. *Journal of Chemical Ecology*, *34*, 1331–1340.

Basu, S. K., Acharya, S. N., & Thomas, J. E. (2008). Genetic improvement of fenugreek (*Trigonella foenum-graecum* L.) through EMS induced mutation breeding for higher seed yield under western Canada prairie conditions. *Euphytica*, *160*, 249–258.

Bilgin, D. D., Zavala, J. A., Zhu, J. I. N., Clough, S. J., Ort, D. R., & DeLUCIA, E. H. (2010). Biotic stress globally downregulates photosynthesis genes. *Plant, Cell & Environment, 33*(10), 1597–1613.

Boege, K., & Marquis, R. J. (2005). Facing herbivory as you grow up: The ontogeny of resistance in plants. *Trends in Ecology & Evolution, 20*(8), 441–448.

Boller, T., & He, S. Y. (2009). Innate immunity in plants: An arms race between pattern recognition receptors in plants and effectors in microbial pathogens. *Science, 324*(5928), 742–744.

Bourgaud, F., Gravot, A., Milesi, S., & Gontier, E. (2001). Production of plant secondary metabolites: A historical perspective. *Plant Science, 161*(5), 839–851.

Brooker, R. W., Maestre, F. T., Callaway, R. M., Lortie, C. L., Cavieres, L. A., Kunstler, G., & Michalet, R. (2008). Facilitation in plant communities: The past, the present, and the future. *Journal of Ecology, 96*(1), 18–34.

Caarls, L., Pieterse, C. M., & Van Wees, S. C. (2015). How salicylic acid takes transcriptional control over jasmonic acid signaling. *Frontiers in Plant Science, 6*, 170.

Carrière, Y., Crickmore, N., & Tabashnik, B. E. (2015). Optimizing pyramided transgenic Bt crops for sustainable pest management. *Nature Biotechnology, 33*(2), 161–168.

Chamarthi, S. K., Kumar, K., Gunnaiah, R., Kushalappa, A. C., Dion, Y., & Choo, T. M. (2014). Identification of fusarium head blight resistance related metabolites specific to doubled-haploid lines in barley. *European Journal of Plant Pathology, 138*, 67–78.

Chen, J., Wu, L., Lu, M., Lu, S., Li, Z., & Ding, W. (2020). Comparative study on the fungicidal activity of metallic MgO nanoparticles and macroscale MgO against soilborne fungal phytopathogens. *Frontiers in Microbiology, 11*, 365.

Chen, R., Li, H., Zhang, L., Zhang, J., Xiao, J., & Ye, Z. (2007). CaMi, a root-knot nematode resistance gene from hot pepper (*Capsicum annuum* L.), confers nematode resistance in tomato. *Plant Cell Reports, 26*, 895–905.

Cheval, C., Aldon, D., Galaud, J. P., & Ranty, B. (2013). Calcium/calmodulin-mediated regulation of plant immunity. *Biochimica et Biophysica Acta (BBA)–Molecular Cell Research, 1833*(7), 1766–1771.

Chisholm, S. T., Coaker, G., Day, B., & Staskawicz, B. J. (2006). Host-microbe interactions: Shaping the evolution of the plant immune response. *Cell, 124*(4), 803–814.

Collins, N. C., Thordal-Christensen, H., Lipka, V., Bau, S., Kombrink, E., Qiu, J. L., & Schulze-Lefert, P. (2003). SNARE-protein-mediated disease resistance at the plant cell wall. *Nature, 425*(6961), 973–977.

da Silva, E. M., Roel, A. R. R. R., Porto, K. R. A., Falco, M. E. F. E., & Matias, R. M. (2017). Insecticidal effect of the ethanol extract of *Baccharis dracunculifolia* (Asterales: Asteraceae). *Revista de Biología Tropical, 65*(2), 517–523.

Dangi, R. S., Lagu, M. D., Choudhary, L. B., Ranjekar, P. K., & Gupta, V. S. (2004). Assessment of genetic diversity in *Trigonella foenum-graecum* and *Trigonella caerulea* using ISSR and RAPD markers. *BMC Plant Biology, 4*(1), 1–11.

David, S., & Asamoah, C. (2011). Farmer knowledge as an early indicator of IPM adoption: A case study from cocoa farmer field schools in Ghana. *Journal of Sustainable Development in Africa, 13*(4), 213–224.

Dixon, M. S., Hatzixanthis, K., Jones, D. A., Harrison, K., & Jones, J. D. (1998). The tomato Cf-5 disease resistance gene and six homologs show pronounced allelic variation in leucine-rich repeat copy number. *The Plant Cell, 10*(11), 1915–1925.

Dizaj, S. M., Lotfipour, F., Barzegar-Jalali, M., Zarrintan, M. H., & Adibkia, K. (2014). Antimicrobial activity of the metals and metal oxide nanoparticles. *Materials Science and Engineering: C*, *44*, 278–284.

Dudareva, N., Negre, F., Nagegowda, D. A., & Orlova, I. (2006). Plant volatiles: Recent advances and future perspectives. *Critical Reviews in Plant Sciences*, *25*(5), 417–440.

Durairaj, C., Sambathkumar, S., & Mohankumar, S. (2016). Integrated pest management of cruciferous vegetables. In *Integrated Pest Management of Tropical Vegetable Crops* (pp. 149–165).

Durner, J., Shah, J., & Klessig, D. F. (1997). Salicylic acid and disease resistance in plants. *Trends in Plant Science*, *2*(7), 266–274.

Durrant, W. E., & Dong, X. (2004). Systemic acquired resistance. *Annual Review of Phytopathology*, *42*, 185–209.

Eneh, F. U., Nwankwo, E. N., & Okwuenu, P. (2016). Phytochemical analysis and toxicity of *Casuarina equisetifola* (whistling pine) to *Sitophilus zeamais* Motschulsky (Coleoptera: Curculionidae). *Phytochemical Analysis*, *1*(3), 233–239.

Falk, K. L., Kästner, J., Bodenhausen, N., Schramm, K., Paetz, C., Vassão, D. G., & Meldau, S. (2014). The role of glucosinolates and the jasmonic acid pathway in resistance of *Arabidopsis thaliana* against molluscan herbivores. *Molecular Ecology*, *23*(5), 1188–1203.

Farhana, Munis, M. F. H., Alamer, K. H., Althobaiti, A. T., Kamal, A., Liaquat, F., & Attia, H. (2022). ZnO nanoparticle-mediated seed priming induces biochemical and antioxidant changes in chickpea to alleviate Fusarium wilt. *Journal of Fungi*, *8*(7), 753.

Fatope, M. O., Takeda, Y., Yamashita, H., Okabe, H., & Yamauchi, T. (1990). New cucurbitane triterpenoids from *Momordica charantia*. *Journal of Natural Products*, *53*(6), 1491–1497.

Ghasemzadeh, A., Jaafar, H. Z., & Rahmat, A. (2011). Effects of solvent type on phenolics and flavonoids content and antioxidant activities in two varieties of young ginger (*Zingiber officinale* Roscoe) extracts. *Journal of Medicinal Plants Research*, *5*(7), 1147–1154.

Gillman, J. H., Zlesak, D. C., & Smith, J. A. (2003). Applications of potassium silicate decrease black spot infection in *Rosa hybrida* "Meipelta" (Fuschia Meidiland™). *HortScience*, *38*(6), 1144–1147.

Glazebrook, J. (2005). Contrasting mechanisms of defense against biotrophic and necrotrophic pathogens. *Annual Review of Phytopathology*, *43*, 205–227.

Grant, M., & Lamb, C. (2006). Systemic immunity. *Current Opinion in Plant Biology*, *9*(4), 414–420.

Gribble, G. W. (2003). The diversity of naturally produced organohalogens. *Chemosphere*, *52*(2), 289–297.

Gu, Y. Q., Wildermuth, M. C., Chakravarthy, S., Loh, Y. T., Yang, C., He, X., & Martin, G. B. (2002). Tomato transcription factors Pti4, Pti5, and Pti6 activate defense responses when expressed in *Arabidopsis*. *The Plant Cell*, *14*(4), 817–831.

Gururani, M. A., Venkatesh, J., Upadhyaya, C. P., Nookaraju, A., Pandey, S. K., & Park, S. W. (2012). Plant disease resistance genes: Current status and future directions. *Physiological and Molecular Plant Pathology*, *78*, 51–65.

Haldar, A. K., Foltz, C., Finethy, R., Piro, A. S., Feeley, E. M., Pilla-Moffett, D. M., & Coers, J. (2015). Ubiquitin systems mark pathogen-containing vacuoles as targets for host defense by guanylate binding proteins. *Proceedings of the National Academy of Sciences*, *112*(41), E5628–E5637.

Halkier, B. A., & Gershenzon, J. (2006). Biology and biochemistry of glucosinolates. *Annual Review of Plant Biology*, *57*, 303–333.

Hanley, M. E., Lamont, B. B., Fairbanks, M. M., & Rafferty, C. M. (2007). Plant structural traits and their role in anti-herbivore defense. *Perspectives in Plant Ecology, Evolution and Systematics*, *8*(4), 157–178.

Hardie, D. G., Schaffer, B. E., & Brunet, A. (2016). AMPK: An energy-sensing pathway with multiple inputs and outputs. *Trends in Cell Biology*, *26*(3), 190–201.

Hare, P. (2011). Antimalarial fungal pesticide. *Nature Biotechnology*, *29*(4), 330–331.

Haynes, P. A., & Roberts, T. H. (2007). Subcellular shotgun proteomics in plants: Looking beyond the usual suspects. *Proteomics*, *7*(16), 2963–2975.

He, J., Chen, F., Chen, S., Lv, G., Deng, Y., Fang, W., & He, C. (2011). Chrysanthemum leaf epidermal surface morphology and antioxidant and defense enzyme activity in response to aphid infestation. *Journal of Plant Physiology*, *168*(7), 687–693.

Hettenhausen, C., Heinrich, M., Baldwin, I. T., & Wu, J. (2014). Fatty acid-amino acid conjugates are essential for systemic activation of salicylic acid-induced protein kinase and accumulation of jasmonic acid in *Nicotiana attenuata*. *BMC Plant Biology*, *14*(1), 1–12.

Hooda, S., & Jood, S. (2003). Effect of soaking and germination on nutrient and antinutrient contents of fenugreek (*Trigonella foenum-graecum* L.). *Journal of Food Biochemistry*, *27*(2), 165–176.

Horiuchi, K., Matsuzaki, H., Kobayashi, K., Goldberg, E. L., & Shibata, Y. (2003). 10Be record and magnetostratigraphy of a Miocene section from Lake Baikal: Re-examination of the age model and its implication for climatic changes in continental Asia. *Geophysical research letters*, *30*(12).

Howe, G. A., & Jander, G. (2008). Plant immunity to insect herbivores. *Annual Review of Plant Biology*, *59*, 41–66.

Jeandet, P., Clément, C., Courot, E., & Cordelier, S. (2013). Modulation of phytoalexin biosynthesis in engineered plants for disease resistance. *International Journal of Molecular Sciences*, *14*(7), 14136–14170.

Kah, M., Tufenkji, N., & White, J. C. (2019). Nano-enabled strategies to enhance crop nutrition and protection. *Nature Nanotechnology*, *14*(6), 532–540.

Kanchiswamy, C. N., Takahashi, H., Quadro, S., Maffei, M. E., Bossi, S., Bertea, C., & Arimura, G. I. (2010). Regulation of Arabidopsis defense responses against Spodoptera littoralis by CPK-mediated calcium signaling. *BMC plant biology*, *10*, 1–10.

Kang, K. W., Lee, S. J., & Kim, S. G. (2005). Molecular mechanism of Nrf2 activation by oxidative stress. *Antioxidants & Redox Signaling*, *7*(11, 12), 1664–1673.

Kanhed, P., Birla, S., Gaikwad, S., Gade, A., Seabra, A. B., Rubilar, O., & Rai, M. (2014). *In vitro* antifungal efficacy of copper nanoparticles against selected crop pathogenic fungi. *Materials Letters*, *115*, 13–17.

Karban, R. (2011). The ecology and evolution of induced resistance against herbivores. *Functional Ecology*, *25*(2), 339–347.

Kavina, J., Ranjith, V., & Sathya, B. (2020). Effect of EMS on chlorophyll mutagen in fenugreek (*Trigonella foenum-graecum* L.). *Journal of Medicinal Plants*, *8*(2), 1–5.

Khan, M., Khan, A. U., Hasan, M. A., Yadav, K. K., Pinto, M. M., Malik, N., & Sharma, G. K. (2021). Agro-nanotechnology as an emerging field: A novel sustainable approach for improving plant growth by reducing biotic stress. *Applied Sciences*, *11*(5), 2282.

Khan, R. A. A., Tang, Y., Naz, I., Alam, S. S., Wang, W., Ahmad, M., & Li, Y. (2021). Management of *Ralstonia solanacearum* in tomato using ZnO nanoparticles synthesized through *Matricaria chamomilla*. *Plant Disease*, *105*(10), 3224–3230.

Kranner, I., Minibayeva, F. V., Beckett, R. P., & Seal, C. E. (2010). What is stress? Concepts, definitions and applications in seed science. *New Phytologist*, *188*(3), 655–673.

Kumar, S., Bhanjana, G., Sharma, A., Sidhu, M. C., & Dilbaghi, N. (2014). Synthesis, characterization, and on-field evaluation of pesticide-loaded sodium alginate nanoparticles. *Carbohydrate Polymers*, *101*, 1061–1067.

Lal, G. (2018). Scenario, importance and prospects of seed spices: A review. *Current Investigations in Agriculture and Current Research*, *4*(2), 491–498.

Lee, C. W., Efetova, M., Engelmann, J. C., Kramell, R., Wasternack, C., Ludwig-Muller, J., & Deeken, R. (2009). *Agrobacterium tumefaciens* promotes tumor induction by modulating pathogen defense in *Arabidopsis thaliana*. *The Plant Cell*, *21*(9), 2948–2962.

Lichtenthaler, H. K. (1996). Vegetation stress: An introduction to the stress concept in plants. *Journal of Plant Physiology*, *148*(1–2), 4–14.

Liu, S. P., Zhu, Y. P., Xie, C., Jue, D. W., Hong, Y. B., Chen, M., & Yang, Q. (2012). Transgenic potato plants expressing StoVe1 exhibit enhanced resistance to *Verticillium dahliae*. *Plant Molecular Biology Reporter*, *30*, 1032–1039.

López-Vargas, E. R., Ortega-Ortíz, H., Cadenas-Pliego, G., de Alba Romenus, K., Cabrera de la Fuente, M., Benavides-Mendoza, A., & Juárez-Maldonado, A. (2018). Foliar application of copper nanoparticles increases the fruit quality and the content of bioactive compounds in tomatoes. *Applied Sciences*, *8*(7), 1020.

Luksiene, Z., Rasiukeviciute, N., Zudyte, B., & Uselis, N. (2020). Innovative approach to sunlight activated biofungicides for strawberry crop protection: ZnO nanoparticles. *Journal of Photochemistry and Photobiology B: Biology*, *203*, 111656.

Maffei, M. E. (2010). Sites of synthesis, biochemistry and functional role of plant volatiles. *South African Journal of Botany*, *76*(4), 612–631.

Mahalingam, R. (2015). Consideration of combined stress: A crucial paradigm for improving multiple stress tolerance in plants. In *Combined Stresses in Plants: Physiological, Molecular, and Biochemical Aspects* (pp. 1–25).

Malhotra, S. K., & Vashishtha, B. B. (2007). *Production, Development, Quality and Export of Seed Spices*. National Research Centre on Seed Spices.

Meyer, J. H., & Keeping, M. G. (2001). Past, present, and future research of the role of silicon for sugarcane in southern Africa. In *Studies in Plant Science* (Vol. 8, pp. 257–275). Elsevier.

Miedes, E., Vanholme, R., Boerjan, W., & Molina, A. (2014). The role of the secondary cell wall in plant resistance to pathogens. *Frontiers in Plant Science*, *5*, 358.

Mikhailova, E. O. (2020). Silver nanoparticles: Mechanism of action and probable bio-application. *Journal of Functional Biomaterials*, *11*(4), 84.

Miranda, M., Ralph, S. G., Mellway, R., White, R., Heath, M. C., Bohlmann, J., & Constabel, C. P. (2007). The transcriptional response of hybrid poplar (*Populus trichocarpa* x *P. deltoids*) to infection by *Melampsora medusae* leaf rust involves induction of flavonoid pathway genes leading to the accumulation of proanthocyanidins. *Molecular Plant-Microbe Interactions*, *20*(7), 816–831.

Missaoui, W. N., Arnold, R. D., & Cummings, B. S. (2018). Toxicological status of nanoparticles: What we know and what we don't know. *Chemico-Biological Interactions*, *295*, 1–12.

Monaghan, J., & Zipfel, C. (2012). Plant pattern recognition receptor complexes at the plasma membrane. *Current Opinion in Plant Biology*, *15*(4), 349–357.

Monde, K., Taniguchi, T., Miura, N., Nishimura, S. I., Harada, N., Dukor, R. K., & Nafie, L. A. (2003). Preparation of cruciferous phytoalexin related metabolites, (–)-dioxibrassinin and

(–)-3-cyanomethyl-3-hydroxyoxindole, and determination of their absolute configurations by vibrational circular dichroism (VCD). *Tetrahedron Letters*, *44*(32), 6017–6020.

Morant, A. V., Jørgensen, K., Jørgensen, C., Paquette, S. M., Sánchez-Pérez, R., Møller, B. L., & Bak, S. (2008). β-Glucosidases as detonators of plant chemical defense. *Phytochemistry*, *69*(9), 1795–1813.

Mulat, Y. W. (2017). Effects of powdery mildew (*Leveillula taurica* and *Erysiphe polygoni*) on yield and yield components of fenugreek in the mid-altitudes of Bale, South Eastern Ethiopia. *Journal of Plant Sciences*, *5*(2), 65–67.

Nocito, F. F., Pirovano, L., Cocucci, M., & Sacchi, G. A. (2002). Cadmium-induced sulfate uptake in maize roots. *Plant Physiology*, *129*(4), 1872–1879.

Ouzounidou, G., & Ilias, I. (2005). Hormone-induced protection of sunflower photosynthetic apparatus against copper toxicity. *Biologia Plantarum*, *49*, 223–228.

Padulosi, S., Thompson, J., & Rudebjer, P. G. (2013). Fighting poverty, hunger, and malnutrition with neglected and underutilized species: Needs, challenges, and the way forward.

Pauwels, L., Inzé, D., & Goossens, A. (2009). Jasmonate-inducible gene: What does it mean? *Trends in Plant Science*, *14*(2), 87–91.

Perveen, A., & Qaiser, M. (2007). Pollen morphology of family Solanaceae from Pakistan. *Pakistan Journal of Botany*, *39*(7), 2243–2256.

Petropoulos, G. A. (2002). Cultivation. In *Fenugreek: The Genus Trigonella* (pp. 26–72). CRC Press, Boca Raton, FL, USA.

Pineda, M., Sajnani, C., & Barón, M. (2010). Changes induced by the Pepper mild mottle tobamovirus on the chloroplast proteome of *Nicotiana benthamiana*. *Photosynthesis Research*, *103*, 31–45.

Posmyk, M. M., Kontek, R., & Janas, K. M. (2009). Antioxidant enzymes activity and phenolic compounds content in red cabbage seedlings exposed to copper stress. *Ecotoxicology and Environmental Safety*, *72*(2), 596–602.

Raina, A., Laskar, R., Khursheed, S., Amin, R., Tantray, Y., Parveen, K., & Khan, S. (2016). Role of mutation breeding in crop improvement: Past, present, and future. *Asian Research Journal of Agriculture*, *2*(2), 1–13.

Ramegowda, V., & Senthil-Kumar, M. (2015). The interactive effects of simultaneous biotic and abiotic stresses on plants: Mechanistic understanding from drought and pathogen combination. *Journal of Plant Physiology*, *176*, 47–54.

Ribot, C., Hirsch, J., Balzergue, S., Tharreau, D., Nottéghem, J. L., Lebrun, M. H., & Morel, J. B. (2008). Susceptibility of rice to the blast fungus, Magnaporthe grisea. *Journal of Plant Physiology*, *165*(1), 114–124.

Rodríguez-Moreno, L., Pineda, M., Soukupová, J., Macho, A. P., Beuzón, C. R., Barón, M., & Ramos, C. (2008). Early detection of bean infection by *Pseudomonas syringae* in asymptomatic leaf areas using chlorophyll fluorescence imaging. *Photosynthesis Research*, *96*, 27–35.

Roselló, S., Ricarte, B., José Díez, M., & Nuez, F. (2001). Resistance to *Tomato spotted wilt virus* introgressed from *Lycopersicon peruvianum* in line UPV 1 may be allelic to Sw-5 and can be used to enhance the resistance of hybrid cultivars. *Euphytica*, *119*, 357–367.

Roze, L. V., Chanda, A., & Linz, J. E. (2011). Compartmentalization and molecular traffic in secondary metabolism: A new understanding of established cellular processes. *Fungal Genetics and Biology*, *48*(1), 35–48.

Sajjadi, S., Shokoohinia, Y., & Mehramiri, P. (2013). Isolation and characterization of steroids, phthalide, and essential oil of the fruits of *Kelussia odoratissima* Mozaff., an endemic mountain celery. *Research in Pharmaceutical Sciences*, *8*(1), 35–42.

Salman Khan, M., Zaka, M., Haider Abbasi, B., Rahman, L., & Shah, A. (2016). Seed germination and biochemical profile of *Silybum marianum* exposed to monometallic and bimetallic alloy nanoparticles. *IET Nanobiotechnology*, *10*(6), 359–366.

Santamaria, M. E., Martínez, M., Cambra, I., Grbic, V., & Diaz, I. (2013). Understanding plant defense responses against herbivore attacks: An essential first step towards the development of sustainable resistance against pests. *Transgenic Research*, *22*, 697–708.

Sarria, S., Wong, B., Martín, H. G., Keasling, J. D., & Peralta-Yahya, P. (2014). Microbial synthesis of pinene. *ACS Synthetic Biology*, *3*(7), 466–475.

Sato, K., Naito, M., Yukitake, H., Hirakawa, H., Shoji, M., McBride, M. J., & Nakayama, K. (2010). A protein secretion system linked to Bacteroidetes gliding motility and pathogenesis. *Proceedings of the National Academy of Sciences*, *107*(1), 276–281.

Satti, S. H., Raja, N. I., Javed, B., & Akram, A. (2021). Mashwani ZuR, Ahmad MS, Ikram M. Titanium dioxide nanoparticles elicited agro-morphological and physico-chemical modifications in wheat plants to control Bipolaris sorokiniana. *PLoS One*, *16*, e0246880.

Schumann, G. L., & D'Arcy, C. J. (2010). Back matter. In *Essential Plant Pathology* (2nd ed., pp. 317–369). The American Phytopathological Society.

Seebold, K. W., Kucharek, T. A., Datnoff, L. A., Correa-Victoria, F. J., & Marchetti, M. A. (2001). The influence of silicon on components of resistance to blast in susceptible, partially resistant, and resistant cultivars of rice. *Phytopathology*, *91*(1), 63–69.

Serghini, K., de Luque, A. P., Castejón-Muñoz, M., García-Torres, L., & Jorrín, J. V. (2001). Sunflower (*Helianthus annuus* L.) response to broomrape (*Orobanche cernua* Loefl.) parasitism: Induced synthesis and excretion of 7-hydroxylated simple coumarins. *Journal of Experimental Botany*, *52*(364), 2227–2234.

Sharma, H. C., Sujana, G., & Manohar Rao, D. (2009). Morphological and chemical components of resistance to pod borer, *Helicoverpa armigera* in wild relatives of pigeonpea. *Arthropod-Plant Interactions*, *3*, 151–161.

Sharma, P., Jha, A. B., Dubey, R. S., & Pessarakli, M. (2012). Reactive oxygen species, oxidative damage, and antioxidative defense mechanism in plants under stressful conditions. *Journal of Botany*, *2012*, Article ID 132408.

Shende, S., Gaikwad, N., & Bansod, S. (2016). Synthesis and evaluation of antimicrobial potential of copper nanoparticles against agriculturally important phytopathogens. *Synthesis*, *1*(4), 41–47.

Shivaji, R., Camas, A., Ankala, A., Engelberth, J., Tumlinson, J. H., Williams, W. P., & Luthe, D. S. (2010). Plants on constant alert: Elevated levels of jasmonic acid and jasmonate-induced transcripts in caterpillar-resistant maize. *Journal of Chemical Ecology*, *36*, 179–191.

Silvestre, A. J., & Gandini, A. (2008). Terpenes: Major sources, properties and applications. In *Monomers, Polymers and Composites from Renewable Resources* (pp. 17–38). Elsevier.

Song, Y., Zhang, Z., Seidl, M. F., Majer, A., Jakse, J., Javornik, B., & Thomma, B. P. (2017). Broad taxonomic characterization of *Verticillium* wilt resistance genes reveals an ancient origin of the tomato Ve1 immune receptor. *Molecular Plant Pathology*, *18*(2), 195–209.

Soylu, S., Bennett, M. H., & Mansfield, J. W. (2002). Induction of phytoalexin accumulation in broad bean (*Vicia faba* L.) cotyledons following treatments with biotic and abiotic elicitors. *Turkish Journal of Agriculture and Forestry*, *26*(6), 343–348.

Spoel, S. H., & Dong, X. (2012). How do plants achieve immunity? Defense without specialized immune cells. *Nature Reviews Immunology*, *12*(2), 89–100.

Sreevidya, V. S., Srinivasa Rao, C., Sullia, S. B., Ladha, J. K., & Reddy, P. M. (2006). Metabolic engineering of rice with soybean isoflavone synthase for promoting nodulation gene expression in rhizobia. *Journal of Experimental Botany*, *57*(9), 1957–1969.

Sureshkumar, D., Begum, S., Johannah, N. M., Maliakel, B., & Krishnakumar, I. M. (2018). Toxicological evaluation of a saponin-rich standardized extract of fenugreek seeds (FenuSMART®): Acute, sub-chronic, and genotoxicity studies. *Toxicology Reports*, *5*, 1060–1068.

Suzuki, N., Rivero, R. M., Shulaev, V., Blumwald, E., & Mittler, R. (2014). Abiotic and biotic stress combinations. *New Phytologist*, *203*(1), 32–43.

Taft, A. S., Ozawa, M., Fitch, A., Depasse, J. V., Halfmann, P. J., Hill-Batorski, L., & Kawaoka, Y. (2015). Identification of mammalian-adapting mutations in the polymerase complex of an avian H5N1 influenza virus. *Nature Communications*, *6*, Article 7491.

Teferedegne, B. (2000). New perspectives on the use of tropical plants to improve ruminant nutrition. *Proceedings of the Nutrition Society*, *59*(2), 209–214.

Tettelin, H., Nelson, K. E., Paulsen, I. T., Eisen, J. A., Read, T. D., Peterson, S., & Fraser, C. M. (2001). Complete genome sequence of a virulent isolate of *Streptococcus pneumoniae*. *Science*, *293*(5529), 498–506.

Thomas, J. E., Bandara, M., Lee, E. L., Driedger, D., & Acharya, S. (2011). Biochemical monitoring in fenugreek to develop functional food and medicinal plant variants. *New Biotechnology*, *28*(2), 110–117.

Thomas, J. E., Basu, S. K., & Acharya, S. N. (2006). Identification of *Trigonella* accessions which lack antimicrobial activity and are suitable for forage development. *Canadian Journal of Plant Science*, *86*(3), 727–732.

Thomma, B. P., Tierens, K. F., Penninckx, I. A., Mauch-Mani, B., Broekaert, W. F., & Cammue, B. P. (2001). Different microorganisms differentially induce *Arabidopsis* disease response pathways. *Plant Physiology and Biochemistry*, *39*(7, 8), 673–680.

Tripathi, D., Singh, M., & Pandey-Rai, S. (2022). Crosstalk of nanoparticles and phytohormones regulates plant growth and metabolism under abiotic and biotic stress. *Plant Stress, 6, 100107.*

Vaidya, K., Ghosh, A., Kumar, V., Chaudhary, S., Srivastava, N., Katudia, K., & Chikara, S. K. (2013). *De novo* transcriptome sequencing in *Trigonella foenum-graecum* L. to identify genes involved in the biosynthesis of diosgenin. *The Plant Genome*, *6*(2), plantgenome2012-08.

Van Bockhaven, J., De Vleesschauwer, D., & Höfte, M. (2013). Towards establishing broad-spectrum disease resistance in plants: Silicon leads the way. *Journal of Experimental Botany*, *64*(5), 1281–1293.

van Loon, L. C., Rep, M., & Pieterse, C. M. (2006). Significance of inducible defense-related proteins in infected plants. *Annu. Rev. Phytopathol.*, *44*, 135–162.

Verhage, A., van Wees, S. C., & Pieterse, C. M. (2010). Plant immunity: It's the hormones talking, but what do they say? *Plant Physiology*, *154*(2), 536–540.

Verma, N., Goyal, M., & Rawat, I. (2020). *Trigonella foenum-graecum* L–Fenugreek.

Walker, T. S., Bais, H. P., Grotewold, E., & Vivanco, J. M. (2003). Root exudation and rhizosphere biology. *Plant Physiology*, *132*(1), 44–51.

Walling, L. L. (2008). Avoiding effective defenses: Strategies employed by phloem-feeding insects. *Plant Physiology*, *146*(3), 859–866.

Walter, M. G., Wamser, C. C., Ruwitch, J., Zhao, Y., Braden, D., Stevens, M., & Pessiki, P. J. (2007). Syntheses and optoelectronic properties of amino/carboxyphenylporphyrins for potential use in dye-sensitized TiO_2 solar cells. *Journal of Porphyrins and Phthalocyanines, 11*(08), 601–612.

Walters, D. (2011). *Plant defense: Warding Off Attack by Pathogens, Herbivores and Parasitic Plants*. John Wiley & Sons.

War, A. R., Paulraj, M. G., Ahmad, T., Buhroo, A. A., Hussain, B., Ignacimuthu, S., & Sharma, H. C. (2012). Mechanisms of plant defense against insect herbivores. *Plant Signaling & Behavior, 7*(10), 1306–1320.

War, A. R., Sharma, H. C., Paulraj, M. G., War, M. Y., & Ignacimuthu, S. (2011). Herbivore induced plant volatiles: Their role in plant defense for pest management. *Plant Signaling & Behavior, 6*(12), 1973–1978.

Wu, Q., Luo, Y., Lu, R., Lau, N., Lai, E. C., Li, W. X., & Ding, S. W. (2010). Virus discovery by deep sequencing and assembly of virus-derived small silencing RNAs. *Proceedings of the National Academy of Sciences, 107*(4), 1606–1611.

Wuyts, N., Swennen, R., & De Waele, D. (2006). Effects of plant phenylpropanoid pathway products and selected terpenoids and alkaloids on the behavior of the plant-parasitic nematodes *Radopholus similis*, *Pratylenchus penetrans*, and *Meloidogyne incognita. Nematology, 8*(1), 89–101.

Yao, L., Jiang, Y., Singanusong, R., D'Arcy, B., Datta, N., Caffin, N., & Raymont, K. (2004). Flavonoids in Australian *Melaleuca*, *Guioa*, *Lophostemon*, *Banksia*, and *Helianthus* honeys and their potential for floral authentication. *Food Research International, 37*(2), 166–174.

Young, N. D., Mudge, J., & Ellis, T. N. (2003). Legume genomes: More than peas in a pod. *Current Opinion in Plant Biology, 6*(2), 199–204.

Zandi, P., Basu, S. K., Cetzal-Ix, W., Kordrostami, M., Chalaras, S. K., & Khatibai, L. B. (2017). Fenugreek (*Trigonella foenum-graecum* L.): An important medicinal and aromatic crop. In *Active Ingredients from Aromatic and Medicinal Plants* (pp. 207–224).

Zaynab, M., Fatima, M., Abbas, S., Sharif, Y., Umair, M., Zafar, M. H., & Bahadar, K. (2018). Role of secondary metabolites in plant defense against pathogens. *Microbial Pathogenesis, 124*, 198–202.

Zhao, Y., Wei, W., Lee, I. M., Shao, J., Suo, X., & Davis, R. E. (2009). Construction of an interactive online phytoplasma classification tool, i PhyClassifier, and its application in analysis of the peach X-disease phytoplasma group (16SrIII). *International journal of systematic and Evolutionary Microbiology, 59*(10), 2582–2593.

Zhou, X., Jia, X., Zhang, Z., Chen, K., Wang, L., Chen, H., & Zhao, L. (2022). AgNPs seed priming accelerated germination speed and altered nutritional profile of Chinese cabbage. *Science of the Total Environment, 808*, 151896.

Zientz, E., Dandekar, T., & Gross, R. (2004). Metabolic interdependence of obligate intracellular bacteria and their insect hosts. *Microbiology and Molecular Biology Reviews, 68*(4), 745–770.

CHAPTER 9

Salinity Stress Alleviation in Fenugreek

AB WAHEED WANI,[1] HARJINDER KAUR,[1] SANJEEV KUMAR,[2] SHAKIR ALI,[3] RUPESH KAUSHIK,[1] ZARINA,[4] KONDLE RAVI,[1] AJAY SAROHA,[1] KOUNSER JAVID,[5] and BATULLA RASHID[5]

[1]*Department of Horticulture, School of Agriculture, Lovely Professional University, Phagwara, Punjab, India*

[2]*Department of Fruit Science, Dr. Yashwant Singh Parmar University of Horticulture and Forestry, Nauni, Solan, Himachal Pradesh, India*

[3]*Division of Fruit Science, Sher-e Kashmir University of Agricultural Science and Technology, Kashmir, India*

[4]*Division of Vegetable Science, Sher-e Kashmir University of Agricultural Science and Technology, Jammu, India*

[5]*Sher-e Kashmir University of Agricultural Science and Technology, Kashmir, India*

ABSTRACT

On a global scale, salinity stress, a significant abiotic factor, severely impacts agricultural productivity, particularly that of fenugreek (*Trigonella foenumgraecum* L.), an important herb known for its culinary and medicinal value due to its diverse therapeutic and nutritional advantages. Unfortunately, the excessive salinity in the soil adversely affects fenugreek's growth and development. The primary objective of this research is to comprehensively explore the physiological, biochemical, and molecular mechanisms that enable fenugreek to withstand salt stress. Salinity stress imposes multiple challenges on fenugreek, leading to oxidative damage, ion toxicity, hindered growth, and osmotic stress. This study delves into how fenugreek employs

Fenugreek: Improvement in Yield Potential Through Plant Growth Regulators (PGRs) and Depolymerized Polysaccharides. Tariq Ahmad Dar, Bilal Ahmad, Moinuddin, and Khalid Rehman Hakeem (Eds.)

various adaptive mechanisms to mitigate the negative effects of salt stress, such as osmotic regulation, ion homeostasis, antioxidant defense, and the activation of stress-responsive genes. Several strategies are involved in this process, including the use of external chemicals like osmoprotectants and plant growth regulators, genetic engineering to enhance stress-related gene expression or RNA interference, and agronomic techniques such as controlled irrigation and soil amendments. The ultimate goal is to gain a comprehensive understanding of fenugreek's salt tolerance, paving the way for improved agricultural practices under saline conditions. How well these methods work at increasing salinity tolerance to boost fenugreek growth and yield. Improvements in cultivars and management techniques that can lessen the impact of salt stress are needed to maintain the viability of fenugreek production and the valuable products it yields.

9.1 INTRODUCTION

Any external constraint, known as stress, slows down photosynthesis and reduces a plant's ability to convert energy into biomass. To feed a projected global population of 9.1 billion by 2050, the Food and Agriculture Organization of the United Nations (FAO) estimates that worldwide food production must increase by 70% between 2005–2007 and 2050. Production in developing countries would need to nearly triple to meet this demand. Reducing agricultural losses due to environmental stresses is a major focus for meeting the rising demand for food (Shanker and Venkateswarlu, 2011). Key abiotic stresses—such as excessive salt content, extreme temperatures, and both high and low temperature extremes—can adversely affect crop productivity, with losses exceeding 70% in some cases (Ahmad et al., 2012; Kaur et al., 2008; Mantri et al., 2012; Thakur et al., 2010; Vorasoot et al., 2003). The negative impact of high nutrient levels, including sodium and calcium ions, on plants is referred to as salt stress (Munns, 2005). While soil salinity existed prior to the advent of agriculture, the issue has recently become more pressing due to irrigation, which is essential for crop growth (Zhu, 2001).

Salinity is one of the major limitations to crop development, yield, and growth. Soils affected by salinity are classified into two main types based on their characteristics and their effects on plant growth: (i) Saline soils, which primarily contain sodium chloride and sodium sulfate, occasionally with high levels of calcium and magnesium chlorides and sulfates; these soils have sufficient soluble neutral salts to hinder the development of agricultural crops; and (ii) sodic soils, which are characterized by the presence of Na^+ ions.

Trigonella foenum-graecum L., commonly known as fenugreek, is an annual dicot in the Papilionaceae subfamily of the Leguminaceae family. *Trigonella foenum-graecum* L. exhibits an unpredictable growth pattern, continuing to develop until hindered by severe cold or desiccation (Acharya et al., 2006). In North America, fenugreek plants typically reach a height of 40 to 60 cm (Acharya et al., 2006; Slinkard et al., 2006). Seeds of *Trigonella foenum-graecum* L. contain 36% protein, 32% insoluble dietary fiber, 6% oil, 1.6% starch, 13% soluble dietary fiber, 3% ash, and 0.4% sugar (Slinkard et al., 2006).

Fenugreek is associated with various pharmacological effects, including hypoglycemic (Zia et al., 2001), chemopreventive (Amin et al., 2005), gastro-protective, anti-inflammatory, antipyretic, hypocholesterolemic (Srinivasan, 2006), and antioxidant activities. The plant contains compounds such as alkaloids (Jain and Madhu, 1988), salicylate (Swain et al., 1985), flavonoids (Kamal and Yadav, 1991), and nicotinic acid (Rajalakshmi et al., 1964). This chapter discusses the various types of salt stress, its effects on fenugreek, and plant adaptations to saline environments.

9.2 KINDS OF SALINITY AND THEIR CAUSES

9.2.1 NATURAL OR PRIMARY SALINITY

Natural processes cause salts to accumulate in the soil or groundwater over time, resulting in primary salinity. It is the outcome of two natural procedures. The first is the breakdown of the underlying substances which include soluble salt. As rocks deteriorate, a number of soluble salts, primarily calcium, sodium, and mg chlorides, as well as to a lesser amount sulfates and carbonates, seep out. The sodium chloride type of salt has the highest solubility. The deposit of sea salt that rainfall and wind have brought ashore is one of two phenomena. "Cyclic salts" are mostly sodium chloride-based marine salts that are carried onshore by the wind and accumulated by rain.

9.2.2 SECONDARY OR ANTHROPOGENIC SALINITY

Human activities are the primary cause of secondary salinization, disrupting the moisture balance in the soil by affecting the equilibrium between water supply (from irrigation or rainfall) and water absorption by crops through evaporation (Garg and Manchanda, 2008). There are two main reasons for this phenomenon: firstly, the conversion of land for perennial vegetation instead of annually grown plants, and secondly, the utilization of salt-enriched groundwater or poor drainage in irrigation systems. The Food and Agriculture Organization's Land and Plant Nutritional Management Service reports that approximately 6% of the world's land is currently experiencing the negative effects of salinity. Among the land equipped for irrigation, 19.5%, which amounts to 45 million hectares, is affected by salt. Interestingly, around one-third of the global food production comes from only 15% of farmland, which is under irrigation and yields about twice as much as non-irrigated land. Despite seemingly low concentrations of dissolved salt (ranging from 200 to 500 mg per kg) in irrigation water of good quality, the cumulative amount of salt added to the soil is substantial. For instance, in a 1,000 m^3 volume of irrigation water, there would be 0.50 tons of salt at a concentration of 500 mg/l. An acre of soil will annually acquire 6,000–10,000 m^3 (or 3–5 tons) of salt due to the high-water requirements of crops. Salt will build up in the root zone since plants only eliminate a little amount of it, thus it must be removed by providing a greater quantity of water than the plants need. If there is insufficient water flow, the additional water elevates the level of the groundwater, causing salts that accumulate in the area surrounding the roots to be mobilized and forced to the surface. Water stagnation happens when the crop is unable to use the complete amount of supplied water (Garg and Manchanda, 2008).

9.3 IMPACTS OF SALT STRESS ON VEGETATION

For a number of reasons, salts in the soil and water might prevent plants from growing. Plant's capability of water absorption is reduced by the quantity of saltwater in the soil combinations, which in turn slows the rate of growth. This is speaking of how salt affects osmosis or water-deficit mechanisms. Second, if the plant absorbs excessive amounts of salt via its osmosis stream, the mitochondria in the transpiring leaflets are going to be damaged, which may further impede growth. It is also known as excessive ionization or the salty impact of specific salinity (Greenway and Munns, 1980). Salinity has three distinctive consequences. As a result of the disrupted ion homeostasis and toxicity, the decreased potential of water, and ion imbalance, initial growth declines down, and plant production is restricted. The negative impact manifests itself at the scale of the whole plant as plant death or a reduction in output. All crucial processes, including germination, are impacted by salt stress. Photosynthesis, oxidative stress, photosynthetic pigments, yield, and nutritional imbalance are all interconnected. These are covered in a number of places.

9.3.1 GERMINATION/GERMINATION PERCENTAGE

Seed germination is a fundamental phase in a plant's life cycle that significantly affects production. Salinity, however, has been shown to have detrimental effects on the germination process in various plants, including rice (Khodarahmpour et al., 2012; Xu et al., 2011), *Brassica* spp., and other species (Fernández-Torquemada and Sánchez-Lizaso, 2013). Ibrar et al. (2003) observed that salinity influences seed germination in multiple ways (Gomes-Filho et al., 2008), including reducing water intake capacity due to lowered osmotic potential during germination. Germination rates and percentages vary significantly across plant species and cultivars.

Lu et al. (2013) proposed a general relationship between germination percentage and time after water addition at various salt concentrations. Their study showed that increasing salt stress (50–200 mM NaCl) gradually decreased germination percentages compared to control conditions. Lower salt concentration (50 mM) resulted in maximum germination (100%), but as salt concentration increased to 100, 150, and 200 mM, germination percentages dropped to 97%, 41%, and 30%, respectively (Kapoor and Pande, 2015). In the fenugreek cultivar Pusa Early Bunching (PEB), Ratnakar and Rai (2013) found that NaCl concentrations of 80 mM and above significantly delayed germination. Germination was substantially delayed beyond 100 mM, and at 150 and 200 mM, seeds failed to germinate even after 24 and 36 hours, respectively. These findings also linked salt stress to delayed maximum germination of seedlings compared to controls, with maximum germination observed at 48, 108, and 120 hours under control, 50 mM, 100 mM, 150 mM, and 200 mM salt concentrations, respectively.

Salt stress negatively impacts embryos, resulting in delayed germination or reduced germination rates, as reported by Kapoor and Pande (2015). Pour et al. (2012) demonstrated that all salt levels suppressed seed germination relative to the control, with the highest salinity level (200 mM) causing a 48.2% reduction in germination. Kaveh et al. (2011) noted delayed germination and lower germination percentages in *Solanum lycopersicum* under salinity. Similarly, Bybordi (2010) observed a drastic reduction in germination percentage of *Brassica napus* between 150 and 200 mM NaCl, with both germination percentage and speed decreasing by 38% and 33%, respectively, at 200 mM NaCl compared to the control.

9.3.2 GROWTH

The first impact of salt stress is a slowdown of development. The plant's capacity to absorb water is initially slowed, which results in a reduced rate of growth. Secondly, by entering the water that evaporates stream and doing so, it may harm the cells of leaves which transpiring, slowing down growth even more.

As a result of the two influences, Munns (1993) identified a two-phase growth response to salt.

1. **Phase 1:** The foremost stage of the response of growth is brought on by the impact of salt on the outside of the plant. Although less so, presence of salts in solution of soil slows down the development of both roots and leaves (Munns, 1993). Plants impacted by drought experience similar physiological and metabolic activities. Because tissues that are meristematic are mainly fulfilled by the phloem, which salt is completely removed from, and as very quickly expanding cells are able to tolerate the mineral salt that comes in the xylem within their growing vacuoles, sodium, and chlorine ion cannot build up in tissues that are developing at the amount that restrict growth.
2. **Phase 2:** The second phase of the growth response is triggered by the detrimental effects of salt within the plant. High concentrations of sodium and chloride ions accumulate in transpiring leaves, causing older leaves to wither and die as absorbed salts concentrate in them. This damage likely occurs because the salt load exceeds the cell's capacity to compartmentalize salts within the vacuole (Tavakkoli et al., 2011).

   At an early stage (15 DAS), shoot length decreased from 6.20 cm in the control to 2.89 cm under 200 mM salt concentration. Salt stress effects intensified in later growth stages, with shoot lengths reaching 39.78 cm in the control and only 9.33 cm at 200 mM salt concentration after 60 DAS (Kapoor and Pande, 2015; Mehrafarin et al., 2011). Root length also showed a decreasing trend with higher salt levels. From 15 to 60 DAS, root length increased from 2.34 to 7.48 cm in the control, while in salt-stressed plants, it rose from 0.89 to 2.44 cm. A substantial reduction in root length was observed, from 7.48 cm in the control to 2.44 cm at 200 mM salt concentration by 60 DAS (Asaadi, 2009; Kapoor and Pande, 2015).

   As salt levels increased, the germination rate significantly declined, starting at a lower salinity concentration (50 mM; Pour et al., 2012). Additionally, hypocotyl length in fenugreek significantly decreased with higher salt levels ($p < 0.0001$). Hypocotyl length decreased by 6.8% at the first salinity level and by 98.3% at the second salinity level (Pour et al., 2012).

9.3.3 SHOOT TO ROOT RATIO, NUMBER OF BRANCHES (NBP), AND NUMBER OF LEAVES (NLP) PER PLANT

At 60 days after sowing within control and 200 mM concentrations of salt, the maximum and lowest shoot/root ratios were found to be 5.19 and 3.87, respectively. Previous studies on fenugreek have noted a similar reduction in growth markers (Niknam et al., 2006; Talukdar, 2012). During the experiment, there are fewer leaves as the salt levels rise. At 60 DAS, the control plant's greatest number of leaves per plant and number of branches per plant were 54.08 and 6.08, respectively. These results support those presented in the study by Aggarwal et al. (2013). With 200 mM salt stress, the vigor index decreased from 584 to 35.4. It is clear that the vigor index showing a positive and significant correlation with the length of root and shoot germination percentage (Niknam et al., 2006; Talukdar, 2012).

9.3.4 IONIC TOXICITY AND SALINITY

Rising soluble salt concentrations in soil hinder plants' ability to absorb and utilize essential mineral elements. Wang et al. (2003) found that proper ion ratios can support a plant's physiological response related to its growth and development. Studies have shown that increased salinity leads to higher ion leakage, with control (0 mM) and 150 mM salinity levels showing the lowest and highest percentages of ion leakage, respectively (seedlings did not survive at 200 mM). This characteristic can result in a 25.2% decrease in ion leakage at a concentration of 15 μM salicylic acid (SA), compared to untreated plants. In contrast, 0 μM (control) and 5 μM SA treatments showed no significant effects. Different salinity and SA levels had statistically significant impacts on ion leakage. However, excessive salt uptake can cause various ion toxicities, reducing the absorption of essential elements such as calcium (Ca^{2+}), potassium (K^+), nitrogen (N), and phosphorus (P) due to elevated levels of sodium (Na^+), chloride (Cl^-), or sulfate (SO_4^{2-}). Additionally, the adverse effects of sodium ions often relate to the structural and functional integrity of cell membranes (Kurth et al., 1986).

In *Atriplex griffithii*, salinity stress raises Na^+ and Cl^- levels in roots, stems, and leaves, with the highest ion accumulation in leaves, followed by stems and roots, indicating a positive association with sodium and calcium ion concentrations. While K^+ levels remained stable in roots, they declined notably in leaves with increased salinity. Similarly, Ca^{2+} levels decreased in shoots and leaves under high salinity, though Mg^{2+} levels in roots, stems, and leaves did not change significantly, except for a more pronounced decline in leaves (Khan et al., 2000). Under salt stress, reductions in Ca^{2+} and Mg^{2+} contents in leaves indicate increased membrane stability and a decrease in chlorophyll content (Parida and Das, 2005).

Chloride (Cl^-) toxicity also significantly contributes to growth suppression, although most plants growing in saline soils accumulate high levels of both Na^+ and Cl^- in their shoot tissues. Tavakkoli et al. (2011) examined the specific ion toxicity of Na^+ and Cl^- in four barley genotypes grown in saline soils. They found that while both ions hindered growth, NaCl-induced stress had a more detrimental impact on growth and photosynthesis. Additionally, they observed distinct Na^+ and Cl^- exclusion mechanisms across barley genotypes. High Cl^- concentrations primarily reduced photosynthesis and chlorophyll content through non-stomatal effects, whereas elevated Na^+ levels decreased K^+ and Ca^{2+} uptake and reduced photosynthesis by limiting stomatal conductance (Tavakkoli et al., 2011).

9.3.5 PHOTOSYNTHESIS AND PHOTOSYNTHETIC PIGMENTS

Photosynthesis is one of the most crucial biochemical reactions for converting sunlight into chemical energy, promoting plant growth. Reduced water potential is the primary reason for the decrease in photosynthesis rates observed in plants under salt-induced stress. Since chlorophyll is essential for photosynthesis and closely linked to plant health, an increase in sodium or chloride buildup in chloroplasts further hinders the photosynthetic rate (Zhang et al., 2005). Chlorophyll concentration has been widely used in research as an indicator of physiological metabolism, as salt stress often reduces it (Chutipaijit et al., 2011). This reduction was observed at 100 mM in the *Trigonella foenum-graecum* L. cultivar RMt-1, with chlorophyll levels at their lowest (200 mM) showing a significant decrease 30 days after planting (Aggarwal et al., 2013). Carotenoids and chlorophyll a and b content in the control group reached maximum levels of 1.65, 0.58, and 0.37 mg/g, respectively, after 45 days after sowing (DAS). Studies have shown that photosynthetic pigment levels, including chlorophyll a, chlorophyll b, and carotenoids, continuously decline with rising salt levels compared to untreated controls, significantly decreasing at 200 mM NaCl by 38.47%, 41.86%, and 48.70%, respectively. However, foliar application of moringa leaf extract (MLE) alone increased chlorophyll a, chlorophyll b, and carotenoid content by 32.85%, 29.06%, and 36.36%, respectively, compared to untreated controls (Saha et al., 2010).

One study on cucumbers demonstrated that total leaf chlorophyll concentrations increased with treatment, while in untreated samples, chlorophyll content fell by 12%, 21%, and 30% at 2, 3, and 5 days after sowing (DAS) as NaCl levels rose (Khan et al., 2013). Along with reduced pigment levels, chlorophyll fluorescence intensity decreased by an average of 16%. Although values for chlorophyll a and b become closer as salt levels rise, chlorophyll a often remains higher than chlorophyll b in plants (Mane et al., 2010). Multiple studies indicate that factors such as membrane damage contribute to chlorophyll loss under stressful conditions (Allakhverdiev et al., 2000; Mane et al., 2010). In *Scenedesmus obliquus*, photosynthetic properties changed, leading to reduced biomass accumulation (Demetriou et al., 2007). López-Climent et al. (2008) found that salt stress impeded citrus growth by decreasing photosystem II (PSII) efficiency, net photosynthetic rate, and overall photosynthetic efficiency. Salt stress also affects the PSII photosynthetic rate, electron transport rate, and D1 protein, thereby impacting *Brassica juncea* growth (Mittal et al., 2012; Iyengar and Reddy, 1996). Dehydration of cell membranes under salt stress also reduces carbon dioxide permeability.

9.3.6 WATER RELATION

Water potential is a crucial physiological indication, according to Bacha et al. (2017) for determining how plants are doing with regard to water. Romero-Aranda et al. (2001) discovered that a reduction in leaf water potential brought on by an elevation in salt in the root media may influence several plant activities. With this condition, plants are unable to sustain turgor at extremely low soil water potentials by taking up water from the soil and absorbing it. Plants, however, adapt osmotically (absorb solutes) for the maintenance of potential gradient for the entry of water at low or moderate salt concentrations (increased soil water potential). As the salt concentration rose, RWC dramatically reduced. The decreases increasingly worsen over time. The RWC fell by 19.87%. At 15, 30, 45, and 60 DAS, there were, respectively 31.05, 35.49, and 52.39% (Kapoor and Pande, 2015; Schonfeld et al., 1988; Talukdar, 2012). In another study, it was discovered that raising the SA concentration resulted in a distinctly greater relative water content. According to Talukdar (2012), the concentration of 15 M SA resulted in the greatest relative water content increase, 8.1% over the control plant. One study using Cucumis sativa discovered that when salt levels increase, the water potential decreases linearly (Khan et al., 2013).

9.3.7 NUTRITIONAL DEFICIENCY

Salinity-induced nutritional issues are known to negatively impact crop production, though the interactions between crop mineral nutrition and salinity are highly complex (Grattan and Grieve, 1998). Salinity affects nutrition by influencing nutrient availability, competitive absorption, transport, and diffusion within the plant. Studies indicate that salinity reduces nutrient uptake and storage in plant tissues (Hu and Schmidhalter, 2005; Rogers et al., 2003). Micronutrient absorption and distribution in saline soils are influenced by factors such as micronutrient solubility, soil pH, redox potential, and binding sites on the outer layers of inorganic and organic particles. Salinity's effects on micronutrient concentrations in plants vary depending on species and salt levels (Oertli, 1991). Micronutrient deficiencies are prevalent in saline soils due to high pH levels (Zhu et al., 2004).

Several studies suggest that salinity can reduce nitrogen uptake by plants. Increased salinity levels correlate with decreased calcium content in plant tissues. Fenugreek plants grown in saline conditions showed reduced leaf, shoot, root, and pod sizes compared to controls. According to Saleh et al. (2009), the lowest calcium (Ca) content was observed at 250 mM NaCl (1.58 mg/g), whereas the highest Ca content was found in control plants (2.90 mg/g). Chlorine (Cl) content in fenugreek plant tissues increased with rising salinity levels. Alhammadi and Edward (2009) reported that control plants had the lowest Cl concentrations, while those treated with 200 mM NaCl had the highest Cl levels. Increasing soil salinity also adversely impacted phosphorus (P) content in the leaves, roots, shoots, and pods, with P concentrations declining as sodium chloride treatments increased. Similar studies found that plant roots accumulated the highest P levels in saline conditions (Sharifi et al., 2007). In fenugreek, iron (Fe) content rose as salt levels increased (Mass et al., 1972). Salinity treatment also led to increased copper (Cu) accumulation in plant tissues, with statistically significant increases in Cu content observed at up to 200 mM NaCl (Tuncturk, 2011).

The interaction between sodium and ammonium cations and/or chlorine and nitrate in saline environments results in reduced nitrogen (N) uptake, affecting crop development and yield. Studies by Lea-Cox and Syvertsen

(1993) and Bar et al. (1997) noted that a decline in nitrate (NO_3^-) uptake often accompanies antagonistic behavior by chloride (Cl^-; Qadir and Schubert, 2002). High sodium chloride concentrations reduce leaf Ca^{2+}, potassium (K^+), and N levels (Keutgen and Pawelzik, 2009; Tuna et al., 2007). In response to external NaCl salinity, recent research observed temporary drops in Ca^{2+} and magnesium (Mg^{2+}) concentrations across all plant organs (Hussin et al., 2013).

9.3.8 OXIDATIVE STRESS AND SALINITY

Salinity commonly results in too much build-up of ROS, which can lead to protein oxidation, enzyme deactivation, DNA damage, and/or interactions with other crucial plant cell components. Salinity also has an impact on plants directly through its direct impacts on plants. When there are excessive NaCl concentrations present, ROS damage membrane lipids. The concentration of MDA and other aldehydes was utilized to estimate the degree of oxidative damage to tissue lipids. The shoot other aldehydes content in the control group increased with the treatment of 50 mM NaCl. At the maximum NaCl concentration (200 mM), this value increased as the salt level increased, reaching 1.943 mol g^1 FW (Pasandi et al., 2014). Malondialdehyde (MDA) and other aldehydes generation revealed salinity-induced increases in lipid peroxidation in the fenugreek shoot. MDA levels increased at all NaCl concentrations, although those at higher levels had more noticeable effects (Pasandi et al., 2014). When comparing control plants at 100 and 200 mM NaCl, the rate of peroxidation rose by 113%, or by a factor of 373%. Salt stress raised the quantity of MDA by 39.76%, 56%, and 101.47% at 50, 100, and 200 mM NaCl, respectively, compared to the untreated control. When stomata shut as a result of salt stress, there is less carbon dioxide accessible in the leaves, which inhibits carbon fixing (Ahmad et al., 2009, 2010a, b, 2011; Cheeseman, 1988; Greenway and Munns, 1980). Water scarcity leads to the production of reactive oxygen species (Elstner, 1987). ROS are extremely reactive and can damage cells by oxidizing lipids, proteins, and nucleic acids, according to Ahmad et al. (2010a, b). In another study it was found that cellular toxicity brought on by salt in a variety of agricultural plants, including tomato, pea, and mustard, is mostly caused by ROS-mediated membrane damage (Ahmad et al., 2009, 2010b; Mittova et al., 2004). This has been seen in various research on plants. When subjected to prolonged salinity treatments (EC 5.4 and 10.6 dS m^{-1}, 60 days), salt-sensitive cultivars were shown to have greater levels of H_2O_2 and lipid peroxidation than salt-tolerant cultivars (Sairam et al., 2002). Hasanuzzaman and Fujita (2011); Hasanuzzaman et al. (2009) discovered in a recent study that increasing salinity enhanced the levels of H_2O_2 and lipid peroxidation in *Brassica napus* and *Triticum aestivum*. In another study it was found that the production of reactive oxygen species under atmospheric circumstances like salt is one of the main causes of decreases in crop productivity. ROS regulation is therefore a crucial mechanism to prevent undesired toxicity due to damage caused by oxidative stress (Halliwell and Gutteridge, 1989).

9.4 STRATEGIES FOR COPING WITH SALT

9.4.1 PLANT REACTIONS THAT PROMOTE SALINITY TOLERANCE

The plant goes through various biochemical and physiological processes in order to resist salinity. Glycophytes and halophytes are two plant groupings that have evolved to thrive in saltwater. Glycophytes are unable to resist higher salt concentrations, in contradiction to halophytes. Different methods are used by halophytes and glycophytes to deal with salt (Krishnamurthy et al., 2018). Ion homeostasis and subdivision, transport, ion uptake, antioxidant enzyme machinery stimulation, polyamine synthesis, hormone modulation, and specific morphological and anatomical adaptations are only a few of the processes that plants go through (Zhang and Dai, 2019; Table 9.1).

TABLE 9.1 Methods to Control Salinity

Sl. No.	Management Strategies	Application	References
1.	Salt evaporation	Capsicum – the leaching process increases as irrigation water salinity increases.	Ben-Gal et al. (2008)
2.	Plant companions and primer	*Piper nigrum* – *Salsola soda* is grown with other plants.	Colla et al. (2006)
3.	Mulching of soil	*Beta vulgaris* subsp. vulgaris – mulching with rice straw and gravel increases crop production.	Zhang et al. (2008)
4.	Potassium	*Solanum tuberosum* – higher tuber yield	Elkhatib et al. (2004)
5.	Calcium	*Solanum lycopersicum* and *Piper nigrum* – quality of fruits and yield is increased while blossom end rot is decreased.	Rubio et al. (2009)
6.	Nitrogen	1 mM ammonia and 7 mM nitrate – reducing the impact of salinity on fruit yield.	Ben-Oliel et al. (2005)

TABLE 9.1 *(Continued)*

Sl. No.	Management Strategies	Application	References
7.	Phosphorus	Up to 3.5 dS m^{-1}, radish lowered the sensitivity to salinity.	Tari (2002)
8.	Plant growth regulators are used.	*Solanum lycopersicum* – when 0.5 mM salicylic acid is applied, Na can accumulate in plant tissues and serve as osmolytes.	Tari (2002)
9.	Application of fertilizers to the leaves	Fruit output was improved by K_2HPO_4 in brinjal.	Elwan (2010)
10.	Injection of bacteria	The use of artichoke-Bacillus subtilis inoculation reduced the negative impacts of salt stress and enhanced output.	Saleh et al. (2005)
11.	Application of non-nutritional additives	SOD and CAT activity is increased in spinach when it is salinized.	Eraslan et al. (2008)
12.	Elevated CO_2 levels	*Solanum lycopersicum* – enhanced airborne carbon dioxide concentration, reducing the detrimental effects of salt stress.	Takagi et al. (2009)

9.4.2 ION HOMEOSTASIS AND COMPARTMENTALIZATION

High salt levels hinder the absorption of essential minerals such as iron, calcium, potassium, zinc, boron, and magnesium. Salinity also disrupts the plant's mineral composition, lowering nitrogen, potassium, and zinc levels in leaves while increasing phosphorus, potassium, calcium, and magnesium levels in roots (Chrysargyris et al., 2019). Elevated salinity decreases the uptake of macronutrients like potassium, phosphorus, and magnesium, and micronutrients such as iron, zinc, copper, and manganese, leading to ion toxicity and causing nutritional imbalances in various plant organs. High salinity depletes mineral availability and affects the interactions between soil and the plant microbiome (Win et al., 2018).

To thrive under salinity stress, plants must maintain ionic homeostasis and equilibrium through ion absorption and compartmentalization. Because plants cannot tolerate high salt concentrations in their cytoplasm, they either sequester excess salt in older tissues or compartmentalize it within vacuoles. Sodium ions enter the soil in the root zone, pass through cell membranes, and are transported to vacuoles by sodium cation/hydrogen cation antiporters and other pumps (Fahad et al., 2015; Gupta and Huang, 2014; Krishnamurthy et al., 2018; Singh et al., 2015; Zhang and Dai, 2019).

9.4.3 *MORPHOLOGICAL AND ANATOMICAL ADAPTATION*

The plant's morphology, anatomy, and ultrastructure all alter in reaction to salt. Roots are required for salt tolerance. Root length, diameter, and number impact root function and increase plant water and nutrient absorption in stressed plants. Salinity encourages roots to grow and expand. It penetrates deep into the soil to minimize salt. To combat the harmful impacts of salinity, the density and electrical conductivity of the roots are enhanced (El Ghazali, 2020; Kheloufi and Mansouri, 2019). This increases ion retention and prevents harmful ions from accumulating. Arbuscular mycorrhizal fungus (AMF) helps to plant tolerance as a mitigator. As a result, both the roots and the AMF form a symbiotic connection that alleviates stress. Significant differences in root to shoot ratio, root and shoot growth, and negative ion retention from a variety of cellular organelles and plant organs are only a few of the ways that plants respond to salinity tolerance. As a result, the above-ground portion of the plant has correct osmoticum, ionic balance, and nutrient transfer. Plants adjust their leaf area ratio (LAR), leaf weight ratio (LWR) and specific leaf area (SLA) in response to salt. By collecting carotenoid pigments and altering the size of their leaves, certain plants can change the size of their stomata and the color of their leaves. Some plants may even lose their leaves, making them more tolerant by preventing salt from entering the foliage (El Ghazali, 2020; Kheloufi and Mansouri, 2019).

9.4.4 *OSMOTIC ADJUSTMENT*

Plant salinity has a deleterious influence on toxicity due to ions, oxidative stress as well as osmotic stress. Maintaining cell turgor, which is critical for growth, productivity, and yield, necessitates osmotic adaptation. Osmolytes produced by plants include glycine, soluble sugar, betaine, and proline. The amino acid proline (Pro), which is present in plant cytosol, functions as an osmoprotectant, a signaling molecule, and a means of stabilizing and preserving proteins, protein-based enzymes, and membranes, as well as a signaling molecule (Tang et al., 2015; Zhang and Dai, 2019). Betaines that act as osmoprotectants include proline betaine, hydroxyproline betaine, pipecolate betaine, and glycine betaine (GB; Tang et al., 2015). GB is the most potent and effective osmoprotectant for decreasing salinity stress. GB supports osmotic adjustment by calculating the potassium and sodium ions ratio and accumulating within the cell, according to considerable research. As a result, the damaging effects of ions are reduced, resulting in salinity (Singh et al., 2015).

9.5 THE PLANT'S RESPONSE TO SALINITY THROUGH PHYTOHORMONES

Salinity alters the way genes are controlled in response to hormones in a quick and continuing manner. Hormones play a major role in determining how plants respond to salt. Crosstalk between numerous hormone genes and signaling pathways, as well as the plant's response to salt stress, has emerged in recent years. Phytohormones, according to Ali et al. (2020), plays an important role in the development and growth of plants as well as in reducing salinity stress.

9.5.1 *BENZOIC ACID*

Cao et al. (2017) found that ABA caused stomata to shut, delayed senescence, controlled numerous catabolic processes as well as metabolic processes, and decreased transpiration. ABA is an intrinsically signaling chemical that maintains the balance between soil and plant leaves. ABA increases the expression of genes involved in the synthesis of osmoprotectants and stomatal closure. One of the proteins that ABA increased is dehydrin, which minimizes salt stress and provides plants tolerance (Fahad et al., 2015; Martínez-Andújar et al., 2020).

9.5.2 *BRASSINOSTEROIDS*

Brassinosteroids (BRs), which can be found as conjugated, free, or fatty acids, are found in nature phytohormones. The majority of BRs are composed of castasterone (CS), its derivatives, and brassinolide (BL) and their derivatives. Brassinosteroids promote the emergence of seeds, vascular differentiation, growth of pollen tube, photosynthesis, and reproduction in response to abiotic stress, such as salt. BRs boost osmoprotectants and antioxidants. In response to salt stress resistance, plants improve BR function (Nolan et al., 2020). In salinity, BRs play a part in proteostasis. In response to salt, different quantities of Brassinosteroids genes are activated. In contrast to the brassinosteroid sensitivity 1-associated receptors kinases (BAK1) family of genes, the BR-resistant 1/2 genes experience downregulation as a consequence of salinity adapting (Siddiqui et al., 2019).

9.5.3 GIBBERELLINS

Gibberellins (GAs), a plant hormone, aid in the reduction of a number of environmental difficulties. Endogenous GA production rises in plants exposed to salt. It modifies a variety of physiological and metabolic systems in the plant to compensate for salt. The plant's GA also increases the synthesis of osmolytes, sugar signaling, and antioxidant machinery, all of which help to quench ROS and maintain cell turgor in salinity stress conditions. GAs have a role in sustaining water relations between plants and in the distribution of photosynthates from source to sink to reduce the detrimental effects of salt stress and promote plant tolerance. Multiple studies have shown that small amounts of GA lower the stress impacts caused by salt by deteriorating DELLA and enhancing SA production. Particularly GA4, aids in reducing the harmful effects of salt. Genes that produce the DELLA protein operate as a positive regulator of plant responses to salt in addition to increasing ABA synthesis (Cao et al., 2017; Fahad et al., 2015; Rao et al., 2016).

9.5.4 JASMONIC ACID

Jasmonic acid (JA) plays a protective role for plants under both abiotic and biotic stress. JA participates in processes such as sprouting, flowering, callus development, and the growth of bulbs and tubers under both normal and stressful conditions. JA mitigates the detrimental effects of salinity by enhancing the expression of various genes involved in metabolic pathways, including Rubisco, arginine decarboxylase, and invertase. To counteract the harmful effects of excess ions, JA modifies protein synthesis and carbon dioxide fixation. JA also reduces the expression of the gene associated with GA-GID binding while increasing transcription of genes responsible for DELLA protein synthesis. In response to salt stress, plants upregulate *MYC2* and *TGA* genes and downregulate JAZ protein-coding genes to offset the adverse impacts of salinity (Yang et al., 2017).

9.5.5 SALICYLIC ACID

Salicylic acid (SA) is a crucial phytohormone involved in defending plants against various abiotic and biotic stressors. Salinity increases SA levels, which in turn promotes expansion, growth, development, photosynthesis, and pigment synthesis. According to Cao et al. (2017), SA is believed to regulate genes responsible for the production of heat shock proteins, antioxidant enzymes, and various secondary metabolites and their byproducts.

SA enhances the plant's tolerance to salinity by increasing the production of enzymes such as peroxidase (POX), superoxide dismutase (SOD), catalase (CAT), and glutathione peroxidase (GPX). Additionally, SA boosts the production of non-enzymatic antioxidants, including tocopherol, vitamin C, glutathione, and carotenoids, to reduce reactive oxygen species (ROS) and aid in salt adaptation. SA is also reported to activate genes involved in glucose metabolism and glycolysis (Cao et al., 2017; *see* Table 9.2).

TABLE 9.2 Gene's Upregulation and Downregulation in Plants Under Conditions of Salinity

Sl. No.	Name of Gene	Downregulate/ Upregulate	Functions	References
1.	Son of sevenless, voltage-dependent anion channel	Upregulate	Homeostasis, ion homeostasis, and exclusion of salt	Yousefirad et al. (2020)
2.	Na^+/H^+ exchanger 1	Upregulate	The vacuole receives salt from the cytosol	Köster et al. (2019)
3.	Leucine-rich repeat receptor-like kinases	Upregulate	Promote development and growth of plant	Amirbakhtiar et al. (2019)
	Cysteine-rich repeat receptor-like kinases			
	Protein kinase R (PKR)-like endoplasmic reticulum kinase			
	S-locus receptor kinase			
4.	Mitogen-activated protein kinase 3	Downregulate	Homeostasis (Ion) and osmoprotection	Guo et al. (2019)
	Mitogen-activated protein kinase kinase-kinase 2			
5.	*BRI1/2* ion	Downregulate	Increase growth of plant and homeostasis	Siddiqui et al. (2019)
6.	*ETR1/ETR2/EIN4*	Downregulate	Develop a tolerance against salinity	Fahad et al. (2015)
7.	*LHC*	Downregulate	Change the lighting system	Wang et al. (2019)

9.6 APPLICATIONS OF OMICS TO SALINITY STRESS OF PLANTS

Capacity of plant to grow and complete its lifespan on a medium with significant amounts of soluble salt is referred to as salt tolerance. Halophytes are vegetation that can thrive in a condition of salinity in the root zone.

Halophytes can be facultative (found in less salinized habitats along the boundary between saline and non-saline uplands) or obligate. The capacity to tolerate salt is a complex trait with several interconnected traits. Researchers are increasingly interested in researching omics technologies like as transcriptomics, metabolomics, proteomics, and genomics to uncover and grasp the molecular components and mechanisms underpinning salt tolerance (Inan et al., 2004). In this section, we'll go through the roles that genomes, metabolomics, proteomics, and transcriptomics play in tolerance to salinity and how they may be utilized to increase plants' capacity to survive salinity.

9.6.1 GENOMIC SCALE

Some different stress-responsive genes in tolerant plants may be lacking in sensitive plants (differences at the genomic structural level). When thorough genetic research is paired with genome-level profiling, a significant portion of the signaling networks that control stress tolerance may become visible. According to (7), the three main functional groups of salt tolerance genes. Following are some examples of salt-related molecules: (i) the molecules that regulate consumption and transportation; (ii) those with a defensive or osmotic impact; and (iii) compounds that might hasten plant growth in salty soil, and so forth. According to Zhu et al. (1998), numerous A. Thaliana salt excessively responsive mutants with very low tolerance to salinity have been discovered. These mutations are caused by salt consumption and transport-related genes. Zhu et al. (1998) found that SOS mutants exhibit a pronounced hypersensitivity to high exogenous sodium ions or lithium ions levels and are not able to develop in the presence of exceptionally low exogenous K^+ levels. In SOS mutants, allelic testing locates the SOS1, SOS2, and SOS3 loci (Martínez-Atienza et al., 2007). Shi et al. (2000) discovered that SOS1 encodes a plasma membrane sodium or hydrogen ion antiporter that prevents sodium from entering the apoplast (Mahajan et al., 2008; Mendoza et al., 1994). SOS3 physically interacts with and triggers SOS2 (Halfter et al., 2000). The regulatory route for Na^+ and K^+ equilibrium, as well as tolerance to salt in plants, is thus established by SOS2 and SOS3. SOS4 and SOS2 may both work together to control SOS1 activity. The synthesis of pyridoxal-5-phosphate is catalyzed by the SOS4 enzyme. A cofactor that could operate as an SOS1 ligand and has a potential interaction sequence for this component. Out of 3 SOS loci, SOS1 has the most influence on a plant's capacity to handle salt. SOS1 mutant plants are more responsive to lithium or sodium ions stresses than SOS3 and SOS2 mutant plants (Zhu et al., 2004). SOS1 fulfills the same function in both double mutants, according to analysis of two double mutants and SOS3 route (Liu et al., 2000).

9.6.2 GENES THAT REGULATE THE RATE OF CELL AND TISSUE GROWTH

Potential candidates for growth-regulating genes are likely connected to signaling pathways and include hormonal compounds, transcription regulators, protein kinases, and protein phosphatases, as well as other signaling molecules like calmodulin-binding proteins. These genes are likely involved in the response to various stressors, such as drought (Chaves et al., 2003), cold, and soil conditions that limit growth, such as compaction or sodicity. Many of these genes are also activated in response to drought stress, often through the action of abscisic acid (ABA).

When cells contract due to salt, dehydration, or freezing temperatures, metabolite concentrations may shift, and membrane protein configurations may alter. Another potential sensor of stress is a distant communication signal that moves from roots to shoots through the transpirational stream, helping to coordinate the plant's overall stress response.

9.6.3 TRANSCRIPTOMIC LEVEL

Resistant plants have different synchronization of major genes which are responsive to stress than sensitive plants. It exhibits abnormal stress-response gene regulation (Kumari et al., 2009). The transcriptome section explained under subheadings.

9.6.4 TRANSCRIPTIONAL FACTORS

A transcription factor, also known as a sequence-specific DNA-binding factor, binds to specific DNA sequences and regulates DNA transcription. Messenger RNA (mRNA) is responsible for producing stress-activated genes associated with crop adaptation and response, as well as for carrying genetic information from DNA. Transcription factors are grouped according to their DNA-binding homology domains, including bZip (basic region leucine zipper) proteins, zinc finger proteins, homeobox proteins, MYB-like proteins (which contain helix-turn-helix motifs), MADS domain proteins, and others. The basic helix-loop-helix (bHLH) transcription factor, encoded by the *OrbHLH2* gene in *Arabidopsis*, positively regulates salt-stress signaling, enhancing a plant's resistance to osmotic and salt stress (Pitzschke et al., 2009; Zhou et al., 2009).

9.6.5 PROTEOMIC SCALE

In salt-tolerant plants, stress-response proteins exhibit different activity levels compared to those in salt-sensitive plants. Numerous studies have explored the molecular mechanisms controlling plant responses to salinity, focusing on transcriptional modifications (Brumós et al., 2009). However, protein expression and post-translational modifications also impact phenotypic performance, which cannot be fully assessed through transcription analysis alone. In addition to genomic and transcriptomic studies, proteome analysis offers a valuable molecular tool. Simple changes in protein content reveal limited information, so functional protein analysis should be used to understand how proteins operate in saline environments and to verify comparative proteomics. Consequently, further studies are needed to explore how proteins contribute to salt tolerance in plants. These methods include examining post-translational modifications, protein-protein interactions, tissue and intracellular localization, and phenotypic effects through gene silencing or overexpression.

The cytoskeleton, along with cytoskeleton-associated proteins, enzymes linked to pathogenesis, salinity stress, and stress-related proteins, play significant roles in plants under salinity stress. Osmotic stress generally induces increased stomatal closure, reduced CO_2 availability, and decreased CO_2 assimilation in most plants. The levels of oxygen-evolving enhancer (OEE) proteins, the large and small subunits of Rubisco, and the rate of CO_2 assimilation are all affected. Under salt exposure, various glycophytic crops and plants exhibit increased production of Rubisco activase, carbonic anhydrase, and assembly complexes (Aghaei et al., 2008; Bandehagh et al., 2011; Chattopadhyay et al., 2011; Pang et al., 2010; Sobhanian et al., 2010). The primary Rubisco activase may also function in a cost-efficient manner under stress to activate Rubisco (Fatehi et al., 2012; Kim et al., 2005).

9.6.6 METABOLOMICS

For studying the responses of plants to salt stress, metabolomics is a crucial functional genome sequencing approach (25). Because of high-throughput omics approaches like metabolomics, our comprehension of alterations in gene-protein-metabolite interactions brought on by salt stress has increased (Urano et al., 2010). This is currently being used in conjunction with additional platforms to study various physiological processes in plants, such as transcriptome sequencing and proteomics (Wu et al., 2013). Metabolite profiling, targeted examination, and molecular fingerprinting are the three main techniques applied in plant metabolomics. Metabolomics analysis has become a popular approach in recent years for examining abiotic issues such as salt tolerance in plants (Alvarez et al., 2008; Renberg et al., 2010).

9.6.7 *METABOLISM OF OSMOLYTES*

Water with a high concentration of dissolved salt ions has a lower osmotic potential, causing osmotic stress in plant cells. In response to this stress, plant cells accumulate inorganic salt ions as well as hydrophilic compounds, known as osmolytes, and hydrophilic proteins with osmoprotective properties. Proline and glycine betaine (GB) are two osmolytes whose metabolism varies in plants exposed to salt stress. In *Thellungiella*, salt stress was shown to increase the levels of enzymes related to proline biosynthesis, while reducing the levels of enzymes involved in proline hydrolysis (Pang et al., 2010).

9.7 CONCLUSION AND FUTURE PROSPECTIVE

The productivity of crops is harmed by salt stress. This is a result of how salt affects physiological functions like respiration, photosynthesis, hormone imbalance, and nutrient absorption. Salinity indirectly lowers agricultural productivity by causing stressed plants to produce more reactive oxygen species. Following their oxidation, these reactive oxygen species break down macromolecules like proteins, lipids, and amino acids. Understanding the fundamental components of a plant's resistance against salinity network is essential for producing more plants which are resistant to salinity stress. Recently, a variety of characteristics of plant resistance to salt have been successfully and curiously examined using metabolomics, transcriptomics, genomics, and proteomics. A number of genes, protein molecules, and metabolites can be used to alter plants in order to increase food production. This interdisciplinary approach will significantly increase plant and agricultural productivity, helping to fulfil the population's rising food need.

KEYWORDS

- **abiotic issues**
- **fenugreek**
- **genomics**
- **proteomics**
- **salinity**
- **salt stress**
- **sustainability**
- **transcriptomics**

REFERENCES

Acharya, S. N., Thomas, J. E., & Basu, S. K. (2006). Fenugreek: An "old world" crop for the "new world." *Biodiversity*, *7*(3–4), 27–30.

Aggarwal, K. B., Ranjan, J. K., Rathore, S. S., Saxena, S. N., & Mishra, B. K. (2013). Changes in physical and biochemical properties of fenugreek (*Trigonella sp.* L.) leaf during different growth stages. *Int. J. Seed Spices*, *3*(1), 31–35.

Aghaei, K., Ehsanpour, A. A., & Komatsu, S. (2008). Proteome analysis of potato under salt stress. *Journal of Proteome Research*, *7*(11), 4858–4868.

Ahmad, P., Jaleel, C. A., & Sharma, S. (2010a). Antioxidant defense system, lipid peroxidation, proline-metabolizing enzymes, and biochemical activities in two *Morus alba* genotypes subjected to NaCl stress. *Russian Journal of Plant Physiology*, *57*, 509–517.

Ahmad, P., Jaleel, C. A., Azooz, M. M., & Nabi, G. (2009). Generation of ROS and non-enzymatic antioxidants during abiotic stress in plants. *Botany Research International*, *2*(1), 11–20.

Ahmad, P., Jaleel, C. A., Salem, M. A., Nabi, G., & Sharma, S. (2010b). Roles of enzymatic and nonenzymatic antioxidants in plants during abiotic stress. *Critical Reviews in Biotechnology*, *30*(3), 161–175.

Ahmad, P., Kumar, A., Ashraf, M., & Akram, N. A. (2012). Salt-induced changes in photosynthetic activity and oxidative defense system of three cultivars of mustard (*Brassica juncea* L.). *African Journal of Biotechnology*, *11*(11), 2694.

Ahmad, P., Nabi, G., & Ashraf, M. (2011). Cadmium-induced oxidative damage in mustard [*Brassica juncea* (L.) Czern. & Coss.] plants can be alleviated by salicylic acid. *South African Journal of Botany*, *77*(1), 36–44.

Alhammadi, M. S., & Edward, G. P. (2009). Effect of salinity on growth of 12 cultivars of the United Arab Emirates date palm. *Communications in Soil Science and Plant Analysis*, *40*(15, 16), 2372–2388.

Ali, Q., Shahid, S., Nazar, N., Hussain, A. I., Ali, S., Chatha, S. A. S., & Hussain, S. M. (2020). Use of phytohormones in conferring tolerance to environmental stress. In *Plant*

Ecophysiology and Adaptation under Climate Change: Mechanisms and Perspectives II: Mechanisms of Adaptation and Stress Amelioration (pp. 245–355).

Allakhverdiev, S. I., Sakamoto, A., Nishiyama, Y., Inaba, M., & Murata, N. (2000). Ionic and osmotic effects of NaCl-induced inactivation of photosystems I and II in *Synechococcus* sp. *Plant Physiology*, *123*(3), 1047–1056.

Alvarez, S., Marsh, E. L., Schroeder, S. G., & Schachtman, D. P. (2008). Metabolomic and proteomic changes in the xylem sap of maize under drought. *Plant, Cell & Environment*, *31*(3), 325–340.

Amin, A., Alkaabi, A., Al-Falasi, S., & Daoud, S. A. (2005). Chemopreventive activities of *Trigonella foenum-graecum* (Fenugreek) against breast cancer. *Cell Biology International*, *29*(8), 687–694.

Amirbakhtiar, N., Ismaili, A., Ghaffari, M. R., Nazarian Firouzabadi, F., & Shobbar, Z. S. (2019). Transcriptome response of roots to salt stress in a salinity-tolerant bread wheat cultivar. *PLOS ONE*, *14*(3), e0213305.

Asaadi, A. M. (2009). Investigation of salinity stress on seed germination of *Trigonella foenum-graecum*. *Research Journal of Biological Sciences*, *4*(11), 1152–1155.

Bacha, H., Tekaya, M., Drine, S., Guasmi, F., Touil, L., Enneb, H., & Ferchichi, A. (2017). Impact of salt stress on morpho-physiological and biochemical parameters of *Solanum lycopersicum* cv. Microtom leaves. *South African Journal of Botany*, *108*, 364–369.

Bandehagh, A., Salekdeh, G. H., Toorchi, M., Mohammadi, A., & Komatsu, S. (2011). Comparative proteomic analysis of canola leaves under salinity stress. *Proteomics*, *11*(10), 1965–1975.

Bar, Y., Apelbaum, A., Kafkafi, U., & Goren, R. (1997). Relationship between chloride and nitrate and its effect on growth and mineral composition of avocado and citrus plants. *Journal of Plant Nutrition*, *20*(6), 715–731.

Ben-Gal, A., Ityel, E., Dudley, L., Cohen, S., Yermiyahu, U., Presnov, E., & Shani, U. (2008). Effect of irrigation water salinity on transpiration and on leaching requirements: A case study for bell peppers. *Agricultural Water Management*, *95*(5), 587–597.

Ben-Oliel, G., Kant, S., Naim, M., Rabinowitch, H. D., Takeoka, G. R., Buttery, R. G., & Kafkafi, U. (2005). Effects of ammonium to nitrate ratio and salinity on yield and fruit quality of large and small tomato fruit hybrids. *Journal of Plant Nutrition*, *27*(10), 1795–1812.

Brumós, J., Colmenero-Flores, J. M., Conesa, A., Izquierdo, P., Sánchez, G., Iglesias, D. J., & Talón, M. (2009). Membrane transporters and carbon metabolism implicated in chloride homeostasis differentiate salt stress responses in tolerant and sensitive citrus rootstocks. *Functional & Integrative Genomics*, *9*, 293–309.

Bybordi, A. (2010). The influence of salt stress on seed germination, growth and yield of canola cultivars. *Notulae Botanicae Horti Agrobotanici Cluj-Napoca*, *38*(1), 128–133.

Cao, D., Lutz, A., Hill, C. B., Callahan, D. L., & Roessner, U. (2017). A quantitative profiling method of phytohormones and other metabolites applied to barley roots subjected to salinity stress. *Frontiers in Plant Science*, *7*, 2070.

Chattopadhyay, A., Subba, P., Pandey, A., Bhushan, D., Kumar, R., Datta, A., & Chakraborty, N. (2011). Analysis of the grasspea proteome and identification of stress-responsive proteins upon exposure to high salinity, low temperature, and abscisic acid treatment. *Phytochemistry*, *72*(10), 1293–1307.

Chaves, M. M., Maroco, J. P., & Pereira, J. S. (2003). Understanding plant responses to drought—from genes to the whole plant. *Functional Plant Biology*, *30*(3), 239–264.

Cheeseman, J. M. (1988). Mechanisms of salinity tolerance in plants. *Plant Physiology*, *87*(3), 547–550.

Chrysargyris, A., Papakyriakou, E., Petropoulos, S. A., & Tzortzakis, N. (2019). The combined and single effect of salinity and copper stress on growth and quality of *Mentha spicata* plants. *Journal of Hazardous Materials*, *368*, 584–593.

Chutipaijit, S., Cha-um, S., & Sompornpailin, K. (2011). High contents of proline and anthocyanin increase protective response to salinity in *Oryza sativa* L. spp. *indica*. *Australian Journal of Crop Science*, *5*(10), 1191–1198.

Colla, G., Rouphael, Y., Fallovo, C., Cardarelli, M., & Graifenberg, A. (2006). Use of *Salsola soda* as a companion plant to improve greenhouse pepper (*Capsicum annuum*) performance under saline conditions. *New Zealand Journal of Crop and Horticultural Science*, *34*(4), 283–290.

Demetriou, G., Neonaki, C., Navakoudis, E., & Kotzabasis, K. (2007). Salt stress impact on the molecular structure and function of the photosynthetic apparatus—The protective role of polyamines. *Biochimica et Biophysica Acta (BBA)-Bioenergetics*, *1767*(4), 272–280.

El Ghazali, G. E. (2020). *Suaeda vermiculata* Forssk. ex JF Gmel.: Structural characteristics and adaptations to salinity and drought: A review. *International Journal of Science*, *9*(02), 28–33.

Elkhatib, H. A., Elkhatib, E. A., Khalaf Allah, A. M., & El-Sharkawy, A. M. (2004). Yield response of salt-stressed potato to potassium fertilization: A preliminary mathematical model. *Journal of Plant Nutrition*, *27*(1), 111–122.

Elstner, E. F. (1987). Metabolism of activated oxygen species. In *Biochemistry of Metabolism* (pp. 253–315). Academic Press.

Elwan, M. W. (2010). Ameliorative effects of di-potassium hydrogen orthophosphate on salt-stressed eggplant. *Journal of Plant Nutrition*, *33*(11), 1593–1604.

Eraslan, F., Inal, A., Pilbeam, D. J., & Gunes, A. (2008). Interactive effects of salicylic acid and silicon on oxidative damage and antioxidant activity in spinach (*Spinacia oleracea* L. cv. Matador) grown under boron toxicity and salinity. *Plant Growth Regulation*, *55*, 207–219.

Fahad, S., Hussain, S., Matloob, A., Khan, F. A., Khaliq, A., Saud, S., & Huang, J. (2015). Phytohormones and plant responses to salinity stress: A review. *Plant Growth Regulation*, *75*, 391–404.

Fatehi, F., Hosseinzadeh, A., Alizadeh, H., Brimavandi, T., & Struik, P. C. (2012). The proteome response of salt-resistant and salt-sensitive barley genotypes to long-term salinity stress. *Molecular Biology Reports*, *39*, 6387–6397.

Fernández-Torquemada, Y., & Sánchez-Lizaso, J. L. (2013). Effects of salinity on seed germination and early seedling growth of the Mediterranean seagrass *Posidonia oceanica* (L.) Delile. *Estuarine, Coastal and Shelf Science*, *119*, 64–70.

Garg, N., & Manchanda, G. (2008). Salinity and its effects on the functional biology of legumes. *Acta Physiologiae Plantarum*, *30*, 595–618.

Gomes-Filho, E., Lima, C. R. F. M., Costa, J. H., da Silva, A. C. M., da Guia Silva Lima, M., de Lacerda, C. F., & Prisco, J. T. (2008). Cowpea ribonuclease: Properties and effect of NaCl-salinity on its activation during seed germination and seedling establishment. *Plant Cell Reports*, *27*, 147–157.

Grattan, S. R., & Grieve, C. M. (1998). Salinity–Mineral nutrient relations in horticultural crops. *Scientia Horticulturae*, *78*(1–4), 127–157.

Greenway, H., & Munns, R. (1980). Mechanisms of salt tolerance in nonhalophytes. *Annual Review of Plant Physiology*, *31*(1), 149–190.

Guo, S. M., Tan, Y., Chu, H. J., Sun, M. X., & Xing, J. C. (2019). Transcriptome sequencing revealed molecular mechanisms underlying tolerance of *Suaeda salsa* to saline stress. *PLOS ONE*, *14*(7), e0219979.

Gupta, B., & Huang, B. (2014). Mechanism of salinity tolerance in plants: Physiological, biochemical, and molecular characterization. *International Journal of Genomics*, *2014*(1), 701596.

Halfter, U., Ishitani, M., & Zhu, J. K. (2000). The *Arabidopsis* SOS2 protein kinase physically interacts with and is activated by the calcium-binding protein SOS3. *Proceedings of the National Academy of Sciences*, *97*(7), 3735–3740.

Halliwell, B., & Gutteridge, J. M. C. (1989). *Free Radicals in Biology and Medicine* (2nd ed.). Clarendon.

Hasanuzzaman, M., & Fujita, M. (2011). Exogenous silicon treatment alleviates salinity-induced damage in *Brassica napus* L. seedlings by up-regulating the antioxidant defense and methylglyoxal detoxification system. *Plant Biology*, American Society of Plant Biology (Abstract).

Hasanuzzaman, M., Fujita, M., Islam, M. N., Ahamed, K. U., & Nahar, K. (2009). Performance of four irrigated rice varieties under different levels of salinity stress. *International Journal of Integrative Biology*, *6*(2), 85–90.

Hu, Y., & Schmidhalter, U. (2005). Drought and salinity: A comparison of their effects on mineral nutrition of plants. *Journal of Plant Nutrition and Soil Science*, *168*(4), 541–549.

Hussin, S., Geissler, N., & Koyro, H. W. (2013). Effect of NaCl salinity on *Atriplex nummularia* (L.) with special emphasis on carbon and nitrogen metabolism. *Acta Physiologiae Plantarum*, *35*, 1025–1038.

Ibrar, M., Jabeen, M., Tabassum, J., Hussain, F., & Ilahi, I. (2003). Salt tolerance potential of *Brassica juncea* Linn. *Journal of Science and Technology (Peshawar)*, *27*.

Inan, G., Zhang, Q., Li, P., Wang, Z., Cao, Z., Zhang, H., & Galbraith, D. W. (2004). Jenks, M. A., Rhodes, D., Hasegawa, P. M., Bohnert, H. J., Joly, R. J., Bressan, R. A., & Zhu, J. K. Salt cress: A halophyte and cryophyte Arabidopsis relative model system and its applicability to molecular genetic analyses of growth and development of extremophiles. *Plant Physiology*, *135*, 1718–1737.

Iyengar, E. R. R., & Reddy, M. P. (1996). Photosynthesis in highly salt tolerant plants. In *Handbook of Photosynthesis* (pp. 909). Marshal Dekar, Baten Rose, USA.

Jain, S. C., & Madhu, A. (1988). Regulation of trigonellin in *Trigonella* species by chemical mutagenic treatments. *Indian Drugs*, *26*(1), 14–16.

Kamal, R., & Yadav, R. (1991). Flavonoids from *Trigonella polycerta in-vivo* and *in-vitro*. *Journal of Phytological Research*, *4*(2), 161–165.

Kapoor, N., & Pande, V. (2015). Effect of salt stress on growth parameters, moisture content, relative water content and photosynthetic pigments of fenugreek variety RMt-1. *Journal of Plant Sciences*, *10*(6), 210–221.

Kaur, G., Kumar, S., Nayyar, H., & Upadhyaya, H. D. (2008). Cold stress injury during the pod-filling phase in chickpea (*Cicer arietinum* L.): Effects on quantitative and qualitative components of seeds. *Journal of Agronomy and Crop Science*, *194*(6), 457–464.

Kaveh, H., Nemati, H., Farsi, M., & Jartoodeh, S. V. (2011). How salinity affect germination and emergence of tomato lines. *Journal of Biological and Environmental Sciences*, *5*(15).

Keutgen, A. J., & Pawelzik, E. (2009). Impacts of NaCl stress on plant growth and mineral nutrient assimilation in two cultivars of strawberry. *Environmental and Experimental Botany*, *65*(2, 3), 170–176.

Khan, M. A., Ungar, I. A., & Showalter, A. M. (2000). Effects of salinity on growth, water relations and ion accumulation of the subtropical perennial halophyte, Atriplex griffithii var. stocksii. *Annals of Botany*, *85*(2), 225–232.

Khan, M. M., Al-Mas'oudi, R. S., Al-Said, F., & Khan, I. (2013). Salinity effects on growth, electrolyte leakage, chlorophyll content and lipid peroxidation in cucumber (*Cucumis sativus* L.). In *International Conference on Food and Agricultural Sciences Malaysia: IACSIT Press* (Vol. 55, pp. 28–32).

Kheloufi, A., & Mansouri, L. M. (2019). Anatomical changes induced by salinity stress in root and stem of two *Acacia* species (*A. karroo* and *A. saligna*). *Poljoprivreda i Sumarstvo*, *65*(4), 137–150.

Khodarahmpour, Z., Ifar, M., & Motamedi, M. (2012). Effects of NaCl salinity on maize (*Zea mays* L.) at germination and early seedling stage. *African Journal of Biotechnology*, *11*(2), 298–304.

Kim, D. W., Rakwal, R., Agrawal, G. K., Jung, Y. H., Shibato, J., Jwa, N. S., & Usui, K. (2005). A hydroponic rice seedling culture model system for investigating proteome of salt stress in rice leaf. *Electrophoresis*, *26*(23), 4521–4539.

Köster, P., Wallrad, L., Edel, K. H., Faisal, M., Alatar, A. A., & Kudla, J. (2019). The battle of two ions: Ca^{2+} signaling against Na^{+} stress. *Plant Biology*, *21*, 39–48.

Krishnamurthy, P., Qingsong, L., & Kumar, P. P. (2018). Proteomics perspectives in the post-genomic era for producing salinity stress-tolerant crops. In *Salinity Responses and Tolerance in Plants* (Vol. 2, pp. 239–266).

Kumari, S., Panjabi nee Sabharwal, V., Kushwaha, H. R., Sopory, S. K., Singla-Pareek, S. L., & Pareek, A. (2009). Transcriptome map for seedling stage specific salinity stress response indicates a specific set of genes as candidates for saline tolerance in *Oryza sativa* L. *Functional & Integrative Genomics*, *9*, 109–123.

Kurth, E., Cramer, G. R., Läuchli, A., & Epstein, E. (1986). Effects of NaCl and $CaCl_2$ on cell enlargement and cell production in cotton roots. *Plant physiology*, *82*(4), 1102–1106.

Lea-Cox, J. D., & Syvertsen, J. P. (1993). Salinity reduces water use and nitrate-N-use efficiency of citrus. *Annals of Botany*, *72*(1), 47–54.

Liu, J., Ishitani, M., Halfter, U., Kim, C. S., & Zhu, J. K. (2000). The *Arabidopsis thaliana* SOS2 gene encodes a protein kinase that is required for salt tolerance. *Proceedings of the National Academy of Sciences*, *97*(7), 3730–3734.

López-Climent, M. F., Arbona, V., Pérez-Clemente, R. M., & Gómez-Cadenas, A. (2008). Relationship between salt tolerance and photosynthetic machinery performance in citrus. *Environmental and Experimental Botany*, *62*(2), 176–184.

Lu, Y., Lam, H., Pi, E., Zhan, Q., Tsai, S., Wang, C., & Ngai, S. (2013). Comparative metabolomics in Glycine max and Glycine soja under salt stress to reveal the phenotypes of their offspring. *Journal of Agricultural and Food Chemistry*, *61*(36), 8711–8721.

Mahajan, S., Pandey, G. K., & Tuteja, N. (2008). Calcium- and salt-stress signaling in plants: Shedding light on the SOS pathway. *Archives of Biochemistry and Biophysics*, *471*(2), 146–158.

Mane, A. V., Karadge, B. A., & Samant, J. S. (2010). Salinity induced changes in photosynthetic pigments and polyphenols of *Cymbopogon nardus* (L.) Rendle. *Journal of Chemical and Pharmaceutical Research*, *2*(3), 338–347.

Mantri, N., Patade, V., Penna, S., Ford, R., & Pang, E. (2012). Abiotic stress responses in plants: Present and future. In *Abiotic Stress Responses in Plants: Metabolism, Productivity and Sustainability* (pp. 1–19).

Martínez-Andújar, C., Martínez-Pérez, A., Ferrández-Ayela, A., Albacete, A., Martínez-Melgarejo, P. A., Dodd, I. C., & Pérez-Alfocea, F. (2020). Impact of overexpression of 9-cis-epoxycarotenoid dioxygenase on growth and gene expression under salinity stress. *Plant Science*, *295*, 110268.

Martínez-Atienza, J., Jiang, X., Garciadeblas, B., Mendoza, I., Zhu, J. K., Pardo, J. M., & Quintero, F. J. (2007). Conservation of the salt overly sensitive pathway in rice. *Plant Physiology*, *143*(2), 1001–1012.

Mass, E. V., Ogata, G., & Garber, M. J. (1972). Influence of salinity on Fe, Mn, and Zn uptake by plants. *Agronomy Journal*, *64*(6), 793–795.

Mehrafarin, A., Badi, H. N., Noormohammadi, G., Zand, E., Rezazadeh, S., & Qaderi, A. (2011). Effects of environmental factors and methanol on germination and emergence of Persian Fenugreek (*Trigonella foenum-graecum* L.). *African Journal of Agricultural Research*, *6*(19), 4631–4641.

Mendoza, I., Rubio, F., Rodriguez-Navarro, A., & Pardo, J. M. (1994). The protein phosphatase calcineurin is essential for NaCl tolerance of *Saccharomyces cerevisiae*. *Journal of Biological Chemistry*, *269*(12), 8792–8796.

Mittal, S., Kumari, N., & Sharma, V. (2012). Differential response of salt stress on *Brassica juncea*: Photosynthetic performance, pigment, proline, D1 and antioxidant enzymes. *Plant Physiology and Biochemistry*, *54*, 17–26.

Mittova, V., Guy, M., Tal, M., & Volokita, M. (2004). Salinity up-regulates the antioxidative system in root mitochondria and peroxisomes of the wild salt-tolerant tomato species *Lycopersicon pennellii*. *Journal of Experimental Botany*, *55*(399), 1105–1113.

Munns, R. (1993). Physiological processes limiting plant growth in saline soils: Some dogmas and hypotheses. *Plant, Cell & Environment*, *16*(1), 15–24.

Munns, R. (2005). Genes and salt tolerance: Bringing them together. *New Phytologist*, *167*(3), 645–663.

Niknam, V., Razavi, N., Ebrahimzadeh, H., & Sharifizadeh, B. (2006). Effect of NaCl on biomass, protein and proline contents, and antioxidant enzymes in seedlings and calli of two Trigonella species. *Biologia Plantarum*, *50*, 591–596.

Nolan, T. M., Vukašinović, N., Liu, D., Russinova, E., & Yin, Y. (2020). Brassinosteroids: Multidimensional regulators of plant growth, development, and stress responses. *The Plant Cell*, *32*(2), 295–318.

Oertli, J. J. (1991). Nutrient management under water and salinity stress. In *Proceedings of the Symposium on Nutrient Management for Sustained Productivity* (pp. 138–165). Depart. Soils Punjab Agric. Univ. Ludhiana, India.

Pang, Q., Chen, S., Dai, S., Chen, Y., Wang, Y., & Yan, X. (2010). Comparative proteomics of salt tolerance in *Arabidopsis thaliana* and *Thellungiella halophila*. *Journal of Proteome Research*, *9*(5), 2584–2599.

Parida, A. K., & Das, A. B. (2005). Salt tolerance and salinity effects on plants: A review. *Ecotoxicology and Environmental Safety*, *60*(3), 324–349.

Pasandi Pour, A., Farahbakhsh, H., & Saffari, M. (2014). Response of fenugreek plants to short-term salinity stress in relation to lipid peroxidation, antioxidant activity and protein content. *Ethno-Pharmaceutical Products*, *1*(1), 45–52.

Pitzschke, A., Djamei, A., Bitton, F., & Hirt, H. (2009). A major role of the MEKK1–MKK1/2–MPK4 pathway in ROS signaling. *Molecular Plant*, *2*(1), 120–137.

Pour, A. P., Farahbakhsh, H., Saffari, M., & Keramat, B. (2012). Effects of seed priming on germination and seedling growth under salinity stress in fenugreek. *International Journal of Agriculture and Crop Sciences (IJACS)*, *4*(12), 779–786.

Qadir, M., & Schubert, S. J. L. D. (2002). Degradation processes and nutrient constraints in sodic soils. *Land Degradation & Development*, *13*(4), 275–294.

Rajalakshmi, R., Nanavaty, K., & Gumashta, A. (1964). Effect of cooking procedures on the free and total niacin content of certain foodstuffs. *Indian Journal of Nutrition and Dietetics*, *1*, 276–280.

Rao, A. Q., ud Din, S., Akhtar, S., Sarwar, M. B., Ahmed, M., Rashid, B., & Husnain, T. (2016). Genomics of salinity tolerance in plants. *Plant Genomics*, *2016*, 273–299.

Renberg, L., Johansson, A. I., Shutova, T., Stenlund, H., Aksmann, A., Raven, J. A., & Samuelsson, G. (2010). A metabolomic approach to study major metabolite changes during acclimation to limiting CO_2 in *Chlamydomonas reinhardtii*. *Plant Physiology*, *154*(1), 187–196.

Rogers, M. E., Grieve, C. M., & Shannon, M. C. (2003). Plant growth and ion relations in lucerne (*Medicago sativa* L.) in response to the combined effects of NaCl and P. *Plant and Soil*, *253*, 187–194.

Romero-Aranda, R., Soria, T., & Cuartero, J. (2001). Tomato plant-water uptake and plant-water relationships under saline growth conditions. *Plant Science*, *160*(2), 265–272.

Rubio, J. S., Garcia-Sanchez, F., Rubio, F., & Martínez, V. (2009). Yield, blossom-end rot incidence, and fruit quality in pepper plants under moderate salinity are affected by K^+ and Ca^{2+} fertilization. *Scientia Horticulturae*, *119*(2), 79–87.

Saha, P., Chatterjee, P., & Biswas, A. K. (2010). NaCl pretreatment alleviates salt stress by enhancement of antioxidant defense system and osmolyte accumulation in mungbean (*Vigna radiata* L. Wilczek). *Plant Physiology*.

Sairam, R. K., Rao, K. V., & Srivastava, G. C. (2002). Differential response of wheat genotypes to long-term salinity stress in relation to oxidative stress, antioxidant activity, and osmolyte concentration. *Plant Science*, *163*(5), 1037–1046.

Saleh, J., Maftoun, M., Safarzadeh, S., & Gholami, A. (2009). Growth, mineral composition, and biochemical changes of broad bean as affected by sodium chloride and zinc levels and sources. *Communications in Soil Science and Plant Analysis*, *40*(19, 20), 3046–3060.

Saleh, S. A., Heuberger, H., & Schnitzler, W. H. (2005). Alleviation of salinity effect on artichoke productivity by *Bacillus subtilis* FZB24, supplemental Ca, and micronutrients. *Journal of Applied Botany and Food Quality*, *79*, 24–32.

Schonfeld, M. A., Johnson, R. C., Carver, B. F., & Mornhinweg, D. W. (1988). Water relations in winter wheat as drought resistance indicators. *Crop Science*, *28*(3), 526–531.

Shanker, A. K., & Venkateswarlu, B. (2011). Abiotic stress in plants—Mechanisms and adaptations. In *TechJaneza Trdine* (Vol. 9, pp. 1–10). Rijeka, Croatia.

Sharifi, M., Ghorbanli, M., & Ebrahimzadeh, H. (2007). Improved growth of salinity-stressed soybean after inoculation with salt pre-treated mycorrhizal fungi. *Journal of Plant Physiology*, *164*(9), 1144–1151.

Shi, H., Ishitani, M., Kim, C., & Zhu, J. K. (2000). The *Arabidopsis thaliana* salt tolerance gene SOS1 encodes a putative Na^+/H^+ antiporter. *Proceedings of the National Academy of Sciences*, *97*(12), 6896–6901.

Siddiqui, H., Sami, F., Juan, H. F., & Hayat, S. (2019). Brassinosteroid regulated physiological process: An omics perspective. In *Brassinosteroids: Plant Growth and Development* (pp. 297–322).

Singh, M., Kumar, J., Singh, S., Singh, V. P., & Prasad, S. M. (2015). Roles of osmoprotectants in improving salinity and drought tolerance in plants: A review. *Reviews of Environmental Science and Biotechnology*, *14*, 407–426.

Slinkard, A. E., McVicar, R., Brenzil, C., Pearse, P., Panchuk, K., & Hartley, S. (2006). Fenugreek in Saskatchewan. Saskatchewan Agriculture and Food.

Sobhanian, H., Motamed, N., Jazii, F. R., Nakamura, T., & Komatsu, S. (2010). Salt stress induced differential proteome and metabolome response in the shoots of *Aeluropus lagopoides* (Poaceae), a halophyte C4 plant. *Journal of Proteome Research*, *9*(6), 2882–2897.

Srinivasan, K. (2006). Fenugreek (*Trigonella foenum-graecum*): A review of health beneficial physiological effects. *Food Reviews International*, *22*(2), 203–224.

Swain, A. R., Dutton, S. P., & Truswell, A. S. (1985). Salicylates in foods. *Journal of the American Dietetic Association*, *85*(8), 950–960.

Takagi, M., El-Shemy, H. A., Sasaki, S., Toyama, S., Kanai, S., Saneoka, H., & Fujita, K. (2009). Elevated CO_2 concentration alleviates salinity stress in tomato plant. *Acta Agriculturae Scandinavica Section B–Soil and Plant Science*, *59*(1), 87–96.

Talukdar, D. (2012). Modulation of plant growth and leaf biochemical parameters in grass pea (*Lathyrus sativus* L.) and fenugreek (*Trigonella foenum-graecum* L.) exposed to NaCl treatments. *Indian Journal of Fundamental and Applied Life Sciences*, *2*(3), 20–28.

Tang, X., Mu, X., Shao, H., Wang, H., & Brestic, M. (2015). Global plant-responding mechanisms to salt stress: Physiological and molecular levels and implications in biotechnology. *Critical Reviews in Biotechnology*, *35*(4), 425–437.

Tari, I. (2002). Acclimation of tomato plants to salinity stress after a salicylic acid pre-treatment. *Acta Biologica Szegediensis*, *46*(3, 4), 55–56.

Tavakkoli, E., Fatehi, F., Coventry, S., Rengasamy, P., & McDonald, G. K. (2011). Additive effects of Na^+ and Cl^- ions on barley growth under salinity stress. *Journal of Experimental Botany*, *62*(6), 2189–2203.

Thakur, P., Kumar, S., Malik, J. A., Berger, J. D., & Nayyar, H. (2010). Cold stress effects on reproductive development in grain crops: An overview. *Environmental and Experimental Botany*, *67*(3), 429–443.

Tuna, A. L., Kaya, C., Ashraf, M., Altunlu, H., Yokas, I., & Yagmur, B. (2007). The effects of calcium sulfate on growth, membrane stability and nutrient uptake of tomato plants grown under salt stress. *Environmental and Experimental Botany*, *59*(2), 173–178.

Tuncturk, R. (2011). Salinity exposure modifies nutrient concentrations in fenugreek (*Trigonella foenum graecum* L.). *African Journal of Agricultural Research*, *6*(16), 3685–3690.

Urano, K., Kurihara, Y., Seki, M., & Shinozaki, K. (2010). "Omics" analyses of regulatory networks in plant abiotic stress responses. *Current Opinion in Plant Biology*, *13*(2), 132–138.

Vorasoot, N., Songsri, P., Akkasaeng, C., Jogloy, S., & Patanothai, A. (2003). Effect of water stress on yield and agronomic characters of peanut (*Arachis hypogaea* L.). *Songklanakarin Journal of Science and Technology*, *25*(3), 283–288.

Wang, W. C., Lin, T. C., Kieber, J., & Tsai, Y. C. (2019). Response regulators 9 and 10 negatively regulate salinity tolerance in rice. *Plant and Cell Physiology*, *60*(11), 2549–2563.

Wang, W., Vinocur, B., & Altman, A. (2003). Plant responses to drought, salinity and extreme temperatures: Towards genetic engineering for stress tolerance. *Planta*, *218*, 1–14.

Win, K. T., Tanaka, F., Okazaki, K., & Ohwaki, Y. (2018). The ACC deaminase expressing endophyte *Pseudomonas* spp. enhances NaCl stress tolerance by reducing stress-related ethylene production, resulting in improved growth, photosynthetic performance, and ionic balance in tomato plants. *Plant Physiology and Biochemistry*, *127*, 599–607.

Wu, D., Cai, S., Chen, M., Ye, L., Chen, Z., Zhang, H., & Zhang, G. (2013). Tissue metabolic responses to salt stress in wild and cultivated barley. *PLOS ONE*, *8*(1), e55431.

Xu, S., Hu, B., He, Z., Ma, F., Feng, J., Shen, W., & Yang, J. (2011). Enhancement of salinity tolerance during rice seed germination by presoaking with hemoglobin. *International Journal of Molecular Sciences*, *12*(4), 2488–2501.

Yang, Z., Wang, Y., Wei, X., Zhao, X., Wang, B., & Sui, N. (2017). Transcription profiles of genes related to hormonal regulations under salt stress in sweet sorghum. *Plant Molecular Biology Reporter*, *35*, 586–599.

Yousefirad, S., Soltanloo, H., Ramezanpour, S. S., Zaynali Nezhad, K., & Shariati, V. (2020). The RNA-seq transcriptomic analysis reveals genes mediating salt tolerance through rapid triggering of ion transporters in a mutant barley. *PLOS ONE*, *15*(3), e0229513.

Zhang, M., Qin, Z., & Liu, X. (2005). Remote sensed spectral imagery to detect late blight in field tomatoes. *Precision Agriculture*, *6*, 489–508.

Zhang, Q. T., Inoue, M., Inosako, K., Irshad, M., Kondo, K., Qiu, G. Y., & Wang, S. P. (2008). Ameliorative effect of mulching on water use efficiency of Swiss chard and salt accumulation under saline irrigation.

Zhang, Q., & Dai, W. (2019). Plant response to salinity stress. In *Stress Physiology of Woody Plants* (pp. 155–173). CRC Press.

Zhou, J., Li, F., Wang, J. L., Ma, Y., Chong, K., & Xu, Y. Y. (2009). Basic helix-loop-helix transcription factor from wild rice (*OrbHLH2*) improves tolerance to salt- and osmotic stress in *Arabidopsis*. *Journal of Plant Physiology*, *166*(12), 1296–1306.

Zhu, J. K. (2001). Plant salt tolerance. *Trends in Plant Science*, *6*(2), 66–71.

Zhu, J. K., Liu, J., & Xiong, L. (1998). Genetic analysis of salt tolerance in *Arabidopsis*: Evidence for a critical role of potassium nutrition. *The Plant Cell*, *10*(7), 1181–1191.

Zhu, Z., Wei, G., Li, J., Qian, Q., & Yu, J. (2004). Silicon alleviates salt stress and increases antioxidant enzymes activity in leaves of salt-stressed cucumber (*Cucumis sativus* L.). *Plant Science*, *167*(3), 527–533.

Zia, T., Hasnain, S. N., & Hasan, S. K. (2001). Evaluation of the oral hypoglycaemic effect of *Trigonella foenum-graecum* L. (*methi*) in normal mice. *Journal of Ethnopharmacology*, *75*(2, 3), 191–195.

CHAPTER 10

Nanoparticles and Fenugreek

HARJINDER KAUR,[1] RUPESH KAUSHIK,[1] KONDLE RAVI,[1] SHAKIR ALI,[2] AB WAHEED WANI,[1] SANJEEV KUMAR,[3] ZARINA,[4] REHAN,[1] AJAY SAROHA,[1] and AFIYA KHURSHEED[5]

[1]*Department of Horticulture, School of Agriculture, Lovely Professional University, Phagwara, Punjab, India*

[2]*Division of Fruit Science, Sher-e Kashmir University of Agricultural Science and Technology, Kashmir, India*

[3]*Department of Fruit Science, Dr. Yashwant Singh Parmar University of Horticulture and Forestry, Nauni, Solan, Himachal Pradesh, India*

[4]*Division of Vegetable Science, Sher-e Kashmir University of Agricultural Science and Technology, Jammu, India*

[5]*Division of Vegetable Science, Sher-e Kashmir University of Agricultural Science and Technology, Kashmir, India*

ABSTRACT

Nanotechnology has emerged as a promising field for agricultural research, offering innovative solutions to enhance crop productivity, quality, and stress tolerance. Fenugreek (*Trigonella foenum-graecum*) is an economically important medicinal and culinary herb with diverse therapeutic properties. The potential of nanoparticles in fenugreek cultivation, focusing on their effects on plant growth, nutrient uptake, disease management, and stress tolerance. Nanoparticles, such as metal nanoparticles (silver, gold, zinc oxide), metal oxide nanoparticles (iron oxide, titanium dioxide), carbon-based nanoparticles (graphene, carbon nanotubes), and polymeric nanoparticles, have shown significant potential in enhancing fenugreek growth and productivity.

Fenugreek: Improvement in Yield Potential Through Plant Growth Regulators (PGRs) and Depolymerized Polysaccharides. Tariq Ahmad Dar, Bilal Ahmad, Moinuddin, and Khalid Rehman Hakeem (Eds.)

These nanoparticles can be applied via foliar spraying, seed priming, or soil incorporation, leading to improved seed germination, enhanced plant growth parameters, increased biomass accumulation, and higher yield. Nanoparticles have been found to positively influence nutrient uptake and assimilation in fenugreek, including the uptake of essential elements like nitrogen, phosphorus, and potassium. Additionally, nanoparticles have demonstrated antimicrobial properties, inhibited the growth of pathogenic microorganisms and reducing disease incidence in fenugreek plants.

10.1 INTRODUCTION

Nanotechnology is a useful tool in the fields of agricultural research, the development of chemical biosensors, which water purification, and plant growth, claim (Carmen et al., 2003; Nair et al., 2010). Over the past few years, research into producing nanomaterials has increased. Due to its possible effects on and interactions with living beings, bio-nanotoxicology has recently gained a lot of attention (Carmen et al., 2003). Several nanomaterials have become the subject of in-depth study for a number of years. Five categories of NPs have been the subject of recent research: carbon particles, zero-valued metals, metal oxides, quantum dots, and nanopolymers. ZnO and TiO_2NPs have received a good deal of attention in studies on heterogenous catalysis to provide catalytic assistance of various metals (Oberdorster et al., 2005). Previous research has shown that nanoparticles of titanium dioxide (TiO_2NPs) are thermally stable and water insoluble. They are made using minerals like rutile, anatase, or brookite. The majority of chemical fertilizers are used in Indian agriculture to encourage plant development, but because they leach, hydrolyze, and degrade organic molecules, they impose adverse impact on the soil's attributes. To reduce nutrient loss from soil and boost productivity, novel nanotechnology and nanomaterial technologies must be used. Nanoparticles offer distinctive physical features and hold the possibility of strengthening the metabolism of plants. Engineered nanoparticles could easily transmit inside the foliage, fragments of DNA, and other macromolecules into the cells of plants, claim (Torney et al., 2007; Galbraith, 2007). This contact with living beings is quite concerning because nanomaterial production has lately expanded (Zhang and Fang, 2010). Engineered nanoparticles (ENPs) are employed extensively in numerous scientific domains, including bioengineering, textiles and the chemical industry, the production of sustainable habitats, and the use of medication delivery systems in military equipment. A significant amount of area of the surface, high surface energy, and excellent quantum confinement are just a few of the unique characteristics of ENPs. We also look at the advantages and disadvantages of ENPs in terms of how they affect biology and the environment. According to Oberdorster et al. (2005) and Maynard et al. (2006), when nanoparticles are subjected to the environment, engineered nanoparticles may experience physical, metabolic, and biological alterations that modify their mobility. Because of this, ecological species are at danger. Numerous studies on phytotoxicity have looked at how field species of plants react to multi-walled nanotubes, but no multigenerational life cycle research have been reported. This emphasizes the need for in-depth investigation into the interactions of MWNTs with commercial vegetable and grain crops such *Amaranthus dubius*, *Lactuca sativa*, *O. sativa*, *Curcumas sativus*, *Capsicum annum*, *Abelmoschus esculentus*, and other plants, as well as the links between carbon nanoparticles and terrestrial plants. The most sensitive plant species were then used in studies on phytotoxicity while MWNTs were present. Fenugreek is a member of Leguminosae family and is mostly grown in Northern Africa, Asia, and the Middle East. Food fiber, flavonoids, and saponins are all abundant in fenugreek seeds. They also include approximately 6% to 7% fat, 23% to 26% protein, and 58% carbohydrates. *Trigonella foenum-graecum* has used as a flavor and in ancient medicine. A few of the advantages that have been reported include anti-inflammatory properties, antipyretic, antioxidant, antibacterial, antidiabetic, cancer prevention, and anti-atherogenic properties. *Cuminum cyminum*, a fragrant shrub from the Umbelliferae family, is used to taste and spice food (Lu et al., 2008).

10.2 NANOMATERIALS

According to Dutschk et al. (2014) and Al-Halafi (2014), nanomaterials are colloidal particle systems with sizes ranging from 1 to 100 nm, exhibiting unique properties such as high surface-to-volume ratios and distinctive optical behaviors. The main categories of nanomaterials include carbon-based materials (Baughman et al., 2002), semiconductors, metal oxides (Lang et al., 2011; Rizzello and Pompa, 2014), zero-valent metals (Diao and Yao, 2009), lipids (Yang and Ma, 2010), quantum dots, dendrimers (Astruc, 2012), and nanopolymers (Ljubimova and Holler, 2012), each characterized by various forms, including microfibers, nanosheets, and nanowires.

Nanoparticles are synthesized and reduced to nanoscale sizes through methods such as crushing or milling, high-pressure homogenization, and sonication (Podsiadlo et al., 2007; Vasquez et al., 2008). In the bottom-up approach, techniques such as reactive precipitation and solvent displacement are used to create nanomaterials (Mehta et al., 2012). It is important to note that nanoparticles may have potential adverse effects due to their increased surface area, which can disrupt agricultural ecosystems (*see* Tables 10.1 and 10.2).

TABLE 10.1 Classification of Nanomaterials

Sl. No.	Nanomaterials	Explanation of NPs	References
1.	NPs	Typically, materials with dimensions (at least two) and a size varies between (1 and 1000 nm) are referred to as NPs.	Mehta et al. (2012)
2.	Nanotubes and nanofibers	A fiber that is smaller than 1 m. Long linear materials of nanometer dimension, optical materials micro-conductors, microfibers, and PEEK, PET, and PTFE nanotubes.	Baughman (2002)
3.	Nanofilm	Nanofilms used as materials for gas catalysts.	Podsiadlo (2007)
4.	Nanoblock	Obtained with great accuracy at the nanometer scale, controlled crystallization or nanoparticles.	Podsiadlo (2007)
5.	Nanocomposites	Nanoscale reinforcements are used in composite nanomaterials instead of traditional fibers or particles.	Podsiadlo (2007)
6.	Nanocrystalline solids	The intrinsic interface between crystals and various orientations makes up polycrystals with sizes between 1 and 10 nm and 50% or more solid content. The clusters that expand by coalescing and incorporating atoms and produced during homogeneous nucleation.	Vasquez (2008)

TABLE 10.2 Plant-Produced Nanoparticles and Their Applications

Nanoparticles	Source Plant	Applications	References
Nanoparticles of copper oxide Fresh leaves of Abutilon indicum.	*Calotropis procera*, *Punica granatum* peels extract	Photo catalytic properties are used in numerous therapies for conditions like spleen illnesses, piles, tumors, and so on.	Ghidan et al. (2016); Ijaz et al. (2017); Reddy (2017)
NPs of Palladium (Pd NPs)	*Filicium decipiens*, *Hippophae rhamnoides*	Medically speaking, antibacterial properties.	Nasrollahzadeh et al. (2015); Sharmila et al. (2017)
Nanoparticles of zinc oxide	Fruit peel extract from tomato, orange, grapefruit, and tomato leaves.	Catalysts, gas sensors, ceramic resistors, super-conducting materials, and biological sectors are just a few of the applications for these materials.	Dhanemozhi et al. (2017); Nava et al. (2017)

TABLE 10.2 *(Continued)*

Nanoparticles	Source Plant	Applications	References
Nanoparticle of cerium oxide	Gloriosa superba leaves	Sunscreen and other beauty products, Bioima Ging, and antimicrobial activities.	Arumugam et al. (2015)
Nanoparticle of silver	Extract of roots of *Zingiber officinale*, extracts of neem, ***Crotalaria retusa***, **mulberry, *Ocimum sanctum*** (Tulsi; root extract)	Antimicrobial agent against the growing threat posed by antibiotic-resistant bacteria Antimicrobial	Ahmad and Ikram (2015); Ahmad et al. (2016); Pirtarighat et al. (2019); Singh and Singh (2017); Singh et al. (2018); Vijaya et al. (2017)
Nanoparticle of gold	Leaf's extract of *Ziziphus zizyphus*	Substances that act as environmental catalysts and as antimicrobials against a variety of microbes.	Aljabali et al. (2018); Balalakshmi et al. (2017)

10.3 NANOTECHNOLOGY TO ENHANCE CROP YIELDS

A wide range of cost-effective nanotech applications to increase crop yield and fertility have advanced thanks to nanoscience. Using nanotechnology for genetic modification and nanodiagnosis instruments may be employed to boost plant health, crop quality, and productivity as well as to help plants recover from various types of stress. Using field-based screenings for a specific transgenic expression, this is easily applicable in real plants. According to Jasrotia et al. (2018), nanoparticles, nanocapsules, and nanofibers can transport and release compounds for gene modifications as well as foreign DNA into cells. By using an ultrasound-mediated method, Zinc sulfate nanoparticles are employed as genome transporters to carry DNA into intact plants (Fu et al., 2012). MSN have the ability to transport chemicals and DNA into separated plant cells of plant and foliage. Gold nanoparticles are placed on the ends of MSN that have been loaded with the desired gene and chemical inducers to prevent the molecules from draining. Conversely, uncapped gold nanoparticles trigger plant gene expression while gradually releasing chemicals. A good illustration of the application of nanotechnology for plant genetic modification is the transport of genes using quantum dots. It has been documented that carbon nanotubes can enter intact plant cell walls and membranes. Abd-Jasrotia et al. (2018) asserts that the inclusion of carbon nanofibers (CNFs) containing foreign DNA can be used to genetically modify golden rice that has been fed with extra vitamin A.

10.4 MODE OF ENTRY OF NANOPARTICLES INTO CELL OF PLANT

Nanoparticles penetrate plants through various entry points, including cuticles, trichomes, stomata, stigma, and hydathodes on leaves, as well as root tips, lateral roots, root system breaks, wounds, and junctions, primarily via foliar spraying. Foliar fertilization with nanomaterials is an efficient method for delivering micronutrients, as it requires minimal amounts and prevents nutrient loss through soil fixation. Engineered nanoparticles (ENPs) must navigate several biological, chemical, and physiological barriers that regulate size exclusion limits (SELs) to be absorbed and transported. The cell walls have an SEL of 5–20 nm, restricting mobility through the apoplastic transport channel (Dietz and Herth, 2011). The Casparian strip, with an SEL of less than 1 nm, further limits entry into the plant's vascular system (Aubert et al., 2010). ENPs enter cells from the cell wall through endocytosis (Etxeberria et al., 2009; Ma et al., 2010; Zhai et al., 2014).

ENPs have been observed in both the apoplast and symplast, although their transport pathways remain somewhat ambiguous. Despite having diameters larger than typical SELs in cell walls, ENPs with diameters between 36 and 50 nm can still be internalized and transported. Dietz and Herth (2011) suggest that this could be due to the dynamic nature of SELs, which may be influenced by factors such as calcium, silicon, proteins, environmental stressors, pathogens, and engineered nanoparticles.

10.5 EFFECTS OF NANOPARTICLES ON PLANT DEVELOPMENT AND GROWTH

Plants are grown with the aid of nanomaterials. In a study, using hydroponics cultivated *Bacopa monnieri* was treated with Ag NPs made from aqueous *Acalypha indica* leaf extracts. Growth enhancement was noted without any adverse effects in the morphological investigations (Krishnaraj et al., 2012). Carbon nanotubes (CNTs) affected *Lycopersicum esculentum* seed growth and germination rates in a different study. According to Khodakovskaya and Biris (2019), CNTs have the ability to penetrate the dense seed coat and assist water absorption in seedlings, a mechanism that can effect on the growth of seeds as well as germination. Broad bean plants responded better to salt environments in terms of development and resilience thanks to titanium dioxide nanoparticles ($nTiO_2$). $nTiO_2$ has pro- and antioxidant properties. Application of $nTiO_2$ increased root dry weight, leaf area, shoot length, chlorophyll b, proline, and soluble sugar levels, as well as the efficiency of antioxidant enzymes. In plants only treated to salinity, the addition of 001% $nTiO_2$ typically increased the antioxidant activities, soluble sugar levels, proline, and amino acid levels (Abdel et al., 2018). In a dose-dependent manner, $nTiO_2$ enhanced the establishment as well as growth of switchgrass, according to Boykov et al. MiRNA expressions were likewise severely constrained. While some miRNAs, like miR390 and miR399, had their expression levels improve as TiO_2 concentrations rose, other miRNAs, like miR169, had their expression levels fall. Different doses of silver nanoparticles (20, 40, and 60 mg/l) applied to the fenugreek plant not only showed improved growth metrics (such as length and shoot's dry weight, and number of leaves/plant), the amount of indole acetic acid (IAA) and other biochemical traits including pigments used for photosynthesis were also increased (Sadak, 2019). Both the quantity and the quality (contents of protein, phenolics, carbohydrates, etc.) of the yield were improved. Treatment of TiO_2 has been demonstrated to accelerate growth of plant, speed up the process of photosynthesis, reduce the severity of disease, and boost output. Corn diseases such as bacterium leaf blight and curvularia leaf spot have been reported to be less common and more severe following the application of TiO_2. Additionally, rice blast and tomato spray mold incidences were also reported to be less common, and grain weight increased due to crop development. The energizing effects of TiO_2 nanoparticles (Qureshi et al., 2018). In order to develop a pure form of plant nano-activator, Gao et al. (2018) enclosed the activator protein Pea T1 inside a monomethoxy-poly-(ethylene glycol)-poly-(lactide-co-glycolide) (mPEG-PLGA) nanoparticle (polymer). The nanoparticles were spherical and had a 200 nm diameter. After being exposed to a nano-activator, the physical and morphological characteristics of wheat seedlings, such as root vitality, seed vigor index, indices related to photosynthetic activity, and chlorophyll content, were improved.

10.5.1 GERMINATION OF SEEDS

Using nano-zinc at 40 ppm, which had the greatest impact on average shoot length (31.33 cm and 29.11 cm in the first and second seasons, respectively), the effects of zinc treatments on seed germination and the physiological growth of fenugreek plants were clearly visible. When the zinc treatments were compared in terms of root length, fresh weight, and dry weight of fenugreek roots, it was found that nano-zinc at a concentration of 40 ppm produced the best results: 10.83 cm and 11.28 cm for root length, 0.172 g and 0.164 g for fresh weight, and 0.127 g and 0.117 g for dry weight, respectively.

Tap water proved to be the most effective treatment for fenugreek roots, resulting in mean root lengths of 10.17 cm and 10.47 cm and fresh weights of 0.148 g and 0.133 g in both seasons, respectively. The highest values for seed germination percentage, average germination time, seedling vigor, and germination index were achieved with the application of 10 g/mL of nano-silver, as reported by Ibrahim and Faryal (2014) and Puvanitha and Mahendran (2017). Similar findings were observed by Hojjat and Hojjat (2015).

Khodakovskaya et al. (2009) established that *multi-walled carbon nanotubes (MWCNTs)* could penetrate tomato seeds and increase germination percentage by enhancing water uptake. More than 90% of the seeds germinated after 20 days of MWCNT treatments, compared to 71% for the untreated control. In accordance with Bris and Khodakovskaya (2011); and Anita and Rao (2014), MWCNTs had a favorable impact on seed germination for various plants, including *Lycopersicon esculentum*, fenugreek, and other vegetables. Carbon nanotubes (CNTs) have been shown by Haghighi and Teixeira da Silva (2014) to influence the growth of turnip, tomato, radish, and onion seeds. Four distinct CNT concentrations (0, 10, 20, and 40 mg L^{-1}) were tested in a laboratory setting. Tomato and onion germination were more affected by CNTs at concentrations of 10–40 mg L^{-1} than radish and turnip growth. CNTs were toxic at a concentration of 40 mg L^{-1} and hindered the germination of *Allium cepa* and *Raphanus sativus* seeds. This suggests that the efficiency of CNTs may depend on the plant species and the location of CNTs on the surface of the testa and roots.

Similar findings were reported by de la Rosa et al. (2013), who found that cucumber seeds were more likely to germinate when exposed to zinc oxide nanoparticles (ZnO-NPs) at concentrations of 400 and 1600 mg L^{-1}. Raskar and Laware (2014) demonstrated that ZnO-NPs improved onion seed germination by 96.52% at lower doses (20 g mL^{-1}), but this benefit diminished at higher concentrations (40 g mL^{-1}). In contrast, titanium dioxide nanoparticles (TiO_2-NPs) at a concentration of 60 mg L^{-1} showed the highest fennel seed germination rate (approximately 76%), while bulk TiO_2 particles reduced the percentage by about 41% at the same concentration. However, the vegetative development rate of seedlings treated with TiO_2-NPs increased by 83% and

39.5%, respectively, compared to bulk TiO_2 particle levels and the control (Khot et al., 2012).

According to Singh et al. (2013), tomato seedlings treated with ZnO-NPs at a dosage of 9 M showed the highest germination rates compared to control and bulk ZnO (4.5 M)-treated seeds. The ZnO-NPs also produced the fastest-growing seedlings in terms of plumule and radicle length. These results suggest that NP levels can influence processes such as plant growth and germination (Zheng et al., 2005). Furthermore, nanoparticles can penetrate the seed coat to enhance water uptake, activating the biochemical systems that aid in seedling germination and development.

However, the exact mechanism of water absorption facilitated by nanoparticles through the seed is still not fully understood (Banerjee and Kole, 2016; Lu et al., 2002). In the cultivar Super Strain B of *Lycopersicon esculentum* Mill., Siddiqui et al. (2014) found that treatment with 8 g/L of silicon dioxide nanoparticles (SiO_2-NPs) with a particle diameter of 12 nm significantly enhanced seed germination and reduced the average time required for germination.

10.5.2 SEEDLING GROWTH

Using nano-zinc at a concentration of 40 ppm, the maximum root length, fresh weight, and dry weight of roots were found to be 12.33 cm, 0.219 g, and 0.141 g, respectively (Swaefy and Abdallh, 2021). The best results for the first season's root length (12.67 cm), fresh weight (0.276 g), and dry weight were observed in plants treated with tap water and 40 ppm zinc sulfate. Plants watered with tap water and treated with 20 ppm zinc sulfate produced the highest number of pods per plant, seeds per pod, and overall pod numbers, with values of 11.27 and 11.37 pods and 15.73 and 15.63 seeds, respectively (Swaefy and Abdallh, 2021).

Fenugreek plants treated with Ag-NPs (silver nanoparticles) exhibited the greatest average leaf number of 1.36 ± 0.23. After treatment, the roots measured 1.73 ± 0.08 cm in length, and shoot length in Ag-NP-treated seedlings was 6.03 ± 0.45 cm compared to 3.09 ± 0.37 cm in the control group, demonstrating Ag-NPs' effectiveness in promoting seedling growth. Wet weight of Ag-NP-treated seedlings was 0.38 ± 0.089 g, compared to 0.12 ± 0.072 g for control plants (Jasim et al., 2017). When fenugreek is exposed to salt stress, an appropriate dose of AgNPs may improve seed germination, promote seedling growth, and positively affect crop yield (Haghighi and Teixeira da Silva, 2014; Hojjat and Kamyab, 2017).

Increasing carbon concentrations resulted in decreased fresh weight of radish seedlings. However, raising carbon nanotube (CNT) concentration from 10 mg/L to 40 mg/L increased the dry weight of tomato and radish shoots, though CNTs had no effect on rapeseed growth. Begum et al. (2011) examined the effects of different graphene concentrations on the growth of fenugreek, cabbage, red spinach, and tomato seedlings. While a 50 mg/L dose enhanced tomato seedling growth without visible damage, a 2000 mg/L dose significantly impacted growth in red spinach, cabbage, and tomato seedlings. Singh et al. (2013) reported mean growth rates and heights of *Brassica oleracea* var. *brassica* seedlings, noting that higher concentrations improved growth parameters, though the highest levels of zinc oxide and lowest levels of zinc dioxide nanoparticles had minimal impact on capitata. Laware and Raskar (2014) demonstrated that TiO_2 nanoparticles (TiO_2–NPs) had a significant negative effect on onion seedling growth at concentrations up to 30 mg/L.

The improvement in seedling growth parameters, such as dry and fresh weights and root and shoot lengths, may be attributed to nanoparticles' role in reactive oxygen species (ROS) generation, rapid uptake of H_2O and O_2, and the induction of oxidative stress.

10.5.3 PLANT GROWTH

The longest roots were observed in fenugreek plants treated with ZnO nanoparticles (ZnO-NPs) at 50 mg/L, reaching 7 cm after 15 days of exposure, representing a significant root growth increase of approximately 66% compared to the control. Zinc oxide nanoparticles also significantly impacted stem development. In fenugreek, treatment with ZnO-NPs (50 mg/L) notably reduced leaf area compared to the control (Chemingui et al., 2019; Monica and Cremonini, 2009). According to Zheng et al. (2005), lower concentrations of TiO_2 nanoparticles (TiO_2-NPs) greatly enhanced spinach growth compared to higher concentrations. These outcomes may be due to nanoparticles enhancing nitrogen metabolism and photosynthesis (Hong et al., 2005).

Mn-NPs (manganese nanoparticles) increased bean growth by an average of 52% in stems and 38% in roots (Pradhan et al., 2013). In another study, Mahmoud et al. (2018) examined the effects of different doses of chitosan nanoparticles as a foliar spray, with a maximum dosage of 60 mg/L resulting in improved plant growth, including increased plant height, number of branches and leaves per plant, and both fresh and dry weights (Bonilla-Bird et al., 2020). These growth enhancements may be due to chitosan's biochemical properties and application frequency, which stimulate various activities in plants (Malerba and Cerana, 2016).

Salama et al. (2019a) reported that ZnO-NPs at a dosage of 30 mg/L yielded optimal growth parameters, including the average number of branches and leaves per plant, as well as fresh and dry masses of leaves and branches. Mankad et al. (2017) suggested that the release of Zn^{2+} ions from ZnO-NPs improved morphological, physical, and metabolic properties, likely contributing to the observed growth enhancement.

10.5.4 THE PRODUCTIVITY OF PLANTS

Hojjat and Khmyab (2017) found that the addition of 30 g mL^{-1} of silver nanoparticles (AgNPs) to a salt-stressed environment impacted the agricultural output of fenugreek. Elmer and White (2016) reported that copper oxide and zinc sulfate nanoparticles were superior to bulk forms of the metals in enhancing the yield of tomato, fenugreek, and aubergine. Diosgenin levels in treated seedlings were 214.06 ± 17.07 µg/mL, compared to 164.49 ± 7.67 µg/mL in control plants (Li et al., 2011).

Using sulfur-coated nano-nitrogen chelate boosted potato output by 56.10%, 59.61%, and 49.76% compared to urea fertilizer, according to Zare-abyanel et al. (2015). This result was attributed to reduced nitrate leaching by 35.72%, 41.56%, and 9.94% compared to urea-based fertilizer. Hernández-Hernández et al. (2019) found that 10 mg L^{-1} of selenium nanoparticles (Se-NPs) increased tomato fruit output by 21%.

Copper and selenium nanoparticle interactions, at levels of 20 mg L^{-1} and 10 mg L^{-1}, respectively, increased the average weight of *Lycopersicon esculentum* fruit by approximately 25% compared to the control. Ponce-Garcia et al. (2019) found that zinc oxide nanoparticles (ZnO-NPs) at a concentration of 25 mg L^{-1} improved the maximum yield of green beans. Another study by Salama et al. (2019a) reported that ZnO-NPs at a dose of 30 mg L^{-1} increased seed production in common bean plants (Abdel Wahab et al., 2019).

Osman et al. (2020) found that *Phaseolus* plants treated with 40 mg L^{-1} of molybdenum nanoparticles (Mo-NPs) produced more seeds per acre. This outcome is explained by the ability of nano-fertilizers, with sizes smaller than 100 nm, to be more easily absorbed by plants (Naresh et al., 2018).

10.5.5 SOIL IMPROVEMENT

Major issues in the area of sustainable soil use are the preservation of microbial biomass in the soil and diversification (Torsvik and Vres, 2002). The amount of NP's, the soil type, and the enzyme activity of the land all affect how nanoparticles affect soil. High nanoparticle concentrations had a deleterious impact on the dehydrogenase enzymatic activities (Joko et al., 2014). Two other negative consequences of nanomaterials are their impact on the soil's capacity for self-clean and the nutrients the formula, which forms the basis for structuring the processes for boosting the productivity of soil and plant nutrition (Janvier et al., 2007; Suresh et al., 2013). Hänsch and Emmerling (2010) found that Silver-nanoparticles effected the biomass of soil microorganisms. According to Asadishad et al. (2018), the addition of zinc oxide and nanoparticles of copper oxide can improve soil's overall enzymatic activity with no impact on the microbial ecology. ZnO-NPs are able to contact microorganisms more efficiently than large zinc molecules due to their modest dimensions that are below 100 nanometer and the height ratios from the outside to the volume (Xie et al., 2011). Because of nanoparticles, pollutants in soil may travel differently. The removal of Cr (VI) from water and soil has also been widely investigated and evaluated, and more recently, reclaimed water has been used for irrigation (Brasili et al., 2020). Injection of nZVI at a low concentration (5 mg L^{-1}) accelerated the transformation of the least dangerous component Cr (VI) into Cr (III). The development of Lycopersicum esculentus plants was sped up and both the vitamin A and chlorophyll levels of the 457 leaves were increased with a dose of 5 mg L^{-1} of nZVI. However, tomato seed germination was decreased at the maximum dose levels of nZVI (100 and 1000 mg L^{-1}). These findings suggested that nZVI nano-remediation should be employed as the restorative strategy, that it's critical to take into account nZVI's remedial benefits as well as its expected fate and toxicity, and that the degree of toxicity is influenced by the specific type of plant. The microbial communities in the soils that have been polluted are altered by the chemistry and size dispersion of the organic debris. By deliberately modifying the features, structure, and smoothness of the soil with other materials, the outcomes of utilizing nanoparticles can be altered. A soilless supplement called biochar enhanced the productivity as well as quality of the soil. Servin and White (2016) found that selenium oxide nanoparticles in soil containing biochar seemed to have only marginal impacts on plants (Das et al., 2016). Therefore, it can be concluded that, provided the nanoparticles do not build up in the soil, nano-fertilization has impact on the state of soil in regard to microbial development and its physiological attributes and the capacity for contaminants to transform into poisonous to soil microbes). As a result, soil fertility improves, and crop yields rise.

10.5.6 RESISTANCE TO BIOTIC AND ABIOTIC STRESS

Micronutrients are essential for preventing crop diseases. Fenugreek germination rate was significantly influenced by both salinity and silver nanoparticles (AgNPs); as salinity increased, germination rate (GR) decreased, while AgNPs increased it by over 20%. AgNPs also impacted fresh weight, with significant differences observed between high AgNP treatments and the control. When 10 g/mL of AgNPs was applied, fresh weight decreased by more than 35% (Hojjat and Kamyab, 2017). The results indicate that treatment with 30 g/mL of AgNPs improved seed germination and seedling growth parameters while mitigating the adverse effects of salinity. AgNP application to fenugreek under saline stress conditions substantially impacted all parameters, including germination percentage, shoot and root lengths, fresh and dry weights, total germination, and germination speed index (Hojjat and Kamyab, 2017).

Elmer and White (2016) reported that copper oxide nanoparticles (CuO NPs), when sprayed on the leaves of tomato and brinjal seedlings, were highly effective at reducing disease and significantly increasing crop yield. Plants assimilate iron, copper, silicon, and zinc nanoparticles at a much higher rate compared to their bulk counterparts. Stampoulis et al. (2009) found that zucchini plants absorbed AgNPs 4.7 times more efficiently than seedlings supplied with conventional fertilizers. He et al. (2011) demonstrated that zinc oxide nanoparticles (ZnO NPs) deformed fungal structures and inhibited the growth of hyphae, conidia, and conidiophores. Additionally, silver nanoparticles at 15 mg/L reduced fungal infections, including those caused by *Botrytis cinerea* (Khan et al., 2019).

To control late blight disease caused by *Phytophthora infestans*, Giannousi et al. (2013) showed that spraying copper oxide (CuO) nanoparticles and copper oxide composites with particle sizes of 11–14 nm was more effective than traditional Cu fungicides. Copper-based nanoparticles (NPs) also outperformed copper-based agrochemicals in protecting tomato plants. Deepak et al. (2013) found that $CuSO_4$ and $Na_2B_4O_7$ nanoparticles were more effective at preventing rust disease in pea crops than 15 other nanoscale micronutrients.

10.6 POST-HARVEST

Food preservation requires food packaging, especially for items that aren't eaten right away. Packaging safeguards food against contamination, bacteria, and chemical deterioration and is utilized particularly during transportation as well as storage. Food packaging materials need to be secure, portable, affordable, recyclable, and resilient to both physical and environmental challenges. Additionally, the ability of gases and liquids to pass through packaging materials is crucial, especially for produce like vegetables and fruits whose shelf lives depend on the availability of oxygen. Different polymeric materials can be used to create safe packaging, but biopolymer-based packaging is particularly advantageous due to its affordable, friendly to environment, and biodegradable (Shukla et al., 2019). Food has been protected and kept fresh for a long time with NT. Nanocomposites, which have promising properties and might to enhance the features of food packaging and quality, are created when nanomaterials are combined with polymeric packaging materials (Youssef et al., 2019b).

10.7 TOXICITY

Nanotechnology has various applications in precision agriculture; however, toxicity is a significant concern due to the unique properties of nanoparticles. The phytotoxicity profile of nanoparticles is still in the early stages of research. While nanoparticles offer beneficial properties, their potential negative impacts, particularly on commercial food crops, require further investigation. Not all nanoparticles harm plants—some have shown positive effects. Among the carbon nanomaterials used in agriculture are natural compounds like fullerene carbon nanoparticles, fullerol, and single- and multi-walled carbon nanotubes. These materials can have both positive and negative effects.

For example, Tripathi and Sarkar (2014) reported that wheat shoots and roots treated with 150 mg/L carbon nano-dots dispersed in 200 μL of water grew 10 times faster than controls after 10 days. In contrast, zinc oxide nanoparticles inhibited root length and number, while titanium dioxide nanoparticles had no significant effect on root length. This study suggests that certain nanomaterials may cause severe phytotoxicity when absorbed by plants, underscoring the need for environmentally responsible disposal of nanoparticles and careful consideration of their future applications in agriculture and ecology.

In 2005, Chai et al. investigated the impact of metallic oxide nanoparticles (ZnO, SiO_2, TiO_2, and CeO_2) on soil bacteria and their metabolic activity in agricultural soil. These nanoparticles were applied at a concentration of 1.0 mg/g. Cytotoxic effects were evaluated by measuring thermogenic metabolism, bacterial viability, and enzyme production. ZnO and CeO_2 nanoparticles were shown to inhibit thermogenic metabolism, reduce the abundance of soil-based *Azotobacter*, phosphorus- and potassium-solubilizing bacteria, and suppress enzyme activity. Conversely, TiO_2 nanoparticles decreased bacterial diversity and enzyme activity (Boonyanitipong et al., 2011).

10.8 CONCLUSION AND FUTURE PROSPECTIVE

Nanotechnology has advanced significantly in the field of plant sciences. Nanomaterials are employed throughout the process, from their synthesis to their application in boosting and safeguarding different parts of plant life. The rapidly rising world population necessitates higher agricultural output. It is crucial to make use of modern technologies that can raise agricultural productivity. The positive effects of nanoparticles on the development and growth of plants, agricultural improvement, pest control, and PHT are well known. More study into other plant species utilizing various nanomaterial combinations and doses is required because different plant species show varying degrees of resistance to nanoparticles. Due to NPs' efficiency, a new, less dangerous "green revolution" will be encouraged. In spite of this, there remains a number of issues that still need to be researched, including those related to the production and functionality of nanomaterials, absorption by plants techniques, permissible limitations, and the ecotoxicity of certain nanomaterials. A major aspect of plant physiology must be included in the expansion of the usage of nanomaterials. The study of how plants interact with their environment and develop may advance with the use of nanobiosensors for real-time secondary metabolism or Phyto regulator identification. There has been extensive study into the creation of tools for editing the genome like zinc finger nucleases and transcription activator-like effector nucleases. To develop plant genetics, more effort must be done to successfully integrate CRISPR with nanoparticle technology. Through integrative approaches for the synthesis of intelligent nanomaterials, the ensuing delay in plant nanotechnology can be resolved. The main barriers to the application of nanotechnology are high processing costs, public perceptions of the environment, the production of industrial goods, and so forth. Governments all across the world must set stringent regulations before these nanoparticles are sold and utilized extensively.

KEYWORDS

- **abiotic stress**
- **biotic stress**
- **crop diseases**
- **dry weight**
- **fenugreek**
- **germination rate**
- **germination speed index**
- **micronutrients**
- **nanoparticles**
- **salt stress**
- **silver nanoparticles**

REFERENCES

Abdel Latef, A. A. H., Srivastava, A. K., El-Sadek, M. S. A., Kordrostami, M., & Tran, L. S. P. (2018). Titanium dioxide nanoparticles improve growth and enhance tolerance of broad bean plants under saline soil conditions. *Land Degradation & Development, 29*(4), 1065–1073.

Ahmed, H. B., Zahran, M. K., & Emam, H. E. (2016). Heatless synthesis of well-dispersible Au nanoparticles using pectin biopolymer. *International Journal of Biological Macromolecules, 91,* 208–219.

Ahmed, S., & Ikram, S. (2015). Silver nanoparticles: One pot green synthesis using *Terminalia arjuna* extract for biological application. *Journal of Nanomedicine & Nanotechnology, 6*(4), 309.

Ahmed, S., Manzoor, K., & Ikram, S. (2016). Synthesis of silver nanoparticles using leaf extract of *Crotolaria retusa* as antimicrobial green catalyst. *Journal of Bionanoscience, 10*(4), 282–287.

Al-Halafi, A. M. (2014). Nanocarriers of nanotechnology in retinal diseases. *Saudi Journal of Ophthalmology, 28*(4), 304–309.

Aljabali, A. A., Akkam, Y., Al Zoubi, M. S., Al-Batayneh, K. M., Al-Trad, B., Abo Alrob, O., & Evans, D. J. (2018). Synthesis of gold nanoparticles using leaf extract of *Ziziphus zizyphus* and their antimicrobial activity. *Nanomaterials, 8*(3), 174.

Arumugam, A., Karthikeyan, C., Hameed, A. S. H., Gopinath, K., Gowri, S., & Karthika, V. (2015). Synthesis of cerium oxide nanoparticles using *Gloriosa superba* L. leaf extract and their structural, optical, and antibacterial properties. *Materials Science and Engineering: C, 49,* 408–415.

Asadishad, B., Chahal, S., Akbari, A., Cianciarelli, V., Azodi, M., Ghoshal, S., & Tufenkji, N. (2018). Amendment of agricultural soil with metal nanoparticles: Effects on soil enzyme activity and microbial community composition. *Environmental Science & Technology, 52*(4), 1908–1918.

Astruc, D. (2012). Electron-transfer processes in dendrimers and their implication in biology, catalysis, sensing, and nanotechnology. *Nature Chemistry, 4*(4), 255–267.

Aubert, T., Esnault, M. A., Cordier, S., Grasset, F., & Cabello-Hurtado, F. (2010). Phytotoxicity of nanosized molybdenum octahedral clusters. In *XVII Congress of the FESPB*.

Balalakshmi, C., Gopinath, K., Govindarajan, M., Lokesh, R., Arumugam, A., Alharbi, N. S., & Benelli, G. (2017). Green synthesis of gold nanoparticles using a cheap *Sphaeranthus indicus* extract: Impact on plant cells and the aquatic crustacean *Artemia* nauplii. *Journal of Photochemistry and Photobiology B: Biology, 173,* 598–605.

Banerjee, J., & Kole, C. (2016). Plant nanotechnology: An overview on concepts, strategies, and tools. In C. Kole (Ed.), *Plant Nanotechnology: Principles and Practices* (pp. 1–14). Springer.

Baughman, R. H., Zakhidov, A. A., & De Heer, W. A. (2002). Carbon nanotubes—The route toward applications. *Science, 297*(5582), 787–792.

Begum, P., Ikhtiari, R., & Fugetsu, B. (2011). Graphene phytotoxicity in the seedling stage of cabbage, tomato, red spinach, and lettuce. *Carbon, 49*(12), 3907–3919.

Biris, A., & Khodakovskaya, M. (2011). Method of using carbon nanotubes to affect seed germination and plant growth. (Patent No. US20120233725A1). United States Patent and Trademark Office.

Bonilla-Bird, N. J., Ye, Y., Akter, T., Valdes-Bracamontes, C., Darrouzet-Nardi, A. J., Saupe, G. B., & Gardea-Torresdey, J. L. (2020). Effect of copper oxide nanoparticles on two varieties of sweetpotato plants. *Plant Physiology and Biochemistry, 154,* 277–286.

Boonyanitipong, P., Kositsup, B., Kumar, P., Baruah, S., & Dutta, J. (2011). Toxicity of ZnO and TiO_2 nanoparticles on germinating rice seed *Oryza sativa* L. *International Journal of Bioscience, Biochemistry and Bioinformatics, 1*(4), 282.

Boykov, I. N., Shuford, E., & Zhang, B. (2019). Nanoparticle titanium dioxide affects the growth and microRNA expression of switchgrass (*Panicum virgatum*). *Genomics, 111*(3), 450–456.

Brasili, E., Bavasso, I., Petruccelli, V., Vilardi, G., Valletta, A., Dal Bosco, C., & Di Palma, L. (2020). Remediation of hexavalent chromium contaminated water through zero-valent iron nanoparticles and effects on tomato plant growth performance. *Scientific Reports, 10*(1), 1920.

Chai, H., Yao, J., Sun, J., Zhang, C., Liu, W., Zhu, M., & Ceccanti, B. (2015). The effect of metal oxide nanoparticles on functional bacteria and metabolic profiles in agricultural soil. *Bulletin of Environmental Contamination and Toxicology, 94,* 490–495.

Changmei, L., Chaoying, Z., Junqiang, W., Guorong, W., & Mingxuan, T. (2002). Research of the effect of nanometer materials on germination and growth enhancement of *Glycine max* and its mechanism. *Soybean Science, 21*(3), 168–171.

Chemingui, H., Smiri, M., Missaoui, T., & Hafiane, A. (2019). Zinc oxide nanoparticles induced oxidative stress and changes in the photosynthetic apparatus in fenugreek

(*Trigonella foenum graecum* L.). *Bulletin of Environmental Contamination and Toxicology, 102*(4), 477–485.

Das, P., Sarmah, K., Hussain, N., Pratihar, S., Das, S., Bhattacharyya, P., & Bhattacharya, S. S. (2016). Novel synthesis of an iron oxalate capped iron oxide nanomaterial: A unique soil conditioner and slow-release eco-friendly source of iron sustenance in plants. *RSC Advances, 6*(105), 103012–103025.

de la Rosa, G., López-Moreno, M. L., de Haro, D., Botez, C. E., Peralta-Videa, J. R., & Gardea-Torresdey, J. L. (2013). Effects of ZnO nanoparticles in alfalfa, tomato, and cucumber at the germination stage: Root development and X-ray absorption spectroscopy studies. *Pure and Applied Chemistry, 85*(12), 2161–2174.

Deepak, S., Ashish, K., Singh, A. K., & Tripathi, H. S. (2013). Induction of resistance in field pea against rust disease through various chemicals/micronutrients and their impact on growth and yield. *Plant Pathology Journal (Faisalabad), 12*(2), 36–49.

Dhanemozhi, A. C., Rajeswari, V., & Sathyajothi, S. (2017). Green synthesis of zinc oxide nanoparticle using green tea leaf extract for supercapacitor application. *Materials Today: Proceedings, 4*(2), 660–667.

Diao, M., & Yao, M. (2009). Use of zero-valent iron nanoparticles in inactivating microbes. *Water Research, 43*(20), 5243–5251.

Dietz, K. J., & Herth, S. (2011). Plant nanotoxicology. *Trends in Plant Science, 16*(11), 582–589.

Dutschk, V., Karapantsios, T., Liggieri, L., McMillan, N., Miller, R., & Starov, V. M. (2014). Smart and green interfaces: From single bubbles/drops to industrial environmental and biomedical applications. *Advances in Colloid and Interface Science, 209,* 109–126.

Elmer, W. H., & White, J. C. (2016). The use of metallic oxide nanoparticles to enhance growth of tomatoes and eggplants in disease-infested soil or soilless medium. *Environmental Science: Nano, 3*(5), 1072–1079.

Etxeberria, E., Gonzalez, P., & Pozueta, J. (2009). Evidence for two endocytic transport pathways in plant cells. *Plant Science, 177*(4), 341–348.

Feizi, H., Kamali, M., Jafari, L., & Moghaddam, P. R. (2013). Phytotoxicity and stimulatory impacts of nanosized and bulk titanium dioxide on fennel (*Foeniculum vulgare* Mill). *Chemosphere, 91*(4), 506–511.

Galbraith, D. W. (2007). Silica breaks through in plants. *Nature Nanotechnology, 2*(5), 272–273.

Gao, H., Qin, Y., Guo, R., Wu, Y., Qiu, D., & Fu, Y. (2018). Enhanced plant growth promoting role of mPEG-PLGA-based nanoparticles as an activator protein PeaT1 carrier in wheat (*Triticum aestivum* L.). *Journal of Chemical Technology & Biotechnology, 93*(11), 3143–3151.

García-Gómez, C., García, S., Obrador, A. F., González, D., Babín, M., & Fernández, M. D. (2018). Effects of aged ZnO NPs and soil type on Zn availability, accumulation and toxicity to pea and beet in a greenhouse experiment. *Ecotoxicology and Environmental Safety, 160,* 222–230.

Ghidan, A. Y., Al-Antary, T. M., & Awwad, A. M. (2016). Green synthesis of copper oxide nanoparticles using *Punica granatum* peels extract: Effect on green peach aphid. *Environmental Nanotechnology, Monitoring & Management, 6,* 95–98.

Giannousi, K., Avramidis, I., & Dendrinou-Samara, C. (2013). Synthesis, characterization and evaluation of copper-based nanoparticles as agrochemicals against *Phytophthora infestans*. *RSC Advances, 3*(44), 21743–21752.

Haghighi, M., & Teixeira da Silva, J. A. (2014). The effect of carbon nanotubes on the seed germination and seedling growth of four vegetable species. *Journal of Crop Science and Biotechnology, 17,* 201–208.

Hänsch, M., & Emmerling, C. (2010). Effects of silver nanoparticles on the microbiota and enzyme activity in soil. *Journal of Plant Nutrition and Soil Science, 173*(4), 554–558.

He, L., Liu, Y., Mustapha, A., & Lin, M. (2011). Antifungal activity of zinc oxide nanoparticles against *Botrytis cinerea* and *Penicillium expansum. Microbiological Research, 166*(3), 207–215.

Hernández-Hernández, H., Quiterio-Gutiérrez, T., Cadenas-Pliego, G., Ortega-Ortiz, H., Hernández-Fuentes, A. D., Cabrera de la Fuente, M., & Juárez-Maldonado, A. (2019). Impact of selenium and copper nanoparticles on yield, antioxidant system, and fruit quality of tomato plants. *Plants, 8*(10), 355.

Hojjat, S. S., & Hojjat, H. (2015). Effect of nano-silver on seed germination and seedling growth in fenugreek seed. *International Journal of Food Engineering, 1*(2), 106–110.

Hojjat, S. S., & Kamyab, M. (2017). The effect of silver nanoparticle on fenugreek seed germination under salinity levels. *Russian Agricultural Sciences, 43*, 61–65.

Hong, F., Zhou, J., Liu, C., Yang, F., Wu, C., Zheng, L., & Yang, P. (2005). Effect of nano-TiO_2 on photochemical reaction of chloroplasts of spinach. *Biological Trace Element Research, 105*, 269–279.

Ijaz, F., Shahid, S., Khan, S. A., Ahmad, W., & Zaman, S. (2017). Green synthesis of copper oxide nanoparticles using *Abutilon indicum* leaf extract: Antimicrobial, antioxidant and photocatalytic dye degradation activities. *Tropical Journal of Pharmaceutical Research, 16*(4), 743–753.

Janvier, C., Villeneuve, F., Alabouvette, C., Edel-Hermann, V., Mateille, T., & Steinberg, C. (2007). Soil health through soil disease suppression: Which strategy from descriptors to indicators? *Soil Biology and Biochemistry, 39*(1), 1–23.

Jasim, B., Thomas, R., Mathew, J., & Radhakrishnan, E. K. (2017). Plant growth and diosgenin enhancement effect of silver nanoparticles in fenugreek (*Trigonella foenum-graecum* L.). *Saudi Pharmaceutical Journal, 25*(3), 443–447.

Jasrotia, P., Kashyap, P. L., Bhardwaj, A. K., Kumar, S., & Singh, G. P. (2018). Scope and applications of nanotechnology for wheat production: A review of recent advances. *Wheat and Barley Research, 10*(1), 1–14.

Jasrotia, P., Kashyap, P. L., Bhardwaj, A. K., Kumar, S., & Singh, G. P. (2018). Nanotechnology scope and applications for wheat production: A review of recent advances.

Jośko, I., Oleszczuk, P., & Futa, B. (2014). The effect of inorganic nanoparticles (ZnO, Cr_2O_3, CuO, and Ni) and their bulk counterparts on enzyme activities in different soils. *Geoderma, 232*, 528–537.

Khan, M. R., Ahamad, F., & Rizvi, T. F. (2019). Effect of nanoparticles on plant pathogens. In *Advances in Phyto Nanotechnology* (pp. 215–240). Academic Press.

Khodakovskaya, M. V., & Biris, A. S. (2019). *U.S. Patent No. 10,244,761*. Washington, DC: U.S. Patent and Trademark Office.

Khodakovskaya, M., Dervishi, E., Mahmood, M., Xu, Y., Li, Z., Watanabe, F., & Biris, A. S. (2009). Carbon nanotubes are able to penetrate plant seed coat and dramatically affect seed germination and plant growth. *ACS Nano, 3*(10), 3221–3227.

Khot, L. R., Sankaran, S., Maja, J. M., Ehsani, R., & Schuster, E. W. (2012). Applications of nanomaterials in agricultural production and crop protection: A review. *Crop Protection, 35*, 64–70.

Krishnaraj, C., Ramachandran, R., Mohan, K., & Kalaichelvan, P. T. (2012). Optimization for rapid synthesis of silver nanoparticles and its effect on phytopathogenic fungi. *Spectrochimica Acta Part A: Molecular and Biomolecular Spectroscopy, 93*, 95–99.

Lang, X., Hirata, A., Fujita, T., & Chen, M. (2011). Nanoporous metal/oxide hybrid electrodes for electrochemical supercapacitors. *Nature Nanotechnology, 6*(4), 232–236.

Laware, S. L., & Raskar, S. (2014). Effect of titanium dioxide nanoparticles on hydrolytic and antioxidant enzymes during seed germination in onion. *International Journal of Current Microbiology and Applied Sciences, 3*(7), 749–760.

Lu, F. R., Shen, L., Qin, Y., Gao, L., Li, H., & Dai, Y. (2008). Clinical observation on *Trigonella foenum-graecum* L. total saponins in combination with sulfonylureas in the treatment of type 2 diabetes mellitus. *Chinese Journal of Integrative Medicine, 14*, 56–60.

Ma, Y., Kuang, L., He, X., Bai, W., Ding, Y., Zhang, Z., & Chai, Z. (2010). Effects of rare earth oxide nanoparticles on root elongation of plants. *Chemosphere, 78*(3), 273–279.

Mahmoud, A. W. M., Abdelaziz, S. M., El-Mogy, M. M., & Abdeldaym, E. A. (2019). Effect of foliar ZnO and FeO nanoparticles application on growth and nutritional quality of red radish and assessment of their accumulation on human health. *Agriculture/Pol'nohospodárstvo, 65*(1), 16–29.

Mahmoud, S. H., Salama, D. M., & Abd El-Aziz, M. E. (2018). Effect of chitosan and chitosan nanoparticles on growth, productivity, and chemical quality of green snap bean. *Bioscience Research, 15*(4), 4307–4321.

Malerba, M., & Cerana, R. (2016). Chitosan effects on plant systems. *International Journal of Molecular Sciences, 17*(7), 996.

Malvadkar, N. A., Hancock, M. J., Sekeroglu, K., Dressick, W. J., & Demirel, M. C. (2010). An engineered anisotropic nanofilm with unidirectional wetting properties. *Nature Materials, 9*(12), 1023–1028.

Mankad, M., Fougat, R. S., Patel, A., Mankad, P., Patil, G., & Subhash, N. (2017). Physiological and biochemical effects of zinc oxide nanoparticles on rice (*Oryza sativa* L.). *Advances in Bioresearch, 8*(6).

Maynard, A. D., Aitken, R. J., Butz, T., Colvin, V., Donaldson, K., Oberdörster, G., & Warheit, D. B. (2006). Safe handling of nanotechnology. *Nature, 444*(7117), 267–269.

Mehta, R. J., Zhang, Y., Karthik, C., Singh, B., Siegel, R. W., Borca-Tasciuc, T., & Ramanath, G. (2012). A new class of doped nanobulk high-figure-of-merit thermoelectrics by scalable bottom-up assembly. *Nature Materials, 11*(3), 233–240.

Monica, R. C., & Cremonini, R. (2009). Nanoparticles and higher plants. *Caryologia, 62*(2), 161–165.

Moraru, C. I., Panchapakesan, C. P., Huang, Q., Takhistov, P., Liu, S., & Kokini, J. L. (2003). Nanotechnology: A new frontier in food science understanding the special properties of materials of nanometer size will allow food scientists to design new, healthier, tastier, and safer foods. *Nanotechnology, 57*(12), 24–29.

Naresh, R. K., Kumar, M., Singh, S. P., Dwivedi, A., & Kumar, R. (2018). Enhancing nutrient use efficiency through next generation fertilizers in vegetable and field crops. *Agricultural-Based Interventions for Sustainable Food Security & Climate Change, 69*.

Nasrollahzadeh, M., Sajadi, S. M., & Maham, M. (2015). Green synthesis of palladium nanoparticles using *Hippophae rhamnoides* Linn leaf extract and their catalytic activity for the Suzuki–Miyaura coupling in water. *Journal of Molecular Catalysis A: Chemical, 396*, 297–303.

Nava, O. J., Soto-Robles, C. A., Gómez-Gutiérrez, C. M., Vilchis-Nestor, A. R., Castro-Beltrán, A., Olivas, A., & Luque, P. A. (2017). Fruit peel extract mediated green synthesis of zinc oxide nanoparticles. *Journal of Molecular Structure, 1147*, 1–6.

Osman, S. A., Salama, D. M., Abd El-Aziz, M. E., Shaaban, E. A., & Abd Elwahed, M. S. (2020). The influence of MoO_3-NPs on agro-morphological criteria, genomic stability of DNA, biochemical assay, and production of common dry bean (*Phaseolus vulgaris* L.). *Plant Physiology and Biochemistry, 151*, 77–87.

Pirtarighat, S., Ghannadnia, M., & Baghshahi, S. (2019). Green synthesis of silver nanoparticles using the plant extract of *Salvia spinosa* grown *in vitro* and their antibacterial activity assessment. *Journal of Nanostructure in Chemistry, 9*, 1–9.

Podsiadlo, P., Kaushik, A. K., Arruda, E. M., Waas, A. M., Shim, B. S., Xu, J., & Kotov, N. A. (2007). Ultrastrong and stiff layered polymer nanocomposites. *Science, 318*(5847), 80–83.

Ponce-García, C. O., Soto-Parra, J. M., Sánchez, E., Muñoz-Márquez, E., Piña-Ramírez, F. J., Flores-Córdova, M. A., & Yáñez Muñoz, R. M. (2019). Efficiency of nanoparticle, sulfate, and zinc-chelate use on biomass, yield, and nitrogen assimilation in green beans. *Agronomy, 9*(3), 128.

Pradhan, S., Patra, P., Das, S., Chandra, S., Mitra, S., Dey, K. K., & Goswami, A. (2013). Photochemical modulation of biosafe manganese nanoparticles on *Vigna radiata*: A detailed molecular, biochemical, and biophysical study. *Environmental Science & Technology, 47*(22), 13122–13131.

Qureshi, A., Singh, D. K., & Dwivedi, S. (2018). Nano-fertilizers: A novel way for enhancing nutrient use efficiency and crop productivity. *International Journal of Current Microbiology and Applied Sciences, 7*(2), 3325–3335.

Raskar, S. V., & Laware, S. L. (2014). Effect of zinc oxide nanoparticles on cytology and seed germination in onion. *International Journal of Current Microbiology and Applied Sciences, 3*(2), 467–473.

Reddy, K. R. (2017). Green synthesis, morphological and optical studies of CuO nanoparticles. *Journal of Molecular Structure, 1150*, 553–557.

Rizzello, L., & Pompa, P. P. (2014). Nanosilver-based antibacterial drugs and devices: Mechanisms, methodological drawbacks, and guidelines. *Chemical Society Reviews, 43*(5), 1501–1518.

Sadak, M. S. (2019). Impact of silver nanoparticles on plant growth, some biochemical aspects, and yield of fenugreek plant (*Trigonella foenum-graecum*). *Bulletin of the National Research Centre, 43*(1), 1–6.

Salama, D. M., Osman, S. A., Abd El-Aziz, M. E., Abd Elwahed, M. S., & Shaaban, E. A. (2019). Effect of zinc oxide nanoparticles on the growth, genomic DNA, production, and quality of common dry bean (*Phaseolus vulgaris*). *Biocatalysis and Agricultural Biotechnology, 18*, 101083.

Servin, A. D., & White, J. C. (2016). Nanotechnology in agriculture: Next steps for understanding engineered nanoparticle exposure and risk. *NanoImpact, 1*, 9–12.

Sharmila, G., Fathima, M. F., Haries, S., Geetha, S., Kumar, N. M., & Muthukumaran, C. (2017). Green synthesis, characterization, and antibacterial efficacy of palladium nanoparticles synthesized using *Filicium decipiens* leaf extract. *Journal of Molecular Structure, 1138*, 35–40.

Shukla, P., Chaurasia, P., Younis, K., Qadri, O. S., Faridi, S. A., & Srivastava, G. (2019). Nanotechnology in sustainable agriculture: Studies from seed priming to post-harvest management. *Nanotechnology for Environmental Engineering, 4*, 1–15.

Siddiqui, M. H., Al-Whaibi, M. H., Faisal, M., & Al Sahli, A. A. (2014). Nano-silicon dioxide mitigates the adverse effects of salt stress on *Cucurbita pepo* L. *Environmental Toxicology and Chemistry, 33*(11), 2429–2437.

Singh, J., Mehta, A., Rawat, M., & Basu, S. (2018). Green synthesis of silver nanoparticles using sun-dried tulsi leaves and its catalytic application for 4-nitrophenol reduction. *Journal of Environmental Chemical Engineering, 6*(1), 1468–1474.

Singh, J., Singh, N., Rathi, A., Kukkar, D., & Rawat, M. (2017). Facile approach to synthesize and characterization of silver nanoparticles by using mulberry leaves extract in aqueous medium and its application in antimicrobial activity. *Journal of Nanostructures, 7*(2), 134–140.

Singh, N. B., Amist, N., Yadav, K., Singh, D., Pandey, J. K., & Singh, S. C. (2013). Zinc oxide nanoparticles as fertilizer for the germination, growth, and metabolism of vegetable crops. *Journal of Nanoengineering and Nanomanufacturing, 3*(4), 353–364.

Srivastava, A., & Rao, D. P. (2014). Enhancement of seed germination and plant growth of wheat, maize, peanut, and garlic using multiwalled carbon nanotubes. *European Chemical Bulletin, 3*(5), 502–504.

Stampoulis, D., Sinha, S. K., & White, J. C. (2009). Assay-dependent phytotoxicity of nanoparticles to plants. *Environmental Science & Technology, 43*(24), 9473–9479.

Suresh, Y., Annapurna, S., Bhikshamaiah, G., & Singh, A. K. (2013). Characterization of green synthesized copper nanoparticles: A novel approach. In *International Conference on Advanced Nanomaterials & Emerging Engineering Technologies* (pp. 63–67). IEEE.

Swaefy, H. M., & Abdallh, A. M. (2021). Mitigation of salinity stress in fenugreek plants using zinc oxide nanoparticles and zinc sulfate. *Journal of Plant Production, 12*(12), 1367–1374.

Torney, F., Trewyn, B. G., Lin, V. S. Y., & Wang, K. (2007). Mesoporous silica nanoparticles deliver DNA and chemicals into plants. *Nature Nanotechnology, 2*(5), 295–300.

Torsvik, V., & Øvreås, L. (2002). Microbial diversity and function in soil: From genes to ecosystems. *Current Opinion in Microbiology, 5*(3), 240–245.

Tripathi, S., & Sarkar, S. (2015). Influence of water-soluble carbon dots on the growth of wheat plant. *Applied Nanoscience, 5*, 609–616.

Vasquez, Y., Henkes, A. E., Bauer, J. C., & Schaak, R. E. (2008). Nanocrystal conversion chemistry: A unified and materials-general strategy for the template-based synthesis of nanocrystalline solids. *Journal of Solid-State Chemistry, 181*(7), 1509–1523.

Vijaya, J. J., Jayaprakash, N., Kombaiah, K., Kaviyarasu, K., Kennedy, L. J., Ramalingam, R. J., & Maaza, M. (2017). Bioreduction potentials of dried root of *Zingiber officinale* for a simple green synthesis of silver nanoparticles: Antibacterial studies. *Journal of Photochemistry and Photobiology B: Biology, 177*, 62–68.

Xie, Y., He, Y., Irwin, P. L., Jin, T., & Shi, X. (2011). Antibacterial activity and mechanism of action of zinc oxide nanoparticles against *Campylobacter jejuni*. *Applied and Environmental Microbiology, 77*(7), 2325–2331.

Yang, K., & Ma, Y. Q. (2010). Computer simulation of the translocation of nanoparticles with different shapes across a lipid bilayer. *Nature Nanotechnology, 5*(8), 579–583.

Youssef, A. M., Assem, F. M., Abdel-Aziz, M. E., Elaaser, M., Ibrahim, O. A., Mahmoud, M., & Abd El-Salam, M. H. (2019). Development of bio-nanocomposite materials and its use in coating of Ras cheese. *Food Chemistry, 270*, 467–475.

Yu-Qin, F., Lu-Hua, L., Pi-Wu, W., Jing, Q., Yong-Ping, F., Hui, W., & Chang-li, L. (2012). Delivering DNA into plant cells by gene carriers of ZnS nanoparticles. *Chemical Research in Chinese Universities, 28*(4), 672–676.

Zareabyaneh, H., & Bayatvarkeshi, M. (2015). Effects of slow-release fertilizers on nitrate leaching, its distribution in the soil profile, N-use efficiency, and yield in potato crop. *Environmental Earth Sciences, 74*, 3385–3393.

Zhai, G., Walters, K. S., Peate, D. W., Alvarez, P. J., & Schnoor, J. L. (2014). Transport of gold nanoparticles through plasmodesmata and precipitation of gold ions in woody poplar. *Environmental Science & Technology Letters, 1*(2), 146–151.

Zhang, L., & Fang, M. (2010). Nanomaterials in pollution trace detection and environmental improvement. *Nano-Today, 5*(2), 128–142.

Zheng, L., Hong, F., Lu, S., & Liu, C. (2005). Effect of nano-TiO_2 on strength of naturally aged seeds and growth of spinach. *Biological Trace Element Research, 104*, 83–91.

Zimmermann, P., & Zentgraf, U. (2005). The correlation between oxidative stress and leaf senescence during plant development. *Cellular and Molecular Biology Letters, 10*(3), 515.

CHAPTER 11

OMICS and Fenugreek

M. NITHYA KRUTHI, PRIYANKA SHARMA, TUSHAR GUPTA, NAINA SHARMA, SHRUTI PANDEY, ABHISHEK SENGUPTA, and PRIYANKA NARAD

Amity Institute of Biotechnology, Amity University, Noida, Uttar Pradesh, India

ABSTRACT

This chapter delves into the utilization of OMICS technologies in the investigation of fenugreek, an extensively employed medicinal plant with a variety of pharmacological characteristics. OMICS is an encompassing term used to describe a collection of advanced molecular biology techniques that facilitate the parallel examination of numerous biological molecules, including DNA, RNA, proteins, and metabolites. These technological advancements have brought about a paradigm shift in the realm of biology, facilitating an all-encompassing comprehension of intricate biological systems at the molecular scale. Fenugreek has undergone comprehensive scientific investigation regarding its therapeutic potential, revealing compelling evidence supporting its efficacy in the management of diabetes, hyperlipidemia, inflammation, and cancer. This chapter delves into the scientific advancements of OMICS technologies in enhancing our understanding of the therapeutic potential and mechanisms of action of fenugreek.

11.1 INTRODUCTION

Trigonella foenum-graecum L., commonly known as Fenugreek, is a significant botanical species widely cultivated for its therapeutic properties and culinary uses. It has been traditionally recognized for its health-promoting benefits and is rich in bioactive compounds such as alkaloids, flavonoids,

Fenugreek: Improvement in Yield Potential Through Plant Growth Regulators (PGRs) and Depolymerized Polysaccharides. Tariq Ahmad Dar, Bilal Ahmad, Moinuddin, and Khalid Rehman Hakeem (Eds.)

saponins, and steroidal compound (Kumar et al., 2018). With the increasing demand for natural products and alternative medicines, fenugreek has gained significant attention in recent years. Fenugreek, Traditionally, a plant with documented therapeutic uses has been utilized to treat a wide range of illnesses (Agaoglu et al., 2022). The recent advancements in high-throughput molecular biology technologies, collectively known as OMICS technologies, have been instrumental in dissecting the functional biology of medicinal plants like fenugreek.

OMICS encompasses a diverse array of high-throughput molecular biology methodologies that enable the thorough investigation of biological molecules, including DNA, RNA, proteins, and metabolites. These technological advancements have significantly transformed the domain of biology by facilitating researchers in acquiring a comprehensive perspective of intricate biological systems. OMICS approaches offer significant insights into the mechanisms of action that underlie the therapeutic effects of fenugreek by concurrently examining multiple molecular components. The emergence of omics technologies has resulted in a fundamental change in the field of plant biology facilitating exhaustive examination of diverse molecular constituents including genomics, transcriptomics, proteomics, and metabolomics. These instruments have yielded valuable observations regarding the intricate regulatory networks and biochemical pathways linked to plant growth, development, and reactions to environmental stimuli. Omics methodologies have been utilized in the study of fenugreek to uncover the molecular mechanisms that contribute to its therapeutic properties and to improve its yield capacity.

OMICS technologies allow for the concurrent analysis of large quantities of biological molecules, such as DNA, RNA, proteins, and metabolites (Li et al., 2017). Such comprehensive molecular profiling has revolutionized the way we understand biological systems, offering a more holistic view of an organism's function and state (Karczewski and Snyder, 2018). Applying these technologies to fenugreek has allowed researchers to delve deeper into the plant's medicinal properties and potential therapeutic applications (Gupta et al., 2023).

A particular scientific focus revolves around the application of plant growth regulators (PGRs) in the cultivation of fenugreek. Plant growth regulators (PGRs) refer to a group of substances, which can be either artificially produced or found naturally, that regulate the growth and development of plants by imitating or modifying the impacts of plant hormones present within the organism. Numerous scientific investigations have been conducted to examine the impact of plant growth regulators (PGRs) on the growth, productivity, and concentration of bioactive compounds in fenugreek. According to a study by Jasim et al. (2019), applying gibberellic acid (GA3) to the leaves had a significant effect on the fenugreek yield. This outcome was attained through the stimulation of plant development, as evidenced by the observed enhancements in plant height, leaf area, and seed weight. A study conducted by Younis et al. (2017) revealed that the application of cytokinins exerts a beneficial influence on the germination process and overall growth of fenugreek.

Another approach to enhancing fenugreek yield and bioactivity is the utilization of depolymerized polysaccharides. Polysaccharides are complex carbohydrates composed of repeating sugar units. Depolymerization, which involves breaking down large polysaccharide molecules into smaller fragments, can improve bioavailability and biological activity. Studies have explored the potential of depolymerized fenugreek polysaccharides in promoting plant growth and stress tolerance. For example, Gupta et al. (2020) reported that fenugreek plants under drought stress showed increased root development, nitrogen uptake, and antioxidant enzyme activity in response to the application of depolymerized fenugreek polysaccharides.

Numerous scientific investigations have elucidated the therapeutic capacity of fenugreek in the management of diabetes, hyperlipidemia, inflammation, and potentially cancerous conditions (Kakani et al., 2011). The genomics research conducted on fenugreek has provided valuable insights into the genetic mechanisms responsible for the production of bioactive compounds, thereby advancing our comprehension of its therapeutic capabilities (Das et al., 2022; Gupta et al., 2023). On the contrary, transcriptomics has provided valuable insights into the patterns of gene expression and regulatory networks that contribute to the medicinal benefits of fenugreek (Chaudhary et al., 2021). Research conducted in the disciplines of metabolomics and proteomics has significantly enhanced our comprehension of the metabolic pathways and protein profiles that could potentially be linked to the therapeutic characteristics of fenugreek (Khan et al., 2023; Khare et al., 2020). Furthermore, the incorporation of these OMICS data using systems biology methodologies has yielded a comprehensive understanding of fenugreek's intricate molecular interactions and its potential as a target for drug development (Guenin et al., 2009; Li et al., 2019). Thus, the use of OMICS technologies in the study of fenugreek is a rapidly evolving field that promises to significantly contribute to our understanding of the medicinal properties of this plant.

This chapter provides a comprehensive review of the application of omics technologies, plant growth regulators (PGRs), and depolymerized polysaccharides in enhancing the yield potential of fenugreek. Our focus will be on elucidating the molecular mechanisms that contribute to the therapeutic properties and medicinal effects of fenugreek. We will discuss strategies to optimize plant growth, yield, and bioactive compound content using these methods. Additionally, we will highlight the challenges and future prospects of integrating omics approaches with PGRs and depolymerized polysaccharides for sustainable fenugreek production.

11.2 GENOMICS IN FENUGREEK

Trigonella foenum-graecum, commonly referred to as Fenugreek, is a member of the Fabaceae family and is highly valued for its fragrant seeds and therapeutic properties. The plant in question has been employed for various purposes for thousands of years in both gastronomic and conventional medicinal applications. In recent years, genomics has made noteworthy advancements in elucidating the genetic components and properties of fenugreek. The aforementioned findings have provided crucial perspectives for enhancing agriculture, generating nutraceuticals, and conducting pharmaceutical investigations (Figure 11.1).

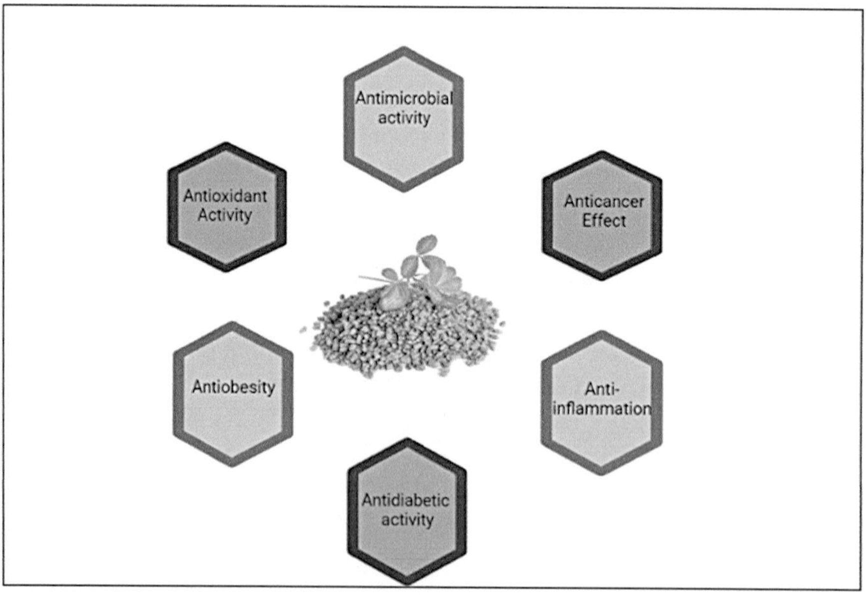

FIGURE 11.1 The AFG-1 variety of *Trigonella foenum-graecum* is depicted below as a representative example of the secondary metabolites that can be produced by this plant. Because it contains such a high concentration of these compounds, fenugreek has a wide range of nutraceutical and medicinal applications.

11.2.1 GENETIC DIVERSITY AND BREEDING

A comprehensive comprehension of the genetic diversity of fenugreek is imperative for the effective advancement of crop development. The genetic diversity of fenugreek was investigated (Shabir and Anjum, 2017), who emphasized the importance of preserving this diversity and utilizing it in breeding programs. The authors emphasized the capacity of genomics to expedite the development of superior fenugreek cultivars that exhibit augmented yields, heightened disease resistance, and enhanced nutritional profiles. In prior studies Gupta et al. (2019) discussed the accessibility of genomic resources that facilitate the identification of genes associated with significant agronomic traits and the subsequent characterization of these genes. The utilization of molecular markers and genetic mapping tools has facilitated the accessibility of marker-assisted breeding and genomic selection in fenugreek.

11.2.2 FUNCTIONAL CHARACTERIZATION AND NUTRACEUTICALS

Genomic research has facilitated the elucidation of the molecular pathways accountable for the synthesis of advantageous compounds in fenugreek. A study presented a scholarly examination of the molecular biology and genomics of fenugreek (Misra and Verma, 2019). The researchers directed their attention towards the identification and delineation of indispensable genes implicated in the biosynthesis of secondary metabolites. Fenugreek is known to possess therapeutic characteristics and nutraceutical potential due to the presence of metabolites such as saponins, alkaloids, flavonoids, and galactomannan. The study emphasized the importance of fenugreek genomics in identifying genes associated with the production of bioactive compounds (Chittora et al., 2020). The utilization of these genes has the capacity to enhance the nutraceutical properties of fenugreek-based products.

11.2.3 PHARMACOLOGICAL APPLICATIONS

The genomic investigation carried out on fenugreek has also been utilized to examine its therapeutic attributes. The genomic and transcriptome methodologies utilized in the identification of genes responsible for the pharmacological properties of fenugreek were examined (Shabir and Anjum, 2017). The aforementioned methodologies facilitated the identification of genetic elements implicated in the pharmacodynamics of fenugreek. The aforementioned activities encompass properties that exhibit anticancer, anti-inflammatory, anti-diabetic, and antioxidant effects. The comprehension of the genetic foundation of the characteristics under investigation enables researchers to devise targeted strategies for drug development and therapeutic interventions.

11.2.4 GENOME-WIDE ASSOCIATION STUDIES IN FENUGREEK

Genome-wide association studies (GWAS), commonly referred to as GWAS, have recently emerged as valuable methodologies for comprehending the genetic underpinnings of intricate traits in diverse agricultural commodities. In recent years, fenugreek, also known as *Trigonella foenum-graecum*, has gained popularity due to the nutraceutical and therapeutic benefits associated with its seeds. Fenugreek is a multipurpose herbaceous plant. The results of genome-wide association studies (GWAS) conducted on fenugreek have yielded vital new information on the genetic variables underlying essential features, paving the path for focused breeding and crop development initiatives. This chapter investigates the use of genome-wide association studies (GWAS) in fenugreek and summarizes essential discoveries made on the topic.

Table 11.1 conducted in fenugreek. It includes information on the study, the traits investigated, key findings, and the corresponding references. Each study focused on different traits of interest in fenugreek and identified significant genomic regions or candidate genes associated with those traits.

TABLE 11.1 Summary of Selected Genome-Wide Association Studies (GWAS)

Study	Trait(s) Investigated	Key Findings	References
Chittora et al. (2020)	Seed yield and yield components	Identified significant SNPs associated with seed yield, seed weight, and pod length.	Chittora et al. (2020); Khan et al. (2021)
Khan et al. (2021)	Seed oil content and fatty acid composition	Identified genomic regions and candidate genes associated with oil content and fatty acid profiles.	Khan et al. (2021); Ulas et al. (2021)
Ulas et al. (2021)	Antioxidant activity and total phenolic content in leaves	Identified SNPs associated with antioxidant activity and phenolic content.	Yadav et al. (2015); Ulas et al. (2021)

11.2.5 OVERVIEW OF GENOME-WIDE ASSOCIATION STUDIES

The widely recognized approach referred to as a "genome-wide association study" (GWAS) entails conducting comprehensive investigations on the entire population of a particular community in order to ascertain genetic variations that exhibit correlation with specific phenotypic traits or characteristics. Scientists possess the capacity to detect genomic regions or single nucleotide polymorphisms (SNPs) that display a noteworthy correlation with particular areas of interest by analyzing numerous genetic markers dispersed throughout the entire genome. The procedure in question is formally referred to as a genome-wide association study (GWAS) in scientific terminology. The genome-wide association study (GWAS), also known as GWAS, is a rigorous methodology employed to gain understanding of the complex genetic architecture that underlies a wide array of traits. The potential of this technology for application in agricultural development, particularly in crop cultivation, has been established. This conclusion was reached through careful and systematic analysis. Genomic-scale association studies, commonly known as GWAS, have been performed on the botanical species *Trigonella foenum-graecum*, also known as fenugreek. In recent times, there has been a notable increase in the application of genome-wide association studies (GWAS), which seek to investigate the genetic foundations of diverse agronomic and nutraceutical characteristics. These investigations have been conducted using large populations of organisms. Chittora et al. (2020) performed a genome-wide association study (GWAS) to detect specific genomic regions in fenugreek that are linked to seed yield and its component traits. A number of statistically significant single nucleotide polymorphisms (SNPs) have been discovered to be correlated with differences in pod length, seed weight, and seed yield. The Single Nucleotide Polymorphisms (SNPs) identified in this study demonstrate the potential to function as selection markers in breeding programs through the utilization of marker-assisted selection.

Study conducted by Ulas et al. (2021) also carried out a genome-wide association study, often known as a GWAS, in order to gain a better understanding of the genetic basis behind the seed oil content and fatty acid composition in fenugreek. They were able to identify genomic regions and candidate genes related with oil content and diverse fatty acid profiles, which gave insights into the genetic regulation of these important nutritional properties. Breeders now have the opportunity, according to the findings of these sort of research, to select for certain genotypes that have ideal oil content and fatty acid composition. This, in turn, leads to an improvement in the nutritional quality of fenugreek seeds.

In addition, genome-wide association studies (GWAS) have been conducted to investigate the nutraceutical properties of fenugreek. These studies suggest that fenugreek may have health benefits. To assess the genetic basis of antioxidant activity and total phenolic content in fenugreek leaves, Yadav et al. (2015) conducted a GWAS, successfully identifying single nucleotide polymorphisms (SNPs) associated with these traits. This research provides valuable information for selecting fenugreek genotypes with enhanced nutraceutical effects, paving the way for the development of functional foods and nutraceutical products with increased antioxidant capacity.

11.2.6 FUNCTIONAL GENOMICS OF FENUGREEK

11.2.6.1 UNRAVELING MOLECULAR MECHANISMS AND METABOLIC PATHWAYS

The expansive area of research known as functional genomics has as its primary objective the comprehension of the functional components and relationships of genes that are present inside the genome of an organism. Fenugreek (*Trigonella foenum-graecum*) has been the subject of extensive research in functional genomics, which has led to significant breakthroughs in our understanding of the molecular mechanisms and metabolic pathways that underlie a variety of important biological activities. In this academic work, the use of functional genomics to the study of fenugreek is investigated, and a detailed analysis of the key results that have been accomplished in this field to this point is presented.

11.2.7 TRANSCRIPTOMICS AND GENE EXPRESSION PROFILING

A subgenre of functional genomics known as "transcriptomics" focuses on the exhaustive study of all RNA transcripts that are present in a cell or tissue at a particular point in time. Transcriptomics is a subdivision of "functional genomics." Fenugreek has been studied with the intention of understanding the gene expression patterns that are related with the various developmental stages, tissue-specific activities, and reactions to external stimuli. This has been accomplished through the utilization of transcriptomics. An examination of the transcriptome profiles of fenugreek seeds at several stages of development was carried out by Das et al. (2022). The objective of the study was to locate essential genes that were thought to be involved in seed growth, metabolic processes, and the storage of nutrients. The current research intended to discover prospective targets for enhancing seed quality and yield, as well as to elucidate the transcriptional processes that drive the maturity of seeds.

11.2.8 METABOLIC PATHWAY ANALYSIS

The investigation of metabolic pathways is an essential component of functional genomics. This study is conducted with the intention of gaining an understanding of the interconnected network of biochemical events that take place within an organism. The investigation of metabolic pathways in fenugreek has shown to be extremely beneficial in elucidating the biosynthesis routes of bioactive chemicals and secondary metabolites in the plant. These compounds and metabolites exhibit both nutraceutical and medicinal characteristics. In their research by Das (2022) conducted an exhaustive investigation into the metabolic pathway of fenugreek. They isolated the genes and enzymes that were specifically responsible for the manufacture of essential secondary metabolites, such as saponins, alkaloids, and flavonoids. Not only did this work reveal the biosynthetic pathways, but it also discovered possible regulatory genes that may be used for metabolic engineering that is specifically aimed at a certain target.

11.2.9 FUNCTIONAL CHARACTERIZATION OF CANDIDATE GENES

The field of functional genomics enables the functional characterization of candidate genes identified through genomics, transcriptomics, or genome-wide association studies (GWAS). A functional analysis of a gene encoding chalcone isomerase (CHI), a key enzyme in the flavonoid biosynthesis pathway in fenugreek, was conducted by Gupta et al. (2019) and published in *Scientific Reports*, a journal known for its rigorous editorial review process. The study showed that inhibiting the CHI gene significantly reduced flavonoid accumulation, highlighting the gene's essential role in producing these valuable compounds. This research exemplifies the application of functional genomics approaches to validate and understand the roles of candidate genes associated with key traits in fenugreek.

11.2.10 INTEGRATION OF OMICS DATA

To achieve a comprehensive understanding of biological processes, functional genomics often requires the integration of data obtained from various omics platforms, such as genomics, transcriptomics, proteomics, and metabolomics. Ulas et al. (2021) conducted a study on fenugreek, wherein they integrated data from transcriptomic and metabolomic approaches to identify the genes and metabolites associated with the plant's overall phenolic content and antioxidant activity. The incorporation of omics data has facilitated a more extensive understanding of the molecular mechanisms that underlie the nutraceutical characteristics and has yielded valuable information regarding potential regulatory networks. This was facilitated by the prior obscurity of these mechanisms.

11.2.11 APPLICATIONS OF GENOMICS IN FENUGREEK BREEDING AND IMPROVEMENT

Plant breeding has experienced a substantial transformation because of the important resources and information provided by genomics in the field of crop improvement. This is because genomics has allowed for the progress of the field. In addition, the herbaceous plant known as fenugreek (*Trigonella foenum-graecum*), which contains qualities that may be used in cooking as well as medicine, has profited from the development of genomic technology. Fenugreek is a member of the *Trigonella foenum-graecum* family. In the current research, an investigation into the use of genomics in the process of breeding and upgrading fenugreek was carried out. The potential of genomics to improve key agronomic features, nutritional value, and resistance to disease is at the heart of this text.

11.2.12 GENOMIC SELECTION AND MARKER-ASSISTED BREEDING

Genomic selection is a method of breeding that makes use of genomic information to predict the breeding value of individuals based on the genetic markers they contain. This method of breeding was developed by scientists at the University of California, Davis. Genomic breeding is another name for this approach to animal reproduction. Genomic selection offers a considerable amount of promise in terms of speeding up the breeding process in fenugreek and improving traits like yield, seed quality, and resistance to disease.

This potential is provided by the significant amount of potential offered by genomic selection. The potential of genomic selection in fenugreek breeding programs was emphasized in Shabir and Anjum's work which was published in 2017. They brought attention to the application of marker-assisted selection as a strategy for locating and introducing desirable genes or alleles into top breeding lines. Breeders that utilize marker-assisted breeding are able to select for certain attributes in a manner that is not only more efficient but also more precise, which eventually results in superior fenugreek varieties.

11.2.13 TRAIT MAPPING AND QUANTITATIVE TRAIT LOCI (QTL) ANALYSIS

A key development in this area is fenugreek genomics, which enables the identification and mapping of genetic loci related with critical agronomic qualities. This is an important step forward for the sector. Through the implementation of trait mapping investigations and quantitative trait locus (QTL) analysis, scientists are capable of identifying genomic regions or markers that exhibit associations with specific traits of interest. Chittora et al. (2020) performed a genome-wide association study (GWAS) on fenugreek to identify quantitative trait loci (QTLs) associated with seed yield and yield component traits in the plant. The research findings unveiled a substantial quantity of single nucleotide polymorphisms (SNPs) and quantitative trait loci (QTLs) that exhibit a correlation with seed yield. Because of these findings, marker-assisted breeding might be used to develop fenugreek varieties with increased crop yields.

11.2.14 GENOMIC DIVERSITY ANALYSIS

It is essential to have a comprehensive grasp of the genetic variability that is present among fenugreek populations in order to ensure the effective breeding of new varieties and the preservation of germplasm. Genomic technology has made it feasible to analyze not only the genetic diversity of fenugreek but also the population structure of the plant. An amplified fragment length polymorphism (AFLP) study was utilized for research on the genetic diversity of fenugreek. The research came to the conclusion that there is a significant level of genetic variety among the fenugreek accessions. This is beneficial information for selecting genetically diversified parents and designing breeding processes that are effective.

11.2.15 FUNCTIONAL GENOMICS AND TRAIT CHARACTERIZATION

Functional genomics techniques, including as transcriptomics and metabolomics, are able to provide light on the molecular pathways that are responsible for critical properties of fenugreek. This is because these techniques are able to analyze large amounts of data. When researchers examine gene expression profiles in combination with metabolite profiles, they are able to identify significant genes and pathways that are associated with the characteristics that are of interest to them. Gupta and Das (2022) investigated how the use of functional genomics may be implemented in the breeding of fenugreek. The characterization of genes related with nutritional characteristics, the production of secondary metabolites, and reactions to stress was a primary focus for them. This knowledge may be put to use in the process of selecting and altering genes in fenugreek crops in order to increase the crop's nutritional quality, nutraceutical properties, and resilience to abiotic stress.

The genome-wide association studies, often known as GWAS, that were performed on fenugreek gave important new information on the genetic foundations of important traits. These factors cover a wide spectrum, from agronomic considerations to nutraceutical potential. On the other hand, more research is needed to validate and fine-map the probable genomic areas and regions that were found in the reported genomic regions. Attempts to sequence the entire genome can give a more thorough understanding of the underlying biological processes if they are combined with other functional genomics methods like as transcriptomics and proteomics.

The investigation of the genome of the plant known as fenugreek has resulted in substantial new insights into the genetic components and features of this versatile crop. Studies on genetic diversity have opened the way for effective breeding programs, while techniques based on functional genomics have made it simpler to discover and define genes involved in the production of bioactive compounds. Both of these developments have had a positive impact on the field of genomics. Both of these breakthroughs have made important contributions to the study of genomics. In addition, the genetic knowledge of fenugreek's pharmacological features has a tremendous degree of potential for both the creation of novel medications and the use of such pharmaceuticals in therapeutic settings. This is because of the enormous amount of potential that is held by this understanding. Research into the genetics of fenugreek should be maintained since it will undoubtedly assist in the discovery of additional mysteries and contribute to the plant's environmentally sustainable production, the development of nutraceuticals, and advancements in the technological capabilities of the pharmaceutical industry. Investigating the Role of Genes in the Development of Important Traits Using Genome-wide Association Genome-wide association studies (GWAS), which were carried out on fenugreek, have given insight on the genetic foundations of a range of features, including as seed yield, oil content, fatty acid composition, and nutraceutical attributes. These studies were done on fenugreek. These insights make it possible to develop unique varieties of fenugreek and to improve the genetic makeup of agricultural species that are closely linked to fenugreek. The use of GWAS to fenugreek offers a great deal of untapped potential for enhancing crop yield, nutritional quality, and functional features. Continued study in this field will, without a shadow of a doubt, help to the sustainable production and utilization of fenugreek in a range of industries, such as agriculture, nutrition, and health. This is without a doubt the case. The study of functional genomics in fenugreek, which is a field that is quickly emerging, has a significant lot of promise for the manufacture of nutraceuticals, in addition to having applications in agricultural and pharmaceutical research, respectively. The goal of study in the future should be to examine the function of genes by targeted gene manipulation utilizing methods such as gene editing technologies. In addition, the combination of data from many omics and methods from systems biology will lead to a more in-depth appreciation of the complicated biological processes that are involved in fenugreek. The use of functional genomics to fenugreek has fundamentally altered our comprehension of the molecular mechanisms and metabolic pathways that underpin fundamental phenotypes and functions in biology. This is because fenugreek is used as a model organism in the study of these particular elements of biological systems. The investigation of metabolic pathways and transcriptomics has assisted researchers in locating essential genes and enzymes that play a role in seed development, the production of secondary metabolites, and the accumulation of nutrients.

The application of genomics has had a considerable influence on the breeding and enhancement of fenugreek due to the supply of helpful tools and insights by genomics. Genomic selection and marker-assisted breeding make it feasible to choose desired characteristics in an efficient manner, whilst trait mapping and quantitative trait locus analysis find genomic areas connected with critical agronomic qualities. Both of these methods were developed relatively recently. The study of genetic diversity is beneficial to the maintenance of germplasm as well as the development of breeding techniques. In addition, the approaches that are based on functional genomics interpret the molecular mechanisms that are at the root of characteristics, which makes it possible to control traits in a manner that is more focused.

It is important to continue research in genomics because it will inevitably result in the development of new fenugreek varieties that have improved agronomic characteristics, nutritional quality, and resistance to disease.

11.3 TRANSCRIPTOMICS IN FENUGREEK

The field of transcriptomics focuses on studying the transcriptome, which encompasses all RNA transcripts present within a specific cell or tissue at a given time (Branch, 2013). Transcriptomics is a powerful tool for uncovering the regulatory mechanisms and gene expression patterns underlying various biological processes and responses in fenugreek (Branch, 2013). Fenugreek is valued for its numerous culinary, medicinal, and traditional uses (Naika et al., 2022). Transcriptomic research, which examines the transcriptome in different tissues or under specific conditions, enables a deeper understanding of fenugreek's molecular properties and functions (Branch, 2013). Through transcriptomic techniques such as RNA sequencing (RNA-Seq), researchers can obtain a comprehensive view of expressed genes and their relative abundance in fenugreek (Przybylska et al., 2017). This approach allows for the identification and characterization of key genes and pathways involved in fenugreek's development, metabolism, stress responses, and the biosynthesis of bioactive compounds. Transcriptomic research has illuminated numerous aspects of fenugreek, including:

1. **Seed Development and Metabolism:** Transcriptomics has helped researchers comprehend the gene expression patterns, and regulatory networks associated with fenugreek seed development, maturation, and the accumulation of storage compounds such as proteins, lipids, and secondary metabolites (Przybylska et al., 2017).

    - **Example:** A study on fenugreek seed development, for instance, identified differentially expressed genes in association with the seed coat development, which comprise of genes encoding seed coat-specific transcription factors and enzymes involved in flavonoid biosynthesis. Another study aimed to comprehend the molecular mechanisms underlying seed maturation in fenugreek and identified genes involved in lipid metabolism, seed storage protein accumulation, and desiccation tolerance (Dar and Uddin, 2018).

2. **Biosynthesis of Bioactive Compounds:** Fenugreek is renowned for its abundance of bioactive compounds, which include alkaloids,

flavonoids, saponins, and mucilage (Ahsan et al., 2020). The identification of the genes and regulatory mechanisms associated with the processes involving the biosynthesis of these compounds provides a molecular understanding of their production in fenugreek through the use of transcriptomic analysis.

- **Example:** A transcriptomic analysis of fenugreek seeds revealed the presence of genes encoding enzymes implicated in alkaloid biosynthesis, such as cytochrome P450 enzymes and glycosyltransferases (Das et al., 2022). Another study examined the biosynthesis of saponins in fenugreek and identified genes responsible for the synthesis of enzymes involved in saponin biosynthesis pathways, such as beta-amyrin synthase and cytochrome P450 enzymes (Hidangmayum et al., 2019).

3. **Stress Responses and Adaptation:** Transcriptomic studies enable researchers to investigate how fenugreek responds to a variety of environmental stresses, such as drought, salinity, temperature extremes, and pathogen attacks (Bhattacharya et al., 2016). Researchers can identify stress-responsive genes, signaling pathways, and regulatory mechanisms involved in fenugreek's stress tolerance and adaptation by analyzing gene expression changes.

   - **Example:** A study identified genes associated with stress signaling, antioxidant defense, and osmotic regulation in fenugreek seeds subjected to drought stress (Hidangmayum et al., 2019). Another study investigated the transcriptomic responses of fenugreek seeds to fungal infection and identified genes associated with defense-related pathways and secondary metabolite biosynthesis (Vaidya et al., 2013). These results illuminate the molecular basis of stress tolerance in fenugreek seedlings.

Transcriptomics provides insight into the gene expression patterns and regulatory mechanisms underlying fenugreek's nutritional composition, medicinal properties, and prospective health benefits. Researchers can identify genes associated with the synthesis of specific nutrients, antioxidants, or bioactive compounds with potential therapeutic properties by investigating the transcriptome (Przybylska et al., 2017). In addition to this, researchers can identify novel genes, regulatory elements, and metabolic pathways that contribute to the biology and functionality of fenugreek through transcriptomic analysis. This information not only improves our comprehension of fenugreek as a valuable plant resource but also paves the way for future research, crop enhancement, and the creation of fenugreek-based products with enhanced nutritional, medicinal, or agronomic properties (Figure 11.2; Raghavan et al., 2022).

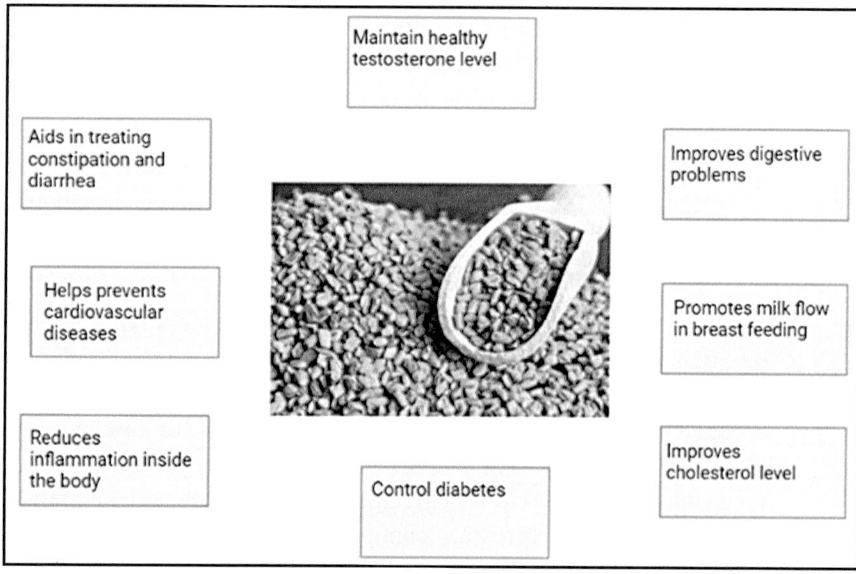

FIGURE 11.2 Health benefits of fenugreek.

11.3.1 TRANSCRIPTOME ASSEMBLY AND ANNOTATION

Fenugreek transcriptome assembly and annotation involves reconstructing and annotating the expressed genes or transcripts in fenugreek tissues or cells (Przybylska et al., 2017). Here is a summary of the processes involved in the assembly and annotation of the transcriptome (Figure 11.3).

1. **RNA Sample Preparation:** RNA must be extracted from fenugreek tissues or cells of interest. For accurate transcriptome analysis, high-quality RNA is required. Commonly used total RNA extraction methods include TRIzol and commercial RNA extraction kits (Evangelistella et al., 2017).
2. **RNA Sequencing:** The extracted RNA is then sequenced using RNA sequencing (RNA-Seq), a high-throughput sequencing technique that generates millions of brief reads representing the transcriptome. Short-read sequencing (e.g., Illumina sequencing) and long-read

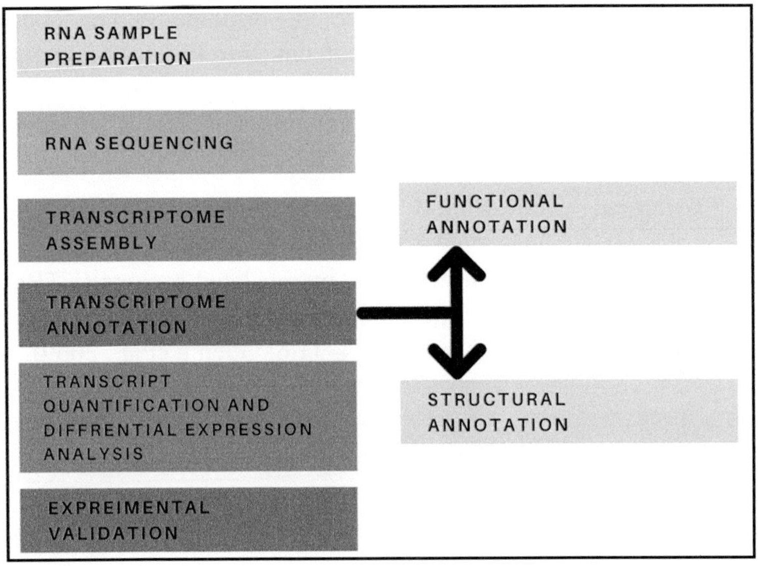

FIGURE 11.3 Flowchart of transcriptome assembly and annotation.

sequencing (e.g., PacBio or Oxford Nanopore sequencing) are the two most prevalent forms of RNA-Seq. Short-read sequencing provides high coverage and is appropriate for quantifying transcripts, whereas long-read sequencing facilitates the resolution of complex transcripts and alternative splicing events (Evangelistella et al., 2017).

3. **Transcriptome Assembly:** The generated RNA-Seq reads are subjected to transcriptome assembly to reconstruct full-length or fragmentary transcripts. De novo assembly, in which transcripts are reconstructed without a reference genome, and guided assembly, in which a reference genome or the genome of a closely related species is used as a scaffold, are the two categories of assembly methods. For transcriptome assembly, various software programs such as Trinity, Velvet, and SPAdes can be utilized (Evangelistella et al., 2017).

4. **Transcriptome Annotation:** The reconstructed transcripts are annotated after assembly to attribute biological information to the assembled sequences. Functional annotation and structural annotation are the two primary phases of annotation.

5. **Functional Annotation:** It is the process of identifying and designating putative functions to reconstructed transcripts based on

sequence similarity to known proteins or domains. This is possible using BLAST (Basic Local Alignment Search Tool) against protein databases such as NCBI's non-redundant (NR) database, UniProt, or domain databases such as Pfam and InterProScan (Guenin et al., 2009).

6. **Structural Annotation:** It is the prediction of the gene structure of the reconstructed transcripts, which includes the identification of open reading frames (ORFs), untranslated regions (UTRs), and splice sites. This can be accomplished using programs like TransDecoder or GeneMark (Figures 11.4–11.6; Guenin et al., 2009).

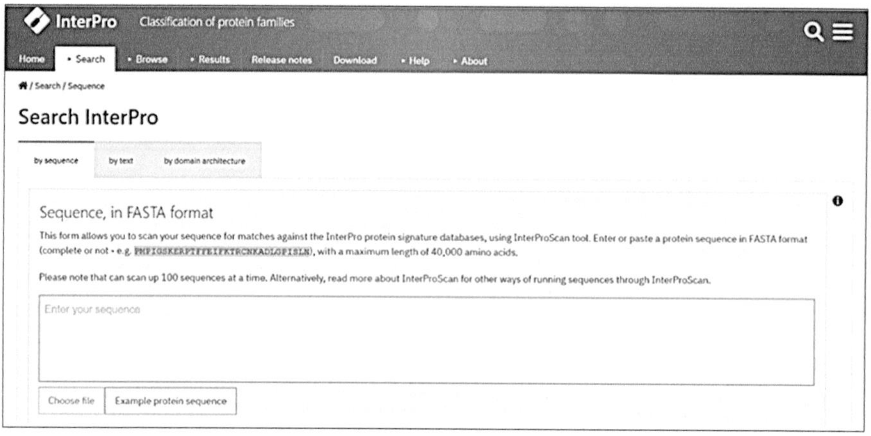

FIGURE 11.4 Homepage of InterProScan database.

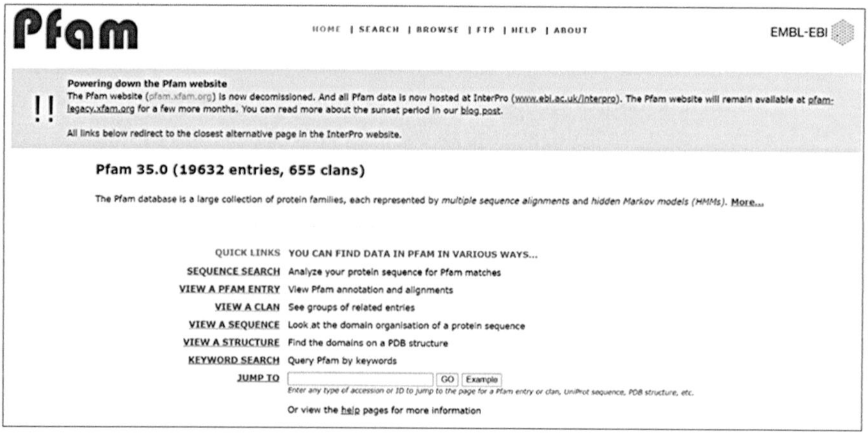

FIGURE 11.5 Homepage of Pfam database.

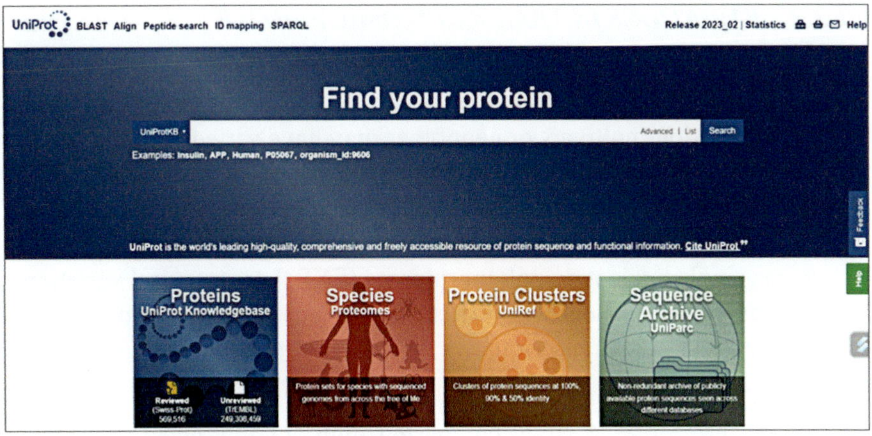

FIGURE 11.6 Homepage of UniProt database.

7. **Transcript Quantification and Differential Expression Analysis:** Following the assembly and annotation of the transcriptome, transcript abundance estimation and differential expression analysis can be conducted. This enables researchers to identify genes or transcripts that are differentially expressed between various samples or conditions, providing insights into the regulation of gene expression in fenugreek.

8. **Experimental Validation:** The techniques such as reverse transcription-PCR (RT-PCR), quantitative real-time PCR (qRT-PCR), and RNA *in situ* hybridization can be used to validate the assembled and annotated transcripts. These techniques help validate the transcriptome assembly and annotation accuracy by confirming the expression patterns (Mahbub et al., 2018). The assembly and annotation of the fenugreek transcriptome provide a comprehensive overview of the expressed genes and their functional properties. This information is valuable for understanding fenugreek's molecular mechanisms and gene expression regulation and identifying potential targets for future functional studies or crop enhancement initiatives (Gupta et al., 2023). However, it is important to be aware that the software and methodologies used for assembling and annotating transcriptomes may vary depending on the research objectives and available resources.

11.3.2 VARIATIONS IN TRANSCRIPTOMIC PROFILE

1. **Genotypic Variations:** Fenugreek demonstrates genetic diversity, and various cultivars or genotypes may have varied gene expression profiles. Comparative transcriptomic analyses of multiple fenugreek varieties or genotypes can reveal the genetic substrate for phenotypic differences. Researchers can identify genetic variations, such as single nucleotide polymorphisms (SNPs) or structural variations, and correlate them with differences in gene expression patterns by analyzing the transcriptomes of different genotypes (Raghavan et al., 2022).

   - **Example:** A study comparing high-yielding and low-yielding fenugreek genotypes identified differentially expressed genes involved in seed development and metabolism, suggesting that genetic differences contribute to yield differences (Ahmed et al., 2020). Comparing fenugreek genotypes with different seed diameters, study identified genotype-specific gene expression patterns associated with seed size determination (Kumar et al., 2016).

2. **Population-Level Variations:** Fenugreek is cultivated in diverse regions around the globe, and natural populations may manifest genetic and transcriptomic variations. Comparative transcriptomic investigations of fenugreek populations or accessions can reveal the genetic diversity within the species. These analyses may reveal population-specific patterns of gene expression, unearth rare alleles, and provide information about the genetic structure and evolutionary history of fenugreek populations.

   - **Example:** A study comparing populations of wild and cultivated fenugreek from various regions revealed population-specific gene expression patterns associated with adaptation and domestication (Kulkarni et al., 2019). These variations can shed light on the genetic structure, diversity, and evolutionary history of populations of fenugreek.

3. **Developmental Stage-Dependent Variations:** Gene expression patterns in fenugreek plants can differ throughout their developmental stages. Transcriptomic analyses of various growth stages, such as seed germination, vegetative growth, flowering, and seed

development, can reveal dynamic alterations in gene expression and regulatory networks. Researchers can acquire insight into the genetic control of phenotypic traits by investigating the transcriptomes at each developmental stage (Raghavan et al., 2022).

- **Example:** A study identified stage-specific gene expression patterns associated with seed maturation, storage protein synthesis, and germination processes by investigating the transcriptomes during seed development (Ghawana et al., 2011). Bhatia et al. (2018) investigated the transcriptomic alterations during flower development in fenugreek and identified genes involved in floral organ formation and secondary metabolite biosynthesis.

4. **Environmental and Stress-Related Variations:** In fenugreek, environmental factors, and stress conditions can affect gene expression patterns. Transcriptomic studies conducted under various environmental signals, such as drought, salinity, temperature stress, or pathogen infection, can aid in the identification of genes implicated in stress responses and adaptive mechanisms (Khare et al., 2020). These studies may reveal differences in the expression of stress-related genes between populations or genotypes of fenugreek, allowing researchers to identify stress-responsive genes and regulatory pathways.

   - **Example:** A study of the transcriptomic response of fenugreek to drought stress identified genes involved in drought tolerance and osmotic regulation (Ahsan et al., 2020). Similarly, genes implicated in ion homeostasis and stress signaling pathways were found to be differentially expressed in fenugreek transcriptomes under salinity stress (Jha et al., 2019).

5. **Tissue-Specific Variations:** Different tissues and organs of the fenugreek plant may express genes differently. Analysis of specific plant tissues, such as leaves, stems, roots, flowers, and seeds, can shed light on tissue-specific gene expression patterns and the functional specialization of various organs. By comparing the transcriptomes of various tissues, researchers can identify genes that are uniquely expressed in certain tissues, identify organ-specific metabolic pathways, and investigate the molecular basis of tissue-specific traits (Bewley and Black, 1994).

- **Example:** A study identified tissue-specific genes involved in photosynthesis, nutrient absorption, and root development by comparing the transcriptomes of fenugreek leaves and roots (Verma et al., 2019). A study uncovered tissue-specific expression of genes involved in secondary metabolite biosynthesis and seed storage by examining the transcriptomic differences between fenugreek leaves and seeds (Kumar et al., 2018).

11.3.3 GENE EXPRESSION PROFILING IN FENUGREEK

Gene expression profiling is a technique for studying the expression of genes in a specific cell or tissue at a certain time. This may be accomplished by monitoring mRNA levels, which are molecules that are transcribed from DNA and subsequently translated into proteins. Gene expression profiling in fenugreek has been used to investigate a range of activities, including seed formation, germination, and stress response (Lockhart and Winzeler, 2000). One researcher discovered, for example, that the expression of genes involved in galactomannan biosynthesis increased during seed development. Galactomannans are a kind of carbohydrate present in fenugreek seeds' endosperm. They're supposed to help in seed growth and germination. Another study discovered that when fenugreek plants were subjected to drought stress, the expression of genes involved in stress response increased (Buckeridge et al., 2000). This shows that gene expression profiling can be used to discover genes implicated in essential metabolic processes. Gene expression profiling is a great approach for studying a wide range of biological processes in fenugreek. It may aid in our understanding of how fenugreek plants form, grow, and adapt to stress. This knowledge might be utilized to improve fenugreek plant farming and breeding, as well as to generate novel fenugreek products (Chaudhary et al., 2021). The following are some of the advantages of employing gene expression profiling in fenugreek:

- It may be used to discover genes involved in critical fenugreek processes like seed formation, germination, and stress response.
- It may be used in investigating the varying expression of genes in response to various environmental situations, including drought stress.
- It may be utilized to create novel fenugreek products, such as higher nutritional value or stress tolerance.

11.3.4 APPLICATIONS OF TRANSCRIPTOMICS IN FENUGREEK

Here are some potential future directions and prospects for fenugreek transcriptomics:

1. **Discovering New Genes and Pathways:** Despite advances in transcriptome studies of fenugreek, there is still more to learn. Future research might concentrate on identifying novel genes and pathways involved in fenugreek biology by studying the transcriptomes of certain tissues, developmental phases, or environmental factors. This leads to understand the molecular processes and regulatory networks of fenugreek.
2. **Gene Functional Characterization:** Transcriptomic data offer a framework for gene functional characterization in fenugreek. To understand the roles of individual genes found by transcriptomics, future research can use techniques such as gene deletion or overexpression. This will assist to confirm the roles of these genes and give insights into their contributions to the growth, development, and productivity of bioactive compounds in fenugreek.
3. **Integration with Other Omics Technologies:** Integrating transcriptomics with other omics technologies, including proteomics, genomics, and metabolomics, can give a more thorough knowledge of the molecular processes of fenugreek. Multi-omics techniques can highlight the connections between gene expression, protein abundance, and metabolite profiles, providing a systems-level view of fenugreek biology. This combination may lead to a more comprehensive knowledge of the complicated biological processes of fenugreek.
4. **Functional Genomics and Gene Editing:** Advances in gene editing technologies, including CRISPR-Cas9, open up intriguing possibilities for improving functional genomics. Transcriptomic data can help researchers choose which genes to target for precision genome editing, allowing them to change particular gene functions and analyze the phenotypic repercussions. This method helps hasten the breeding and development of fenugreek crops with desirable features (Choudhary et al., 2023).
5. **Epigenomics and Gene Regulation:** Epigenetic changes are important in gene regulation and can have long-term consequences

on plant characteristics. Future transcriptome investigations can be combined with epigenomic analysis to further understand the impact of epigenetic changes on gene expression in fenugreek. This will shed light on the epigenetic control of fenugreek growth, development, and responsiveness to environmental cues (Choudhary et al., 2023).

6. **Translational Applications:** Transcriptomic research in fenugreek can be used to develop practical applications. The discovery of important genes and pathways involved in bioactive molecule biosynthesis can be exploited to improve the production of particular compounds with medicinal or nutritional value. Transcriptomics can also help in creation of molecular markers for the marker-assisted breeding projects in fenugreek crops to increase agronomic attributes and production (Choudhary et al., 2023).

These prospective views show the possibility for additional advances in fenugreek transcriptomics research. Transcriptomics will continue to contribute to our understanding and utilization of fenugreek in various fields, including agriculture, biotechnology, and medicine, by exploring novel genes, integrating with other omics technologies, employing gene editing approaches, studying epigenetic regulation, and translating findings into practical applications.

11.4 PROTEOMICS OF FENUGREEK

The total set of proteins produced by an organism, tissue, or cell at a specific time is referred to as its proteome. The term "proteome" was first used in the mid-1990s as an analogue to "genome" (Suna and Mayr, 2018). Proteomics is the study of protein interactions, biological processing, composition, and structure (Al-Amrani et al., 2021). Proteomics technology is well-suited for evaluating phenotypic aspects and holds the potential to identify novel prognostic and diagnostic targets, given that proteins are the primary effector molecules in all biological processes. Proteomics offers new possibilities for improved patient categorization through protein biomarker analysis on an unprecedented scale (Suna and Mayr, 2018).

Protein interaction data can provide insights into protein function and help create intricate interaction networks. Such data can be integrated with additional genetic information to reconstruct complex biochemical pathways by categorizing proteins according to their functional roles. These interactions and shared subnetworks among proteins can be mapped, helping researchers deduce functionality through evolutionary conservation of gene sequences and protein domains across species—a particularly useful approach when genomic and proteomic details are limited (Bertone and Snyder, 2005).

Proteomics can be used to analyze alterations, interactions, and changes in protein expression resulting from fenugreek consumption or treatment, helping researchers understand the molecular mechanisms underlying any potential health benefits of fenugreek. Proteomics studies can, for example, identify specific proteins that are up- or downregulated in response to fenugreek, providing insights into the biological pathways affected by its components. However, three major challenges must be addressed to successfully apply proteomic approaches to plants:

- Thorough recognition of proteins, the isoforms, and how prevalent they are in the tissue;
- Characterization of each protein's biochemical and cellular functions; and
- Examination of regulation of protein and its connections to other systems responsible for regulation (Bertone and Snyder, 2005).

11.4.1 PROTEOMICS WORKFLOW

In proteomics analysis, top down and bottom-up procedures can be used. Peptide-based proteomics is another name for the bottom-up approach. In this, the protein will be broken down by using trypsin and separation will be carried out by the use of a specific column, after this the peptides are analyzed using Mass spectrometry. The fractionation stage allows the bottom-up approach to be divided into two categories. In the first method, 2-DE is used for separating proteins from the gel. The separated proteins are then broken down into peptides which the Mass spectrometry can recognize. The other strategy is referred to as "shotgun" proteomic. In this case, protein digestion takes place without any fractionation, and Liquid Chromatography is employed for separating the peptides found by Mass spectrometry. Workflows for proteomics include preparation of the sample and analytical flow. Protein separation, identification of protein, and validation of protein are included in the latter (Figure 11.7; Al-Amrani et al., 2021).

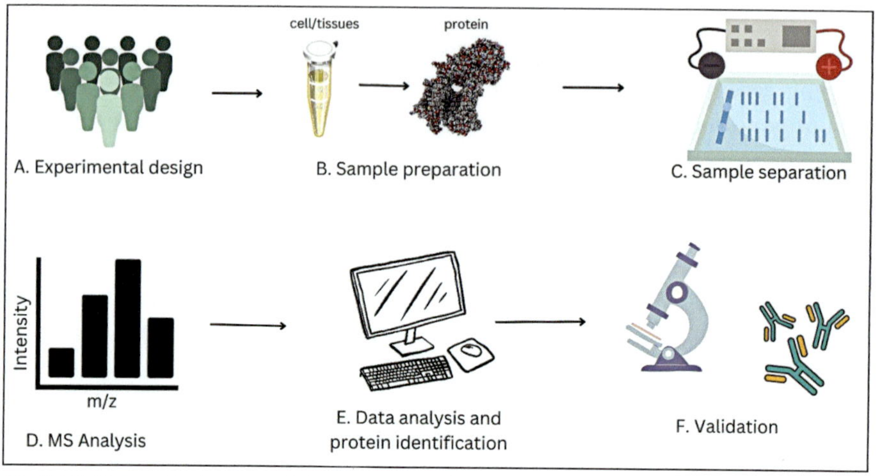

FIGURE 11.7 The basic workflow of proteomics.

Note: (A) Numerous protein comparisons will emerge from a proteomics investigation. Therefore, for accurate data interpretation, a thoroughly planned experimental design with many biological and technical replicates is essential. (B) Protein extraction is crucial for acquiring a representative sample of proteins. Some proteins may be underrepresented in tissue or cell extracts because they are difficult to extract using standard lysis buffers. (C) Before the MS (mass spectrometry) analysis, proteins are separated using gels, or peptides are separated using LC (liquid chromatography). (D) To identify the protein, the acquired MS spectra are programmatically compared to a database of spectra. (E) and (F) Ultimately, independent methods, typically antibody-based, are used for validation of proteins of interest.

The following actions are involved in a typical proteomics study:

1. **Sample Preparation:** For precise and repeatable proteomics analysis, sample preparation is essential. Conditions for sample collection, handling, and storage are included here. To avoid protein breakdown, for instance, it is essential to quickly freeze samples after collection and treat them with protease inhibitors after thawing. A single lysis stage can be used to extract protein from a biological material, or subfractionation techniques can be used. This is done to explore a particular sub-proteome or to lessen sample complexity (Suna and Mayr, 2018).

2. **Separation:** Prior to MS analysis, further separation processes are necessary due to the complexity of the material. To lessen the under-sampling of proteins which are low in abundance, this is important. If not, high-abundant proteins predominate in the MS results. High Performance liquid chromatography (HPLC) or one dimensional or

two-dimensional gel electrophoresis ("1-DE, 2-DE") can be used to accomplish this at the level of protein or at the level of peptides, respectively (Suna and Mayr, 2018).

3. **MS Analysis:** It comes after separation. "MS" is the foundation of contemporary "proteomics" and was responsible for the biggest advance in the field. Attomole concentrations can be detected with the excellent accuracy of "mass, resolution, and sensitivity" of today's MS. The intricacy of the sample, however, affects how sensitive the MS analysis is. High throughput and numerous protein simultaneous analysis are capabilities of Mass spectrometry. In a single run, the mass spectrometer produces thousands of MS spectra in order to calculate the "mass-to-charge ratio (m/z)" of an ionized entity (Suna and Mayr, 2018).
4. **Analyzing Data for Identifying and Quantifying Protein:** The analysis of data for the identification and quantification of proteins is the next phase in the workflow. A peptide's m/z is measured using MS. By comparing the recorded Mass spectrometry spectra with a database through the use of software tools and algorithms for scoring, the real protein is unmistakably identified (Suna and Mayr, 2018).
5. **Validation:** The workflow's final phase comes after choosing the target proteins to be examined further. Realize that additional validation of proteomics data is frequently necessary. Typically, independent methods are used to validate the target proteins, such as immunohistochemistry or Western blotting (Suna and Mayr, 2018).

11.4.2 DATABASE FOR PROTEOMICS DATA

Databases play a crucial role in performing any study. Some databases that can be used to study proteomics of fenugreek are:

1. **LactMed® Database:** It provides information about medications and other compounds to which nursing mothers may be exposed. It offers details on the concentrations of these compounds in breast milk and infant blood, as well as any potential negative effects on nursing infants. Where appropriate, suggested alternative therapies to these medications are provided. All data is sourced from the scientific literature, which is thoroughly cited. To ensure scientific

validity and accuracy, the data is reviewed by a peer review panel (Drugs and Lactation Database (LactMed®), 2006).

2. **UniProt:** This database provides protein sequence and functional data. It contains a substantial amount of information from the scientific literature regarding the biological roles of proteins (https://www.uniprot.org/).
3. **ExPASy**: Using the ProtParam tool on the ExPASy platform, various physical and chemical properties of a specific protein can be computed, whether the protein is present in SwissProt or TrEMBL, or even for a protein sequence entered by a user. This tool can also be used to predict the primary structure of a protein.
4. **LOMETS**: It is a local meta-threading server used for the rapid and automatic prediction of protein tertiary structures (3D) and spatial restraints (https://zhanggroup.org/LOMETS/).
5. **PSIPRED:** It is a fast and accurate secondary structure prediction method that utilizes two feed-forward neural networks to analyze PSI-BLAST (Position-Specific Iterated BLAST) data.
6. **Phyre2:** It predicts tertiary protein structures using HHsearch to align hidden Markov models, significantly improving alignment accuracy and detection rates. Additionally, it features a novel protein folding simulation method called Poing, which models protein regions with no detectable similarity.
7. **PRIDE:** Mass spectrometry (MS)-based proteomics data are stored in the PRIDE (PRoteomics IDEntifications) database, an open-source data repository. The PRIDE database is a structured repository that holds researchers' original experimental data without applying any editorial oversight.

11.4.3 CASE STUDIES WHERE FENUGREEK WAS STUDIED USING PROTEOMICS

1. Seeds from fenugreek are known to possess extremely strong antioxidant activity after phytochemical screening and rudimentarily extracting tannins, saponins, steroid, and cardiac glycosides. The top three phytochemicals (tannins, saponins, and steroid) that were chosen against AKT-1 based on their potent energy of binding

following virtual screening hits. Out of all the compounds, tannins have the best pharmacophore fit score. AKT-1's binding pockets were found to contain some important residues (Glutamine, Tryptophan, Threonine, Valine, Tyrosine, Aspartic Acid) during molecular interactions. It was found that the protein's structure underwent some major conformational changes as the ligand connected to the protein, according to MD simulations performed at 100 ns. These three compounds block the "AKT-1 pathway," which is associated with the inflammatory activity of the cancer cell (Ahmad et al., 2022).

2. Most often, fenugreek allergy is linked to peanut allergy. By consuming meals containing fenugreek, peanut allergic people may get sensitized to it. This is likely due to substantial cross-reactivity, which suggests that both legume plants have homologous IgE-binding epitopes. The identification of potential fenugreek (*T. foenum-graecum*) allergens was done utilizing proteomic and IgE-binding analyses. Significant similarities to the main peanut allergens may be the cause of patients with peanut allergies developing secondary fenugreek allergies due to cross-reactivity (Fæste et al., 2010).

3. A study revealed that fenugreek extract has specific cytotoxic effects *in vitro* on a variety of cancer cell lines, as well as T-cell lymphoma. The protein profile of the specific fenugreek extract utilized differs considerably from other regional sub-types of fenugreek extract, according to the cluster analysis of proteomics data. This implies Fenugreek may be effective in the better preventing and treating of cancer due to its *in vitro* impact as a compound with considerable cyto-toxicity to cancer cells (Alsemari et al., 2014).

4. In a study, the plants of fenugreek were exposed to "methyl Jasmonates," which served as a trigger, and cholesterol, which is a precursor to "steroids and steroidal saponins," in order to assess the proteome's response to stress and evaluate the manufacture of steroidal saponin (diosgenin). There were found to be higher levels of photosynthetically sensitive proteins to both biotic and abiotic stimulus. Additionally, there was an overexpression of proteins involved in stress (particularly oxidative stress) and defense, protein synthesis, and secondary metabolism. According to the proteomic analysis findings, methyl Jasmonates and cholesterol caused a

proteome-level defense response in response to stress. These findings offered fresh perspectives on the proteome of fenugreek's reaction to treatment with methyl Jasmonates and cholesterol (Ciura et al., 2017).

5. Numerous researches have shown that fenugreek extract possesses antioxidant, antiradical, and anti-inflammatory qualities but its anti-aging aspects remain unexplored. In 2022, a study was conducted which showed promising anti-aging capabilities of fenugreek extract and observed Rutin to be the biological marker. Rutin is said to contain several beneficial properties including "antioxidant, cardio-protection, neuroprotection, anticarcinogenic" abilities (Figure 11.8; Eaknai et al., 2022).

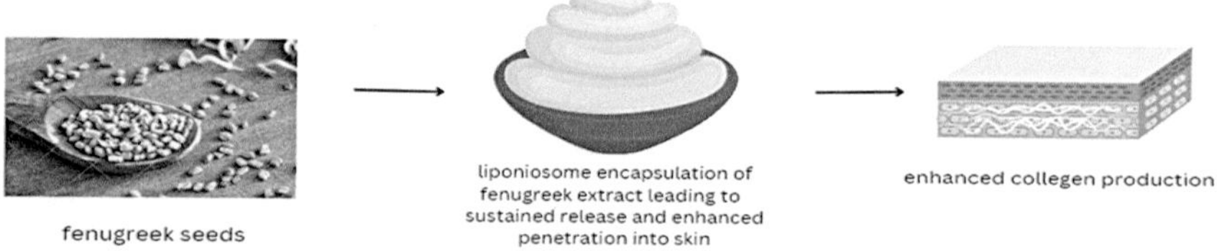

FIGURE 11.8 A diagrammatic representation showing mechanism for fenugreek's promising antiaging capabilities.

11.4.4 APPLICATIONS OF PROTEOMICS

1. **Understanding Molecular Mechanisms:** The molecular mechanisms behind the effects of fenugreek can be investigated using proteomics techniques. Researchers can find variations in protein expression, adjustments in protein interactions, or interactions between proteins by comparing the proteome of cells or tissues before and after fenugreek therapy. This knowledge can shed light on the cellular processes that fenugreek chemicals affect.

2. **Identification of Bioactive Proteins:** Proteomics can be used to locate and describe the proteins in fenugreek that may be involved in the herb's therapeutic effects. Researchers can pinpoint particular proteins with possible bioactive qualities, such as enzymes, antimicrobial peptides, or proteins implicated in antioxidant pathways, by studying the fenugreek proteome.

3. **Biotic Stress Response:** Recent developments in stress proteomics have made it possible for plant biologists to research the molecular processes involved in stress adaptation (Rathi et al., 2016). The study of biological systems and processes in higher plants can be effectively accomplished by proteomics. Identifying those proteins that are relevant to a pathogen's resistance is made possible by revealing the proteomes of plants and disease-causing agents in tissues that are infected. The use of proteomic analysis will also make it easier to understand how the proteome changes in reaction to various biotic environmental challenges in fruit crops (Al-Amrani et al., 2021).
4. **Discovering Biomarkers:** A compound which can indicate whether the body is processing normally or not is known as a biomarker. Discovery of biomarkers utilizes "2D-PAGE." Additionally, it might give a contrast of the protein patterns of healthy/normal and diseased cells, including tissues from the tumor and body fluids (Al-Amrani et al., 2021).
5. **Systems Biology:** Systematic studies of medication action, toxicity, resistance, and efficacy; systemic studies of disease mechanisms and relations between host and pathogen to identify prospective biomarkers and therapeutic targets.
6. **Agriculture:** Studies focus on the interactions between plants and pathogens, as well as crop engineering to enhance resistance to environmental challenges such as flooding and drought.

11.4.5 CHALLENGES IN PROTEOMICS

Proteomics and "tandem mass spectrometry" for identifying proteins are severely hampered by the fact that many alternative splice isoforms are excluded from or only included in text comments in protein databases. Proteomic research might not be able to find actual products of protein of different isoforms since the target-sequence wasn't included in the database that was being searched. The lack of authentic protein sequences in the sequence library may also lead to erroneous peptide and protein identification since degenerate peptides are included that correspond to several similar proteins.

When concerning a long protein list that has been found in various areas, a "lack of standardization" is another prevalent problem. Different proteinIDs and different names could be utilized for the very same proteins if an additional database labeled as target is used for "MS/MS" protein searching. For the interpretation of proteomic data in biological terms, it is laborious to integrate annotations from several diverse sources because there aren't any universal protein identities and naming guidelines. Because of this, the extensive annotations found in many different information resources are frequently used inefficiently and fragmentarily during expression data processing (Hongzhan et al., 2007).

11.5 METABOLOMICS IN FENUGREEK

Metabolomics is a scientific discipline that investigates the composition and behavior of minute and large chemical entities, commonly referred to as metabolites, present in cells, and tissues and also secreted in biofluids, of all organisms undergoing metabolism. The metabolome refers to the entirety of macromolecules and their interplay within a biological environment (Eaknai et al., 2022). Metabolomics has evolved as an excellent approach for comprehending the molecular underpinnings of health and disease. Through the analysis of the metabolome, scholars can acquire a deeper understanding of the fundamental biochemical mechanisms that contribute to illnesses and can devise novel diagnostic methodologies and therapeutic interventions. Metabolomics has been a non-invasive approach that enables the identification and quantification of metabolites in living organisms. Hence this method was used for the exploration of large numbers of samples in a short duration with the maintenance of a high level of resolution.

11.5.1 METABOLITE PROFILING OF VARIOUS METABOLITES OF FENUGREEK

Metabolomics has evolved as an efficient tool in fenugreek research, providing valuable insights into its chemical composition, bioactive compounds, and metabolic pathways. Several metabolites of fenugreek have been examined in various scholarly works, including Trigonelline, Flavonol glycosides, Diosgenin, Galactomannan, and 4-Hydroxyisoleucine. In their research, Riasat et al. (2018) conducted an analysis of multiple metabolites, such as Coumarin, Hesperetin, Hesperidin, Catechin, Trans-ferulic acid, p-Coumaric acid, Quercetin, Chlorogenic acid, and Rutin, in various fenugreek samples obtained from different geographical regions. Figure 11.9 depicts several metabolites resulting from the metabolism of Fenugreek.

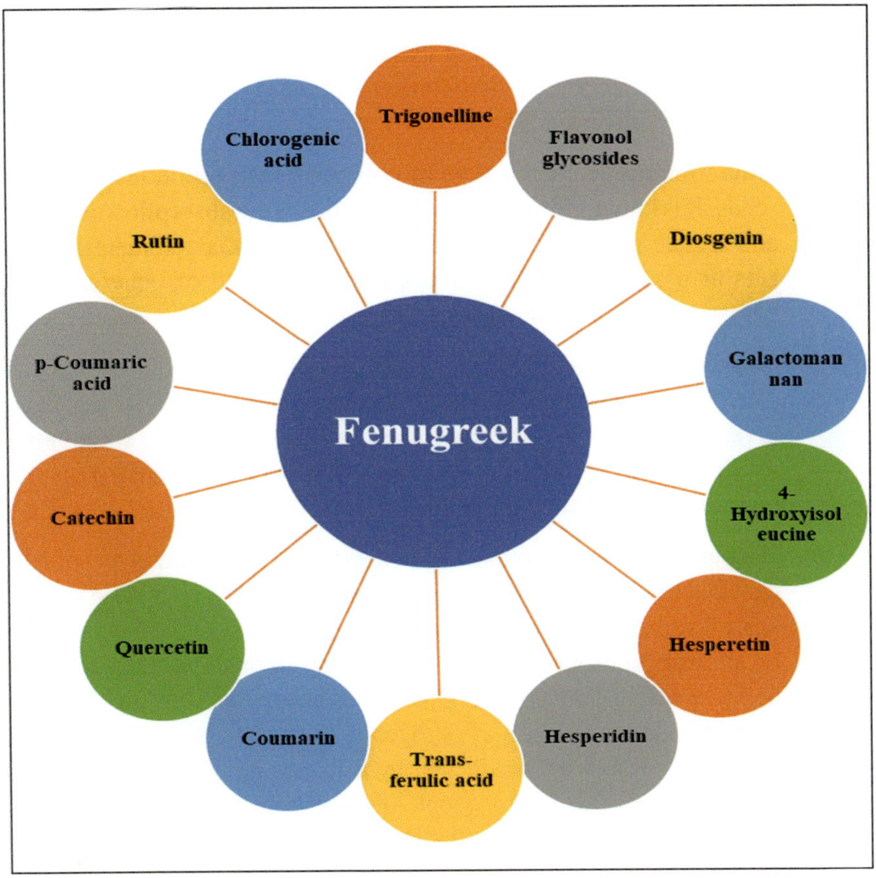

FIGURE 11.9 Illustration of several metabolites of fenugreek metabolism.

1. **Trigonelline:** It is an alkaloid that is found in high concentrations in fenugreek seeds. It has proven to have several beneficial applications, including lowering blood sugar levels by reducing glycosuria in diabetes, reducing inflammation, and improving cognitive function (Srinivasan, 2006).
2. **Flavonol Glycosides:** They are a type of flavonoid that is found in fenugreek. Flavonoids are antioxidants that have exhibited several health benefits, including protecting against cancer, heart disease, and stroke.
3. **Diosgenin:** It is a steroidal saponin found in fenugreek plant seeds. It has been researched for its potential anti-cancer, anti-inflammatory, and anti-diabetic effects. It is also a precursor for the production of

various steroidal hormones (Sharma et al., 1990). It is also used as an estrogen precursor that is used in managing menopause. It also increases the secretory functions and induces uterine contraction in females (Srinivasan, 2006).

4. **Galactomannan:** It is a type of polysaccharide found in fenugreek seeds. It is known for its ability to slow down the absorption of sugars and reduce postprandial blood glucose levels. Galactomannan also acts as a prebiotic and also helps in the growth of beneficial gut bacteria (Sharma et al., 1990).
5. **4-Hydroxyisoleucine:** The amino acid derivative known as 4-hydroxyisoleucine is found in the fenugreek seeds that has been investigated for its potential anti-diabetic effects by stimulating insulin secretion, improving glucose utilization, and reducing insulin resistance (Hannan et al., 2003).
6. **Hesperetin:** It is a flavonoid that possesses the properties of an antioxidant that has several health benefits, including protection against cancer, heart diseases, and strokes. It also has the properties of phytoestrogen, which can be beneficial for women going through menopause, as it helps in easing the symptoms including hot flashes and night sweats by mimicking the effects of estrogen in the body. Hesperetin can also help to increase urine production by acting as a natural diuretic. This flavonoid is also helpful for people who are trying to lose weight or who have a medical condition that requires them to reduce their fluid intake.
7. **Hesperidin:** It is a type of flavonoid glycoside found in many fruits and vegetables, including fenugreek. It is a powerful antioxidant that protects the cell damage caused due to free radicals. Hesperidin also has proven anti-inflammatory, anti-diabetic, and cardiovascular benefits. Although the amount of hesperidin in fenugreek varies in different varieties of fenugreek and environmental conditions, a typical serving of fenugreek seeds or leaves may contain around 100 milligrams of hesperidin.
8. **Trans-Ferulic Acid:** It is a type of antioxidant that protects cells from the damage caused due to free radicals and also promotes the production of other Antioxidants like Vitamin A, Vitamin C, and Vitamin E. Other therapeutic application of Trans-ferulic acid includes Lowering blood sugar level in diabetic patients by stimulating the production of Insulin, reducing inflammation by its

ability to scavenge free radicals, and improving cognitive function in patients suffering from various dementias including Alzheimer's disease. Trans-ferulic acid also helps in boosting the immune system by promoting the production of White Blood cells and inhibiting the growth and development of cancer cells.

9. **Coumarin:** They are derivatives of Benzo-α-pyrone belonging to the class of phytochemicals known as coumarins. These compounds show anticoagulating activity by inhibiting vitamin K epoxide reductase, leading to the depletion of active vitamin K and interfering with the synthesis of clotting factor, it inhibits the production of pro-inflammatory cytokines and enzymes, including tumor necrosis factor-alpha (TNF-α) and cyclooxygenase-2 (COX-2), and modulate immune responses and possess anti-inflammatory activity. Coumarin compounds exhibit antimicrobial and antifungal activities against a vast pool of microorganisms. They can inhibit the growth and viability of bacteria, fungi, and even certain drug-resistant strains. Coumarins may disrupt microbial cell membranes or interfere with essential microbial enzymes.

10. **Quercetin:** It is a flavonoid compound found in various natural sources like fruits and vegetables and plants. Quercetin inhibits the release of inflammatory mediators, such as cytokines and histamines and hence exhibit potential anti-inflammatory properties. It can help in the reduction of inflammation during chronic diseased conditions like asthma, arthritis, and inflammatory bowel disease. It also acts as a powerful antioxidant, protecting cells and tissues from oxidative stress and damage caused by free radicals. Its antioxidant activity contributes to its potential benefits in preventing or managing conditions associated with oxidative stress, including cardiovascular diseases, neurodegenerative disorders, and certain cancers. Quercetin also possesses anti-allergic properties and has been studied for its ability to reduce allergic reactions, such as hay fever, allergic asthma, and allergic rhinitis. It may help stabilize mast cells and inhibit the release of histamine, leading to decreased allergic symptoms. Quercetins have also shown potential neuroprotective effects, including reducing neuroinflammation, protecting against oxidative stress, and modulating signaling pathways involved in diseases like Alzheimer's disease and Parkinson's disease. However, more research is needed to establish its therapeutic potential in humans.

11. **Catechin:** It is a type of flavonoids have been studied for their potential role in weight management. They may help increase fat oxidation, enhance thermogenesis (calorie burning), and reduce fat absorption, thereby assisting in weight loss and weight maintenance. Catechins have also shown potential benefits in maintaining oral health. They may help inhibit the growth of oral bacteria, reduce plaque formation, and prevent tooth decay and gum disease. They are also known for their antimicrobial and anti-inflammatory properties.

12. **p-Coumaric acid:** It has been investigated for its hepatoprotective properties which may help in protecting the liver from damage caused by toxins and oxidative stress. It may also support liver health by modulating enzymes involved in detoxification processes. p-Coumaric acid is also studied for its potential benefits in maintaining skin health including anti-aging properties by reducing oxidative stress and protecting against UV-induced damage. It also helps in the management of skin disorders like dermatitis and psoriasis due to its anti-inflammatory effects.

13. **Rutin:** It is also known as quercetin-3-O-rutinoside or rutoside, is a flavonoid glycoside found in fenugreek seeds and many other natural sources like fruits and vegetables. Rutin has been studied for its ability to strengthen capillaries and blood vessels which can help reduce capillary permeability and fragility, which is beneficial in conditions like chronic venous insufficiency, varicose veins, and hemorrhoids. Rutin has shown potential benefits for cardiovascular health that help in improving endothelial function, reduce LDL cholesterol oxidation, inhibit platelet aggregation, and enhance blood vessel integrity. These effects contribute to its potential in the reduction of the risks associated with cardiovascular diseases including heart disease and strokes. Rutin helps to strengthen and make blood vessels more elastic, which can help to improve blood circulation. This can be beneficial for people with conditions such as varicose veins, hemorrhoids, and high blood pressure. Rutin also helps in the protection of eyes from damage caused due to the action of free radicals. It also helps in reducing the risks of age-related macular degeneration and cataracts.

14. **Chlorogenic Acid:** It is a polyphenolic compound present in many natural sources including plants. The chlorogenic acid possesses several therapeutic properties, Chlorogenic acid exhibits strong antioxidant properties, which neutralizes free radicals and provide protection to cells against oxidative stress. Chlorogenic acid also possesses anti-inflammatory properties by inhibiting the production of mediators, such as cytokines and enzymes involved in the inflammatory response.

Different metabolites along with their chemical structures have been shown in Table 11.2.

TABLE 11.2 Fenugreek Metabolites Along with Their Chemical Structures

Metabolite	Chemical Structure	Metabolite	Chemical Structure
Trigonelline	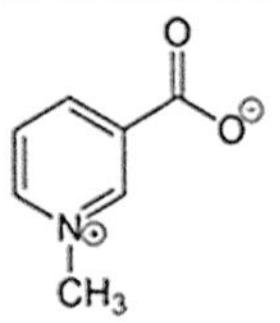	4-Hydroxyisoleucine	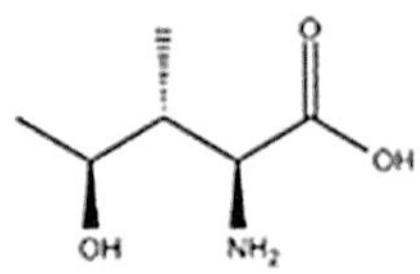
Flavonol glycosides	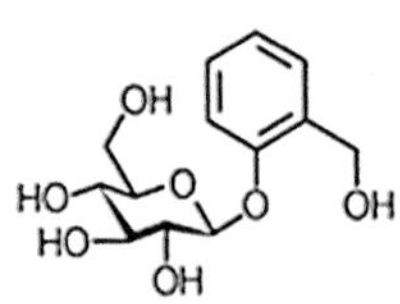	Hesperetin	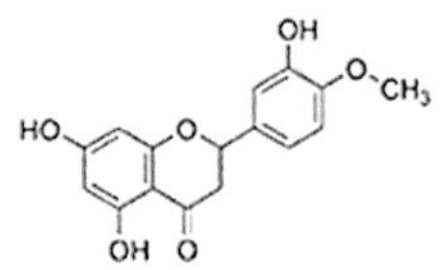
Diosgenin	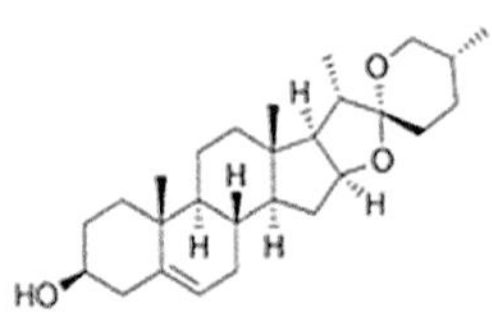	Hesperidin	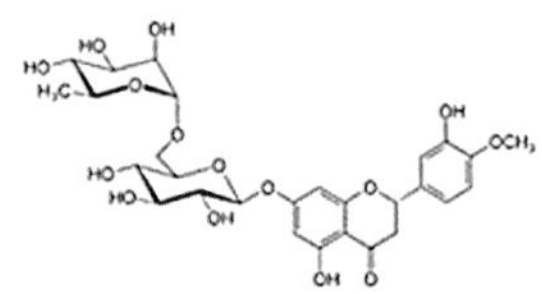
Galactomannan	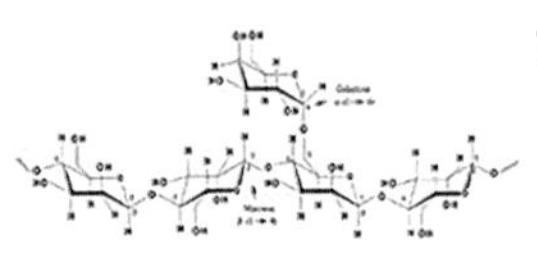	Trans-ferulic acid	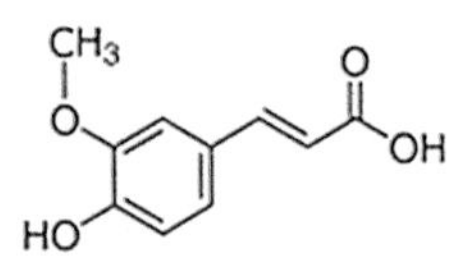
Coumarin	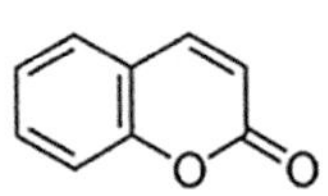	Quercetin	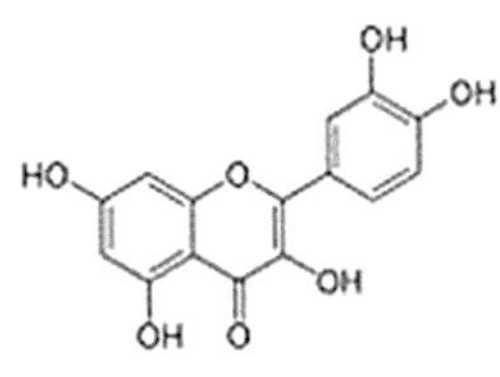

TABLE 11.2 *(Continued)*

Metabolite	Chemical Structure	Metabolite	Chemical Structure
Catechin	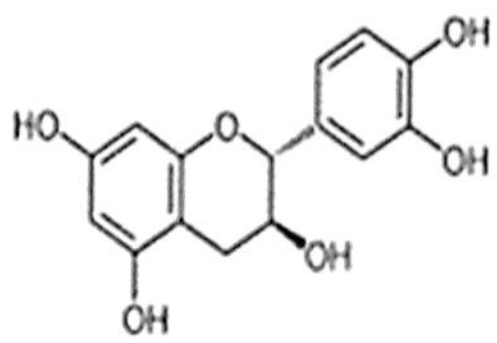	p-Coumaric acid	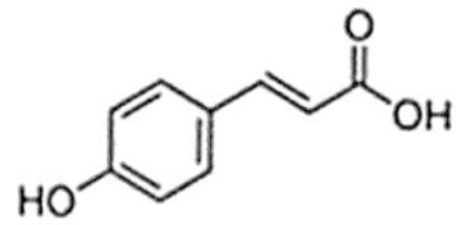
Rutin	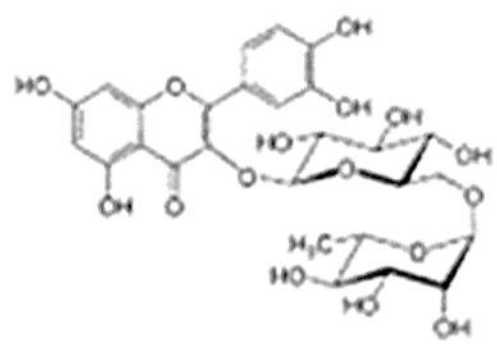	Chlorogenic acid	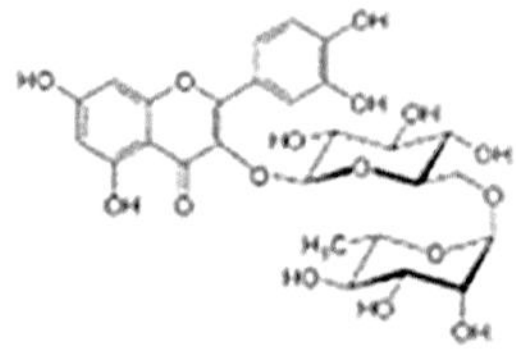

11.5.2 METABOLITE EXTRACTION TECHNIQUES AND INSTRUMENTATION

Different biotechnological extraction and analysis techniques and equipment may be used for the extraction, qualification, and quantification of different metabolites of fenugreek for the metabolomic analysis. A few of the known techniques used for quality control analyses were studied (Kandil et al., 2022). Figure 11.10 illustrates the multiple biotechnological approaches used in the extraction and analysis of metabolite data for the metabolomic studies.

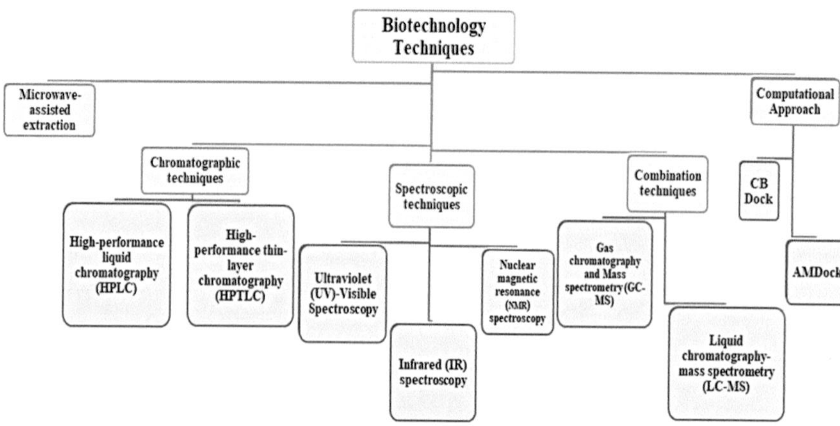

FIGURE 11.10 Biotechnological approaches for metabolite data analysis.

1. **Microwave-Assisted Extraction (MAE):** This method is used for extracting compounds from solid matrices using microwave energy. MAE is a relatively young technique, but due to its increased efficiency and speed, it has quickly become a popular extraction method. It is often used to extract phenolic compounds, flavonoids, and other bioactive compounds from plants. MAE was employed for the extraction and quantification of phenolic and saponin compounds from fenugreek seeds (Akbari et al., 2019).

2. **Chromatographic Techniques:**

   i. **High-Performance Liquid Chromatography (HPLC):** This technique is used for the separation and identification of micro- and macromolecules in a mixture. It is a powerful tool for the extraction and analysis of fenugreek metabolites. HPLC can be used to identify and quantify a wide range of compounds in fenugreek, including flavonoids, alkaloids, and saponins. This information can be used to study the health benefits of fenugreek and develop new fenugreek-based products. HPLC was used for the quantification and analysis of several phenolic and non-phenolic secondary metabolites in *Trigonella foenum-graecum* (herbarium file – 37527) in the study by Riasat et al. (2018). The results of HPLC are accurate and applicable to a broad range of compounds, including flavonoids, alkaloids, and saponins. Additionally, HPLC provides fast and reliable results.

   ii. **High-Performance Thin-Layer Chromatography (HPTLC):** This technique is used for the separation and identification of compounds in mixtures. HPTLC is a more sensitive and accurate method than traditional thin-layer chromatography (TLC). The results obtained from HPTLC can be used for identifying different metabolites in fenugreek extract. They can also be used to separate and quantify the amount of each metabolite in the extract. HPTLC is proven to be a highly sensitive, precise, and efficient analytical technique for the separation of fenugreek metabolites.

3. **Spectroscopic Techniques:**

   i. **Ultraviolet-Visible Spectroscopy:** This technique uses ultraviolet and visible light to measure the absorbance of test and reference extracts. It is used to identify and quantify a large group of metabolites in fenugreek extract by comparing the absorbance of the extract to that of standard solutions. Metabolites such as flavonol glycosides, saponins, and triterpenoids can be identified and quantified using UV spectroscopy. The study by Sahar Idris et al. (2021) used UV-visible spectroscopy to estimate rutin and quercetin from fenugreek extract.

ii. **Infrared (IR) Spectroscopy:** Specifically known as Fourier-transform infrared (FTIR) spectroscopy, this technique is impressive for analyzing chemical compounds, including metabolites, in fenugreek and other plant extracts. It allows for the identification and characterization of various functional groups present in plant extracts and is widely employed for the qualitative and quantitative analysis of metabolites. The study by Sweeta Akbari et al. (2019) used FTIR analysis to evaluate the chemical composition of fenugreek oil.

iii. **Nuclear Magnetic Resonance (NMR) Spectroscopy:** This technique is used to determine and analyze the structure, dynamics, chemical environment, and composition of metabolites in fenugreek and other natural products. NMR provides several advantages over other analytical techniques, including non-destructive sampling, higher sensitivity, and increased specificity. The study by Graziani et al. (2018) applied a rapid NMR-based approach to identify secondary metabolites in plant extracts.

4. **Combination Techniques:** Various combination techniques can be used for more accurate results and higher levels of quantification and analysis.

   i. **Gas Chromatography-Mass Spectrometry (GC-MS):** This technique is used for identifying and quantifying metabolites in biological samples. GC-MS operates on the principle that different compounds vaporize and travel through a column at different rates, depending on their chemical structure. The mass spectrometer detects the compounds and determines their mass-to-charge ratio, which can be used as an identifier to study metabolites through various online databases, such as Metabolic Workbench. The study by Henda Keskes et al. (2018) utilized this technique for analyzing metabolites in fenugreek. The study by Sweeta Akbari et al. used GC-MS for the extraction of chemical compounds from fenugreek oil.

ii. **Liquid Chromatography-Mass Spectrometry (LC-MS):** This powerful analytical technique combines the physical separation capability of liquid chromatography (LC) with the mass-analyzing property of mass spectrometry (MS). LC-MS is a versatile method used to analyze large numbers of samples. It is a highly sensitive and specific technique, capable of detecting even low levels of analytes. LC-MS is also a quantitative method, allowing for the precise quantification of analyte concentrations in a sample. The quantitative results obtained by LC-MS are highly sensitive, specific, versatile, and robust. The study by Sahar Idris et al. (2021) used LC-MS to identify secondary metabolites in fenugreek seeds.

5. **Computational Approach**: Computational tools like CB dock and AMDock can also be employed for the analysis and quantification of several metabolites in fenugreek. A study by Maaz et al. (2022) used these tools to identify anti-inflammatory metabolites, including alkaloids, 4-hydroxy isoleucine, sapogenins, and others.

11.5.3 DATABASES FOR METABOLOMIC DATA

Online Databases have a vital role in understanding and storing the data of metabolic profiles and applications of various metabolites of *Trigonella foenum-graecum* and other plant products. One such database that may be used for the extraction and analyzing the metabolic data of *Trigonella foenum-graecum* is Phytochemicals.info.

Phytochemicals.info (https://phytochemicals.info/)—this database provides comprehensive information on various phytochemicals, including metabolites found in plants. It covers a wide variety of compounds and their properties, along with the metabolic data of *Trigonella foenum-graecum.*

Recent advancements in metabolomics, such as multi-omics integration and data integration approaches, have enhanced our understanding of fenugreek metabolites and their biological activities. However, challenges remain, including standardization of analytical techniques, data analysis, and the need for comprehensive metabolite databases specific to fenugreek. There is a need for the building and updating more databases with the details of plant omics data like the *Trigonella foenum-graecum* organism.

11.5.4 APPLICATION OF METABOLOMICS IN FENUGREEK STUDY

The study of Metabolites known as metabolomics has evolved to play a vital role for the study and research of the Fenugreek. Some of the applications of the metabolomics may be listed as:

1. **Metabolite Profiling:** Metabolomics can be used to profile and quantify the metabolites present in fenugreek. This provides insights into the chemical composition and diversity of fenugreek seeds, leaves, or other plant parts. Metabolite profiling may be helpful in identifying and quantifying bioactive compounds, including alkaloids, flavonoids, saponins, and other phenolic compounds, which contribute to the nutritional and medicinal properties of fenugreek.
2. **Quality Control and Authentication:** Metabolomics study may be utilized for quality control and authentication of fenugreek products. By analyzing the metabolite profiles, variations in chemical composition and the presence of adulterants or contaminants can be detected. Metabolomics-based fingerprinting techniques, such as pattern recognition and chemometrics, can help distinguish genuine fenugreek samples from adulterated or low-quality ones.
3. **Understanding Metabolic Pathways:** Metabolomics allows researchers to study the metabolic pathways and biochemical transformations occurring in fenugreek. By profiling metabolites and their changes under different conditions (e.g., growth stages, stress conditions, or post-harvest treatments), metabolomics provides insights into the biosynthesis, degradation, and regulation of important compounds in fenugreek. This information can aid in understanding the biological processes and regulatory mechanisms involved in fenugreek metabolism.
4. **Functional Genomics and Gene Discovery:** Metabolomics can complement genomics and transcriptomics approaches in fenugreek research. By integrating metabolomic data with genomic and transcriptomic data, correlations between gene expression and metabolite accumulation can be identified. This can lead to the discovery of key genes involved in fenugreek metabolism and the biosynthesis of bioactive compounds. Metabolomics can also help in elucidating the functional roles of specific genes and pathways in fenugreek.

5. **Nutritional and Health Studies:** Metabolomics enables the assessment of nutritional quality and health-related properties of fenugreek. By profiling metabolites, researchers can identify and quantify important nutrients, such as vitamins, minerals, amino acids, and fatty acids. Metabolomics can also provide insights into the bioavailability and metabolism of fenugreek-derived compounds in the human body, contributing to studies on its potential health benefits and therapeutic applications.
6. **Metabolomics-Assisted Breeding:** Metabolomics can support fenugreek breeding programs by identifying and selecting plants with desired metabolite profiles. By comparing metabolomic profiles of different fenugreek genotypes or breeding lines, metabolomics can aid in identifying markers associated with specific metabolites of interest, including those related to flavor, aroma, or medicinal properties. This information can guide the development of improved fenugreek varieties with enhanced nutritional or pharmaceutical characteristics.

11.6 FUTURE DIRECTIONS

1. **Advanced OMICS Technologies:** The development and refinement of next-generation OMICS technologies, such as multi-omics integration and artificial intelligence-based analysis, could provide deeper insights into the molecular mechanisms of fenugreek's pharmacological effects.
2. **Elucidating the Role of PGRs and Depolymerized Polysaccharides:** Future studies could focus on exploring how plant growth regulators (PGRs) and depolymerized polysaccharides impact fenugreek yield, both in quantity and quality (phytochemical composition), at a molecular level.
3. **Genetic Modification and Bioengineering:** Applying genetic modification and bioengineering approaches to enhance fenugreek yield and increase its therapeutic potential could be another direction. This will require a thorough understanding of fenugreek's genomic structure and gene function, which can be achieved using OMICS technologies.

4. **Clinical Trials and Safety Evaluations:** Once the therapeutic benefits and mechanisms of fenugreek are well understood, it's essential to conduct comprehensive clinical trials to confirm these findings in humans. It is also important to study potential side effects or adverse reactions to ensure safety.
5. **Integration with Traditional Knowledge:** Finally, integrating modern scientific discoveries with traditional knowledge about fenugreek can lead to innovative applications in medicine and agriculture. This is an area that has not been fully explored and could offer promising potential for the future.

11.7 CONCLUSION

In conclusion, our in-depth exploration of the intersection between OMICS technologies and the cultivation of fenugreek provides compelling evidence that both Plant Growth Regulators (PGRs) and depolymerized polysaccharides can significantly improve yield potential. The deployment of genomics, proteomics, and metabolomics has allowed us to gain unparalleled insight into the biological processes of fenugreek at a molecular level, paving the way for precision agriculture and targeted interventions.

PGRs have emerged as a promising avenue to enhance growth, yield, and overall performance of fenugreek plants. Their regulatory roles in plant metabolic pathways, influencing various growth and development aspects, have made them instrumental in crop yield enhancement. Furthermore, the application of depolymerized polysaccharides exhibits a significant positive effect on fenugreek crop yield, corroborating their role in improving plant nutrition and overall health.

The integration of these elements into fenugreek cultivation practices can contribute to enhanced yield potential, meeting increasing demand, and ensuring food security. Future work should aim to refine the application of these interventions, considering potential ecological and economic impacts. Furthermore, it is worth investigating the possibility of incorporating other OMICS technologies, such as transcriptomics and phenomics, in order to acquire a more holistic comprehension of fenugreek's biology and its interactions with these interventions aimed at enhancing yield.

Ultimately, this chapter paves the way for harnessing advanced biotechnological tools and bioactive substances in sustainable fenugreek cultivation, optimizing yield while minimizing environmental impact. The findings provide a strong foundation for future research and practical applications in fenugreek farming and may be extrapolated to other crops, fostering a broader revolution in sustainable and efficient agricultural practices.

CONFLICT OF INTEREST

All the authors declare that they have no competing interests.

KEYWORDS

- **bioinformatics**
- **fenugreek**
- **genomics**
- **metabolites**
- **OMICS**
- **transcriptomics**

REFERENCES

Agaoglu, S., D'Onofrio, C., & Brandt, K., & Hawkesford, M. (2022). Plant genomics and proteomics. *Plant Biotechnology Journal, 20*(2), 302–316.

Ahmad, H. M., Abrar, M., Izhar, O., Zafar, I., Rather, M. A., Alanazi, A. M., Malik, A., Rauf, A., Bhat, M. A., Wani, T. A., & Khan, A. A. (2022). Characterization of fenugreek and its natural compounds targeting AKT-1 protein in cancer: Pharmacophore, virtual screening, and MD simulation techniques. *Journal of King Saud University–Science, 34*(6), 102186. https://doi.org/10.1016/J.JKSUS.2022.102186.

Ahmed, A., Yousuf, B., Anwar, S., Aslam, A., & Malik, A. U. (2020). Differential gene expression in high yielding and low yielding genotypes of fenugreek (*Trigonella foenum-graecum* L.). *Gene, 747*, 144649.

Ahsan, N., Mishra, A., Bhattacharya, A., & Ghosh, S. (2020). Transcriptomic analysis reveals drought-responsive genes and molecular pathways in fenugreek (*Trigonella foenum-graecum* L.). *Frontiers in Plant Science, 11*, 198.

Akbari, S., Abdurahman, N. H., Yunus, R. M., Alara, O. R., & Abayomi, O. O. (2019). Extraction, characterization, and antioxidant activity of fenugreek (*Trigonella foenum graecum*) seed oil. *Materials Science for Energy Technologies, 2*(2), 349–355. https://doi.org/10.1016/j.mset.2018.12.001.

Al-Amrani, S., Al-Jabri, Z., Al-Zaabi, A., Alshekaili, J., & Al-Khabori, M. (2021). Proteomics: Concepts and applications in human medicine. *World Journal of Biological Chemistry, 12*(5), 57. https://doi.org/10.4331/WJBC.V12.I5.57.

Alsemari, A., Alkhodairy, F., Aldakan, A., Al-Mohanna, M., Bahoush, E., Shinwari, Z., & Alaiya, A. (2014). The selective cytotoxic anti-cancer properties and proteomic analysis of *Trigonella Foenum-Graecum*. *BMC Complementary and Alternative Medicine, 14*, 114. https://doi.org/10.1186/1472-6882-14-114.

Bertone, P., & Snyder, M. (2005). Prospects and challenges in proteomics. *Plant Physiology, 138*(2), 560–562. https://doi.org/10.1104/PP.104.900154.

Bewley, J. D., & Black, M. (1994). Seed development and maturation. In *Seeds: Physiology of Development and Germination* (pp. 35–115).

Bhatia, R., Singh, K. P., Roy, R., & Agarwal, P. K. (2018). Transcriptome analysis of flower development in fenugreek (*Trigonella foenum-graecum* L.). *Scientific Reports, 8*(1), 14266.

Bhattacharya, A., Saha, G., & Ghosh, S. (2016). Transcriptomic analysis of fenugreek reveals genes involved in saponin biosynthesis. *Planta, 244*(3), 515–531.

Branch, S. (2013). Fenugreek (*Trigonella foenum-graecum* L.) as a valuable medicinal plant. *International Journal of Advanced Biological and Biomedical Research, 1*, 922–931.

Buckeridge, M. S., Dietrich, S. M. C., & De Lima, D. U. (2000). Galactomannans as the reserve carbohydrate in legume seeds. In *Developments in Crop Science* (Vol. 26, pp. 283–316). Elsevier.

Chaudhary, S., Chaudhary, P., & Patel, S. (2021). Genomics, transcriptomics, proteomics, and metabolomics approaches. In *Fenugreek: Biology and Applications* (pp. 355–373).

Chaudhary, S., Chaudhary, P., & Patel, S. (2021). Genomics, transcriptomics, proteomics and metabolomics approaches. In *Fenugreek: Biology and Applications* (pp. 355–373).

Chittora, V., Gupta, N., Sharma, P., & Singh, P. (2020). Genome-wide association study reveals genetic architecture of seed yield and yield component traits in fenugreek (*Trigonella foenum-graecum* L.). *PLOS One, 15*(9), e023.

Choudhary, V., Gangwar, H., & Jaiswal, V. (2023). Application of transcriptomics in exploring important genes in medicinal plants. In *Phytochemical Genomics: Plant Metabolomics and Medicinal Plant Genomics* (pp. 263–284). Singapore: Springer Nature Singapore.

Ciura, J., Bocian, A., Kononiuk, A., Szeliga, M., Jaromin, M., & Tyrka, M. (2017). Proteomic signature of fenugreek treated by methyl jasmonate and cholesterol. *Acta Physiologiae Plantarum, 39*(5), 1–14. https://doi.org/10.1007/S11738-017-2416-7.

Dar, T. A., & Uddin, M. (2018). Fenugreek: A miraculous medicinal herb. *Journal of Complementary Medicine & Alternative Healthcare, 7*(2), 1–3.

Das, A., Singh, V., Yadav, U., Gupta, A., & Patel, S. (2022). Integrating transcriptomics and metabolomics for understanding bioactive compound biosynthesis in fenugreek. *Planta Medica, 88*(5), 375–389.

Dong, Z., & Chen, Y. (2013). Transcriptomics: Advances and approaches. *Science China Life Sciences, 56*, 960–967.

Eaknai, W., Bunwatcharaphansakun, P., Phungbun, C., Jantimaporn, A., Chaisri, S., Boonrungsiman, S., Nimmannit, U., & Khongkow, M. (2022). Ethanolic fenugreek extract: Its molecular mechanisms against skin aging and the enhanced functions by nanoencapsulation. *Pharmaceuticals, 15*(2), 254. https://doi.org/10.3390/PH15020254.

Evangelistella, C., Valentini, A., Ludovisi, R., Firrincieli, A., Fabbrini, F., Scalabrin, S., Cattonaro, F., Morgante, M., Mugnozza, G. S., Keurentjes, J. J., & Harfouche, A. (2017). De novo assembly, functional annotation, and analysis of the giant reed (*Arundo donax* L.) leaf transcriptome provide tools for the development of a biofuel feedstock. *Biotechnology for Biofuels, 10*, 1–24.

Fæste, C. K., Christians, U., Egaas, E., & Jonscher, K. R. (2010). Characterization of potential allergens in fenugreek (*Trigonella foenum-graecum*) using patient sera and MS-based proteomic analysis. *Journal of Proteomics, 73*(7), 1321–1333. https://doi.org/10.1016/J.JPROT.2010.02.011.

Ghawana, S., Paul, A., Kumar, H., Kumar, A., Singh, H., Bhardwaj, P. K., & Kumar, S. (2011). An RNA isolation system for plant tissues rich in secondary metabolites. *BMC Research Notes, 4*(1), 85.

Graziani, V., Scognamiglio, M., Belli, V., Esposito, A., Chambery, A., Russo, R., Panella, M., Russo, A., Ciardiello, F., Troiani, T., Potenza, N., & Fiorentino, A. (2018). Metabolomic approach for a rapid identification of natural products with cytotoxic activity against human colorectal cancer cells. *Scientific Reports, 8*(1), 1–11. https://doi.org/10.1038/s41598-018-23704-9.

Guenin, S., Mauriat, M., Pelloux, J., Van Wuytswinkel, O., Bellini, C., & Gutierrez, L. (2009). Normalization of qRT-PCR data: The necessity of adopting a systematic, experimental conditions-specific, validation of references. *Journal of Experimental Botany, 60*(2), 487–493.

Gupta, A., Mishra, A. P., Kumar, S., & Pandey, A. K. (2020). Depolymerized fenugreek polysaccharides improve drought tolerance by modulating antioxidant defense system and osmolyte accumulation in *Trigonella foenum-graecum* L. plants. *Plants, 9*(1), 68.

Gupta, A., Sharma, T., Singh, S. P., Bhardwaj, A., Srivastava, D., & Kumar, R. (2023). Prospects of microgreens as budding living functional food: Breeding and biofortification through OMICS and other approaches for nutritional security. *Frontiers in Genetics, 14*.

Gupta, P., Das, A., Singh, S., & Srivastava, S. (2022). Genomic insights into the medicinal properties of fenugreek. *Trends in Plant Science, 27*(1), 75–89.

Gupta, S., Kumar, R., Singh, P., Pandey, V., & Sharma, A. (2019). Genomic resources for functional characterization of *Trigonella foenum-graecum* L. plant breeding. *Molecular Biology Reports, 46*(4), 4475–4489.

Hannan, J. M., Rokeya, B., Faruque, O., Nahar, N., Mosihuzzaman, M., & Azad Khan, A. K. (2003). Effect of soluble dietary fiber fraction of *Trigonella foenum graecum* on glycemic, insulinemic, lipidemic and platelet aggregation status of type 2 diabetic model rats. *Journal of Ethnopharmacology, 88*(1), 73–77.

Hidangmayum, A., Dwivedi, P., Katiyar, D., & Hemantaranjan, A. (2019). Application of chitosan on plant responses with special reference to abiotic stress. *Physiology and Molecular Biology of Plants, 25*, 313–326.

Hongzhan, H., Shukla, H. D., Cathy, W., & Satya, S. (2007). Challenges and solutions in proteomics. *Current Genomics, 8*(1), 21. https://doi.org/10.2174/138920207780076910.

Idris, S., Mishra, A., & Khushtar, M. (2021). Phytochemical estimation of germinated *Trigonella foenum-graecum* L. seed extract for better application in phytotherapy. *Annals of Phytomedicine, 10*(2), 213–222. https://doi.org/10.21276/ap.2021.10.2.301.

Jasim, A. T., Al-Dujaili, A. H., Abdulrahman, W. A., & Salman, J. M. (2019). The effect of gibberellic acid on growth and yield of fenugreek (*Trigonella foenum-graecum* L.) in field conditions. *Journal of the Saudi Society of Agricultural Sciences, 18*(3), 336–341.

Jha, U. C., Bohra, A., Singh, N. P., & Mishra, R. K. (2019). RNA-Seq analysis revealed genes associated with drought stress response in kabuli and desi chickpea (*Cicer arietinum* L.). *Functional & Integrative Genomics, 19*(1), 65–80.

Jia, X., Sun, C., Zuo, Y., Li, G., Ren, L., & Chen, G. (2016). Integrating transcriptomics and metabolomics to characterize the response of *Astragalus membranaceus* Bge. var. mongolicus (Bge.) to progressive drought stress. *BMC Genomics, 17*, 1–16.

Kakani, R. K., Singh, S. K., Pancholy, A., Meena, R. S., Pathak, R., & Raturi, A. (2011). Assessment of genetic diversity in *Trigonella foenum-graecum* based on nuclear ribosomal DNA, internal transcribed spacer, and RAPD analysis. *Plant Molecular Biology Reporter, 29*, 315–323.

Kandil, N. H., Ayoub, I. M., El-Ahmady, S. H., & El-Moghazy, S. A. (2022). Advances in the quality control of fenugreek seeds using chromatographic, spectroscopic, and DNA-based techniques: A comprehensive review. *Phytochemical Analysis: PCA, 33*(2), 155–169. https://doi.org/10.1002/pca.30911.

Karczewski, K. J., & Snyder, M. P. (2018). Integrative omics for health and disease. *Nature Reviews Genetics, 19*(5), 299–310.

Keskes, H., Belhadj, S., Jlail, L., El Feki, A., Sayadi, S., & Allouche, N. (2018). LC–MS–MS and GC–MS analyses of biologically active extracts of Tunisian fenugreek (*Trigonella foenum-graecum* L.) seeds. *Journal of Food Measurement and Characterization, 12*, 209–220.

Khan, A., Shukla, S., Sinha, S., & Meeran, S. (2023). Proteomic analysis of fenugreek: Implications for health and disease. *Journal of Proteome Research, 22*(6), 2793–2805.

Khan, M. R., Singh, S. P., Sharma, R., & Gupta, A. (2021). Industrial crops and products. *164*, 113334.

Khare, S., Singh, N. B., Singh, A., Hussain, I., Niharika, K. M., Yadav, V., Bano, C., Yadav, R. K., & Amist, N. (2020). Plant secondary metabolites synthesis and their regulations under biotic and abiotic constraints. *Journal of Plant Biology, 63*, 203–216.

Kulkarni, P. S., Arif, M., & Sainis, J. K. (2019). Comparative transcriptome analysis of wild and cultivated populations of fenugreek provides insights into adaptation and domestication. *Scientific Reports, 9*(1), 1–13.

Kumar, R., Awasthi, R., Padmanabhan, P., & Saha, S. (2016). Comparative transcriptome analysis of differentially expressed genes in foxtail millet (*Setaria italica* L.) during dehydration stress. *Biochemical and Biophysical Research Communications, 470*(2), 470–476.

Kumar, V., Sharma, A., Shukla, R., Pandey, A., Singh, S., & Bhandari, A. (2018). A review on biological and chemical diversity in *Trigonella foenum-graecum*. *World Journal of Pharmaceutical Sciences, 6*(5), 91–102.

Li, J., Zhang, Y., Xu, D., Deng, X., & Li, Y. (2017). Transcriptome analysis of seed coat development in fenugreek (*Trigonella foenum-graecum*). *BMC Genomics, 18*(1), 219.

Li, T., Wang, Y. H., Liu, J. X., Feng, K., Xu, Z. S., & Xiong, A. S. (2019). Advances in genomic, transcriptomic, proteomic, and metabolomic approaches to study biotic stress in fruit crops. *Critical Reviews in Biotechnology, 39*(5), 680–692. https://doi.org/10.1080/07388551.2019.1608153.

Lockhart, D. J., & Winzeler, E. A. (2000). Genomics, gene expression and DNA arrays. *Nature, 405*(6788), 827–836.

Maaz, M., Dilshad, E., Suliman, A., & Ashraf, N. M. (2022). Identification of anti-inflammatory metabolites from *Trigonella foenum-graecum* using computational approaches. *Current Trends in OMICS, 2*(1), 00–00. https://doi.org/10.32350/cto.21.04.

Mahbub, J., Mou, R. A., Sikta, S. A., Rahman, A., & Dash, P. R. (2018). Biological and medicinal significance of *Trigonella foenum-graecum*: A review. *International Journal of Life Sciences Review, 4*(2), 15–26.

Misra, A., & Verma, M. (2019). Molecular biology and genomics of *Trigonella foenum-graecum* L.: An appraisal. *Journal of Applied Biology & Biotechnology, 7*(3), 14–24.

Mulimani, P. (2018). Updates available: Time for orthodontics' research approach to catch up with the rapidly evolving techniques and discoveries in molecular biology research? *APOS Trends in Orthodontics, 8*(4), 179. Gale OneFile: Health and Medicine. https://link.gale.com/apps/doc/A567801881/HRCA?u=anon~bf5600e3&sid=googleScholar&xid=00644bed (accessed on 25 September 2024).

Naika, M. B., Sathyanarayanan, N., Sajeevan, R. S., Bhattacharyya, T., Ghosh, P., Iyer, M. S., Jarjapu, M., Joshi, A. G., Harini, K., Shafi, K. M., & Kalmankar, N. (2022). Exploring the medicinally important secondary metabolites landscape through the lens of transcriptome data in fenugreek (*Trigonella foenum-graecum* L.). *Scientific Reports, 12*(1), 13534.

Przybylska, J., Kaczmarek, M., & Baranski, R. (2017). Seed development and maturation in fenugreek (*Trigonella foenum-graecum* L.): Anatomical and transcriptomic studies. *PLOS ONE, 12*(1), e0169948.

Raghavan, V., Kraft, L., Mesny, F., & Rigerte, L. (2022). A simple guide to de novo transcriptome assembly and annotation. *Briefings in Bioinformatics, 23*(2), bbab563.

Rathi, D., Gayen, D., Gayali, S., Chakraborty, S., & Chakraborty, N. (2016). Legume proteomics: Progress, prospects, and challenges. *Proteomics, 16*(2), 310–327. https://doi.org/10.1002/PMIC.201500257.

Riasat, M., Heidari, B., Pakniyat, H., & Jafari, A. A. (2018). Assessment of variability in secondary metabolites and expected response to genotype selection in fenugreek (*Trigonella* spp.). *Industrial Crops and Products, 123*, 221–231. https://doi.org/10.1016/j.indcrop.2018.06.002.

Shabir, G. A., & Anjum, F. M. (2017). Fenugreek (*Trigonella foenum-graecum* L.): A review of its genetic diversity and prospects for crop improvement through genomics. *Planta, 246*(5), 681–699.

Sharma, R. D., Raghuram, T. C., & Rao, N. S. (1990). Effect of fenugreek seeds on blood glucose and serum lipids in type I diabetes. *European Journal of Clinical Nutrition, 44*(4), 301–306.

Srinivasan, K. (2006). Fenugreek (*Trigonella foenum-graecum*): A review of health beneficial physiological effects. *Food Reviews International, 22*(2), 203–224. https://doi.org/10.1080/87559120600586315.

Suna, G., & Mayr, M. (2018). Proteomics. In *Encyclopedia of Cardiovascular Research and Medicine* (pp. 166–180).

Ulas, M., Yalcin, E., Aykut, A., & Sahin, H. (2021). Industrial crops & products, *164*, 113357.

Vaidya, K., Ghosh, A., Kumar, V., Chaudhary, S., Srivastava, N., Katudia, K., Tiwari, T., & Chikara, S. K. (2013). De novo transcriptome sequencing in *Trigonella foenum-graecum* L. to identify genes involved in the biosynthesis of diosgenin. *The Plant Genome, 6*(2), 2012–08.

Verma, R., Mishra, A., & Ghosh, S. (2019). Comparative transcriptome profiling of fenugreek (*Trigonella foenum-graecum* L.) leaves and roots for tissue-specific expression analysis and SSR marker development. *3 Biotech, 9*(2), 72.

Yadav, V. K., Khan, M. F., Singh, S., & Mishra, A. (2015). Transcriptome analysis reveals genes involved in the biosynthesis of alkaloids, flavonoids, saponins, and mucilage in fenugreek (*Trigonella foenum-graecum* L.) seeds. *Functional & Integrative Genomics, 15*(4), 459–470.

Younis, M. E., Ismail, A. M., Hossain, A. M., & Mahmoud, O. M. (2017). Effect of foliar application of some cytokinins on growth and essential oil production of fenugreek plants (*Trigonella foenum-graecum* L.). *Journal of Medicinal Plants Studies, 5*(5), 155–159.

CHAPTER 12

Influence of Depolymerized Sodium Alginate on Morpho-Physiological and Yield Parameters of Fenugreek

TARIQ AHMAD DAR[1], UNJMAN RASHID[2], BILAL AHMAD[3], BARKET ALI[1] and KHALID REHMAN HAKEEM[4]

[1]*Post Graduate Department of Botany, Sri Pratap College, Cluster University Srinagar, Jammu and Kashmir, India*

[2]*Department of Botany, Jamia Hamdard University, New Delhi, India*

[3]*Department of Botany, Govt Degree College, KhanSahib, Budgam, affliated with University of Kashmir Hazratbal Srinagar, India*

[4]*Department of Biological Sciences, Faculty of Science, King Abdulaziz University, Jeddah, Saudi Arabia*

ABSTRACT

Fenugreek (*Trigonella foenum-graecum*) is a multipurpose medicinal plant belonging to family Fabaceae. It contains active constituents like 4-hydroxy-isoleucine, diosgenin, and trigonelline. The plant is recommended for diabetic patients to take as a vegetable as it is anti-diabetic. Keeping in view the immense importance of fenugreek therefore there is a need to increase the growth and yield of fenugreek by using different techniques. Plant growth regulators are signaling molecules that enhance the overall growth and development of plants. Recently, depolymerized polysaccharides, such as sodium alginate, chitosan, and carrageenan, have been employed as plant growth regulators and have shown promising results in promoting plant growth.

Fenugreek: Improvement in Yield Potential Through Plant Growth Regulators (PGRs) and Depolymerized Polysaccharides. Tariq Ahmad Dar, Bilal Ahmad, Moinuddin, and Khalid Rehman Hakeem (Eds.)

12.1 INTRODUCTION

The plant growth regulators have been found to enhance the production of secondary metabolites in plants. The application of plant growth regulators in the form of foliar sprays enhances productivity and stimulates the formation of active constituents of medicinal and aromatic plants (Ahmad et al., 2017, 2022; Naeem et al., 2020). Radiation processed natural polysaccharides can be utilized in the field of agriculture after converting them to more useful forms (Mukarram et al., 2023; Sabharwal et al., 2004). According to this technique, the natural polymers like chitosan, carrageenan, and sodium alginate are subjected to irradiation by gamma rays, which break them down to lower molecular weight oligomers. The oligomers, obtained by the radiation degradation when applied to plants in different forms like foliar sprays, tissue culture methods or hydroponics, are known to modulate various physiological activities of plants. The oligomers promote seed germination, elongation of shoot and root, initiation of flowers and induction of phytoalexins, mitigation of heavy metal stress and promotion of antimicrobial activity (Aftab et al., 2014; Hu et al., 2004; Kume et al., 2002). The lower weight oligomers of natural polysaccharides significantly reduce the use of insecticides and chemical fertilizers and are also known to shorten the harvesting period of various crops. Alginates are noticeable naturally occurring marine polysaccharides available in nature. They are present mainly in the cell wall of marine brown-algae like *Laminaria hyperborean*, *Macrocystis pyrifera*, *Laminaria digitata*, *Scophyllum nodosum*, *Laminaria japonica*, *Edonia maxima*, *Lessonia nigrescens*, and *Sargassum* spp. Sodium alginate is the sodium salt of alginate consisting of homopolymeric poly-β-(1,4) D-mannuronic acid and poly-α-(1,4) L-guluronic acid.

Irradiated sodium alginate (ISA) enhanced the growth and productivity of Zea (Hegazy et al., 2009). ISA enhances the alkaloid and active constituent production and in different medicinal and aromatic plants like *Artemisia annua*, *Eucalyptus*, *Fenugreek*, *Peppermint*, *Papaver somniferum*, *Foeniculum vulgare*, *Mentha arvensis*, and *Cymbopogon flexuosus* (Ahmad et al., 2017, 2022; Dar et al., 2015; Idrees et al., 2015; Mukarram et al., 2023; Shabbir et al., 2017). The effect of different concentrations of irradiated sodium alginate (ISA 40, 80, and 120 mg L^{-1} or ISA40, ISA80, and ISA120, respectively), was studied on the growth, physiological, biochemical, and yield parameters as well as on the content and yield of seed alkaloids and trigonelline of *Trigonella foenum-graecum*.

Un-irradiated sodium alginate (40 mg L^{-1} or USA40 and deionized water were used as control treatments, which gave statistically equal values and had no significant effect on the parameters studied. Hence, all the treatment-means were compared with only water-spray Control. The growth, physiological and biochemical parameters were determined at 60 and 90 DAS, while the yield parameters were studied at 120 DAS. Among the ISA treatments, ISA 80 mg L^{-1} (ISA80) proved to be most superior for the majority of the attributes studied.

12.2 GROWTH ATTRIBUTES

The growth attributes, namely, shoot and root length, shoot and root fresh mass, shoot and root dry mass, number of leaves and leaf area per plant, recorded at 60 and 90 DAS, were affected significantly by the foliar application of irradiated sodium alginate (ISA). Compared to the control (deionized water spray), ISA80 increased the shoot length per plant by 30 and 36% at 60 and 90 DAS, respectively (Table 12.1). Root length was also affected significantly by ISA-spray. Application of the treatment ISA80 proved the optimum one for this parameter at both the growth stages. As compared to the Control, ISA80 increased the root length per plant by 25 and 30% at 60 and 90 DAS, respectively (Table 12.1). Foliar application of ISA increased the shoot fresh mass significantly at both the growth stages. Treatment ISA80 gave the highest value. This treatment enhanced the fresh mass of plant by 40.3 and 30.0% at 60 and 90 DAS, respectively, over the Control. The values recorded for plants sprayed with deionized water (Control) were recorded to be the lowest at both the stages of growth (Table 12.1). Foliar application of ISA proved statistically significant on root fresh mass at both the growth stages. Treatment ISA80 proved the best, giving 22.44 and 23.83% higher mass of fresh root than the Control at 60 and 90 DAS, respectively (Table 12.1). Foliar spray of ISA was significant on shoot dry mass per plant. Treatment ISA80 proved the best over the other treatments. The Control gave the poorest value at all the growth stages. The increase in shoot dry mass due ISA80 over the Control was 21.85 and 19.67% at 60 and 90 DAS, respectively (Table 12.2). Foliar spray of ISA applied gave significant results regarding root dry mass per plant. The optimum treatment brought out to be ISA80 at both the stages of growth, which maximally enhanced the dry mass of root to the extent of 20 and 29.47% at 60 and 90 DAS, respectively, as compared to the

Control (Table 12.2). There was significant effect of ISA on the number of leaves per plant. The best treatment was noted to be ISA80 at all the growth stages, increasing the number of leaves significantly. Application of ISA80 gave 23.88 and 32.85% higher values at 60 and 90 DAS, respectively, over the Control (Table 12.2). Application of ISA increased the leaf area per plant significantly at both the growth stages. Among the treatments applied ISA80 proved optimum at both the stages. Leaf area per plant, given by ISA80, was 26.85 and 32.85% higher than the Control at 60 and 90 DAS, respectively (Table 12.2).

TABLE 12.1 Effect of Three Levels of Irradiated Sodium Alginate (ISA) (40, 80, and 120 mg/L, referred to as ISA40, ISA80, and ISA120, Respectively) on Shoot and Root Length, Shoot Fresh Mass, and Root Fresh Mass of *Trigonella foenum-graecum* L., Recorded at 60 and 90 Days After Sowing (DAS)*

Treatments	Shoot Length (cm)		Root Length (cm)		Shoot Fresh Mass (g)		Root Fresh Mass (mg)	
	60 DAS	90 DAS	60 DAS	90 DAS	60 DAS	90 DAS	60 DAS	90 DAS
Control	36.60	57.00	6.80	8.20	3.92	8.60	294.00	495.00
USA40	37.20	59.28	7.00	8.60	4.07	8.70	300.00	508.00
ISA40	43.55	70.68	7.95	9.02	4.45	9.71	329.20	574.00
ISA80	47.94	77.52	8.50	10.66	4.80	10.57	360.00	613.00
ISA120	44.65	72.39	8.00	9.75	4.60	9.89	335.00	598.00
LSD at 5%	3.18	3.76	0.54	0.46	0.19	0.24	12.39	16.91

**Data are presented as means of three replicates.*

TABLE 12.2 Effect of Three Levels of Irradiated Sodium Alginate (ISA) (40, 80, and 120 mg/L, designated as ISA40, ISA80, and ISA120, Respectively) on Shoot and Root Dry Mass, Number of Leaves, and Leaf Area per Plant of *Trigonella foenum-graecum* L., Recorded at 60 and 90 Days After Sowing (DAS)*

Treatments	Shoot Length (cm)		Root Length (cm)		Shoot Fresh Mass (g)		Root Fresh Mass (mg)	
	60 DAS	90 DAS	60 DAS	90 DAS	60 DAS	90 DAS	60 DAS	90 DAS
Control	627.00	1220.00	70.00	95.00	18.00	35.00	108.00	210.00
USA40	652.00	1270.00	73.00	99.00	19.00	37.00	114.00	222.00
ISA40	689.70	1360.00	78.40	108.00	20.60	42.70	124.20	256.20
ISA80	764.00	1460.00	84.00	123.00	22.30	46.50	137.16	279.30
ISA120	721.00	1370.00	79.80	111.00	21.24	43.00	126.36	260.40
LSD at 5%	19.03	20.94	5.47	5.49	1.93	2.93	4.07	16.09

**Data are presented as means of three replicates.*

12.3 BIOCHEMICAL PARAMETERS

Biochemical parameters, viz. content of leaf-chlorophyll and -carotenoids, content of leaf-nitrogen, -phosphorus, and -potassium, activities of CA and NR in the leaves, studied at 60 and 90 DAS, were affected significantly by the foliar spray of ISA. Salient features of the data are described in subsections.

12.3.1 LEAF NITRATE REDUCTASE (NR) ACTIVITY

ISA-spray resulted in statistically significant effect on leaf NR activity. Among various treatments applied, ISA80 proved to be the best at both the stages of growth, increasing the leaf-nitrate reductase activity by 14.81 and 20.86% at 60 and 90 DAS, respectively, over the Control (Table 12.3).

12.3.2 LEAF CARBONIC ANHYDRASE (CA) ACTIVITY

Effect of foliar spray of ISA proved significant on leaf CA activity against Control. Of the treatments, ISA80 gave maximum value for CA activity. It surpassed the Control by 16.98 and 17.73% at 60 and 90 DAS, respectively (Table 12.3).

12.3.3 LEAF CHLOROPHYLL CONTENT

The treatments applied gave significantly higher value compared to the Control for the leaf-chlorophyll content at both the stages of s growth. Of all the treatments, ISA80 proved the best. Application of ISA80 gave 21.96 and 22.22% higher values over the Control at 60 and 90 DAS, respectively (Table 12.3).

12.3.4 LEAF CAROTENOIDS CONTENT

The ISA-spray applied, proved significantly effective on leaf carotenoids content as against the control.

Amongst the treatments ISA80 proved the best, exceeding the control by 12.0 and 14.70% at 60 and 90 DAS, respectively (Tables 12.3).

TABLE 12.3 Effect of Three Levels of Irradiated Sodium Alginate (ISA) (40, 80, and 120 mg/L, designated as ISA40, ISA80, and ISA120, Respectively) and Phosphorus (40 kg P/ha, Designated as P40), Applied Alone and in Combination, on Nitrate Reductase (NR) Activity and Carbonic Anhydrase (CA) Activity in *Trigonella foenum-graecum* L. at 60 and 90 Days After Sowing (DAS)*

Treatments	NR Activity (nM NO_2-g^{-1}FWh^{-1})		CA Activity [µmol (CO_2) kg^{-1} (FW) s^{-1}]		Total Chlorophyll (mg g^{-1} FW)		Carotenoids Content (mg g^{-1} FW)	
	60 DAS	90 DAS	60 DAS	90 DAS	60 DAS	90 DAS	60 DAS	90 DAS
Control	405.00	345.00	3.12	2.65	1.73	1.35	0.200	0.170
USA40	412.00	355.00	3.18	2.69	1.75	1.37	0.204	0.170
ISA40	445.50	386.40	3.49	2.96	1.94	1.51	0.210	0.185
ISA80	465.70	417.40	3.65	3.12	2.11	1.65	0.224	0.195
ISA120	453.60	393.30	3.52	3.02	1.97	1.54	0.216	0.187
LSD at 5%	11.31	11.51	0.09	0.12	0.076	0.058	0.007	0.018

**Data are presented as means of three replicates.*

12.3.5 LEAF-PHOSPHORUS CONTENT

Influence of ISA spray on leaves, was significant. ISA80 proved superior to the other treatments applied, giving the maximum value for leaf phosphorus content at both the stages of growth. ISA80 enhanced the leaf phosphorus content by 15.8 and 18.4%, over the Control (Table 12.4).

TABLE 12.4 Effect of Three Levels of Irradiated Sodium Alginate (ISA) (40, 80, and 120 mg/L, Designated as ISA40, ISA80, and ISA120, Respectively) on Leaf Nitrogen, Leaf Phosphorus, and Leaf Potassium Contents in *Trigonella foenum-graecum* L., Recorded at 60 and 90 Days After Sowing (DAS)*

Treatments	Leaf-Nitrogen Content (mg g^{-1})		Leaf-Phosphorus Content (mg g^{-1})		Leaf-Potassium Content (mg g^{-1})	
	60 DAS	90 DAS	60 DAS	90 DAS	60 DAS	90 DAS
Control	31.00	28.50	3.40	2.80	30.50	27.00
USA40	31.30	28.90	3.40	2.80	30.85	27.40
ISA40	33.10	30.70	3.63	3.00	32.00	28.89
ISA80	34.10	32.15	3.77	3.22	32.94	30.24
ISA120	33.40	31.30	3.67	3.13	32.33	29.43
LSD at 5%	0.76	0.77	0.10	0.15	0.44	0.44

**Data are presented as means of three replicates.*

12.3.6 LEAF-NITROGEN CONTENT

Foliar-spray of ISA increased the leaf-N content positively at both the stages of growth. Treatment ISA80 recorded the highest value at both the growth stages, outshining the control by 10 and 12.80% at 60 and 90 DAS, respectively (Table 12.4).

12.3.7 LEAF-POTASSIUM CONTENT

Application of ISA was significantly effective on leaf potassium content in comparison to the Control at both the stages. ISA80 proved the optimum treatment, enhancing the leaf-potassium content over the Control by 8 and 12% at 60 and 90 DAS, respectively (Table 12.4).

12.4 YIELD AND QUALITY PARAMETERS

Application of irradiated sodium alginate (ISA) as a foliar spray significantly increased seed yield and yield parameters, seed alkaloid content and yield, as well as trigonelline content and yield, as recorded at 120 days after sowing (DAS). The ISA80 treatment proved to be optimal. Treatments had a significant effect on the number of pods per plant, with ISA80 yielding the maximum value, increasing pod number by 30% compared to the control (Table 12.5). However, ISA foliar spray had no significant effect on the number of seeds per pod (Table 12.5). The number of seeds per plant was significantly affected by treatments, with ISA80 performing best, increasing seed number by 55.84% over the control (Table 12.5).

The different foliar doses of ISA also significantly influenced pod length, with ISA80 providing the maximum increase of 34.1% over the control (Table 12.5). Treatments significantly affected the 1000-seed weight as well, with ISA80 increasing it by 18.22% compared to the control (Table 12.6). ISA foliar spray was significantly effective in improving seed yield per plant, with ISA80 achieving the highest increase of 84% over the control, while the control (deionized water spray) showed the lowest value (Table 12.6).

Biological yield per plant was positively influenced by ISA treatments, with ISA80 achieving the highest increase of 40.81% over the control (Table 12.6). The harvest index was also significantly increased with ISA treatments; ISA80 surpassed the control by 40% (Table 12.6).

Seed alkaloid content was significantly enhanced by ISA foliar spray, with ISA80 as the best treatment, increasing alkaloid content by 25% over the control. The lowest value was recorded for the control. Seed alkaloid yield was also significantly improved, with ISA80 increasing it by 133.0% over the control. Similarly, ISA foliar spray significantly enhanced trigonelline content, with ISA80 yielding the maximum increase of 13% compared to the control. ISA application significantly increased trigonelline yield per plant, with ISA80 achieving the highest enhancement of 100.0% over the control.

TABLE 12.5 Effect of Three Levels of Irradiated Sodium Alginate (ISA) (40, 80, and 120 mg/L, Designated as ISA40, ISA80, and ISA120, Respectively) on Number of Pods per Plant, Number of Seeds per Plant, and Pod Length in *Trigonella foenum-graecum* L.*

Treatments	1000 Seed Weight (g)	Seed Yield (g Plant^{-1}	Biological Yield (g Plant^{-1})	Harvest Index (%)
Control	9.00	0.69	3.70	18.64
USA40	9.00	0.69	3.88	18.55
ISA40	9.65	0.97	4.62	22.51
ISA80	10.64	1.27	5.21	26.10
ISA120	9.50	1.02	4.88	22.54
LSD at 5%	0.23	0.17	0.24	1.95

**Data are presented as means of three replicates.*

TABLE 12.6 Effect of Three Levels of Irradiated Sodium Alginate (ISA) (40, 80, and 120 mg/L, Designated as ISA40, ISA80, and ISA120, Respectively) on Seed Yield, Biological Yield, and Harvest Index of *Trigonella foenum-graecum* L.*

Treatments	Number of Pods Plant^{-1}	Number of Seeds Pod^{-1}	Number of Seeds Plant^{-1}	Pod Length (cm)
Control	7.00	11.00	77.00	7.80
USA40	7.00	11.00	77.00	8.03
ISA40	8.75	11.55	101.00	9.00
ISA80	10.00	12.00	120.00	10.50
ISA120	9.10	11.88	108.00	9.00
LSD at 5%	0.72	NS	7.57	0.94

**Data are presented as means of three replicates.*

KEYWORDS

- **biochemical parameters**
- **carotenoids**
- **fenugreek**
- **growth stages**
- **irradiated sodium alginate**
- **leaf area per plant**
- **seed alkaloids**
- **sodium alginate**

REFERENCES

Aftab, T., Khan, M. M. A., Naeem, M., Idrees, M., Siddiqi, T. O., & Varshney, L. (2014). Effect of irradiated sodium alginate and phosphorus on biomass and artemisinin production in *Artemisia annua* L. *Carbohydrate Polymers, 110*, 396–404.

Ahmad, B., Dar, T. A., Khan, M., Ahmad, A., Rinklebe, J., Chen, Y., & Ahmad, P. (2022). Oligochitosan fortifies antioxidative and photosynthetic metabolism and enhances secondary metabolite accumulation in arsenic-stressed peppermint. *Frontiers in Plant Science, 13*, 987746.

Ahmad, B., Khan, M., Akhtar, M., Jaleel, H., Sadiq, Y., Shabbir, A., & Uddin, M. (2017). Exogenously sourced γ-irradiated chitosan-mediated regulation of growth, physiology, quality attributes and yield in *Mentha piperita* L. *Turkish Journal of Biology, 41*(2), 388–401.

Dar, T. A., Uddin, M., Khan, M. M. A., Ali, A., Mir, S. R., & Varshney, L. (2015). Effect of Co-60 gamma irradiated chitosan and phosphorus fertilizer on growth, yield and trigonelline content of *Trigonella foenum-graecum* L. *Journal of Radiation Research and Applied Sciences, 8*(3), 446–458.

Hegazy, E. A., Abdel-Rehim, H., Diaa, D. A., & El-Barbary, A. (2009). Controlling of degradation effects in radiation processing of polymers. *Controlling of Degradation Effects in Radiation Processing of Polymers.*

Hu, X., Jiang, X., Hwang, H., Liu, S., & Guan, H. (2004). Promotive effects of alginate-derived oligosaccharide on maize seed germination. *Journal of Applied Phycology, 16*, 73–76.

Idrees, M., Dar, T. A., Naeem, M., Aftab, T., Khan, M. M. A., Ali, A., Uddin, M., & Varshney, L. (2015). Effects of gamma-irradiated sodium alginate on lemongrass: Field trials monitoring production of essential oil. *Industrial Crops and Products, 63*, 269–275.

Kume, T., Nagasawa, N., & Yoshii, F. (2002). Utilization of carbohydrates by radiation processing. *Radiation Physics and Chemistry, 63*(3–6), 625–627.

Mukarram, M., Khan, M. M. A., Kurjak, D., & Corpas, F. J. (2023). Chitosan oligomers (COS) trigger a coordinated biochemical response of lemongrass (*Cymbopogon flexuosus*) plants to palliate salinity-induced oxidative stress. *Scientific Reports, 13*(1), 8636.

Naeem, M., Nabi, A., Aftab, T., & Khan, M. M. A. (2020). Oligomers of carrageenan regulate functional activities and artemisinin production in *Artemisia annua* L. exposed to arsenic stress. *Protoplasma, 257*, 871–887.

Sabharwal, S., Varshney, L., Chaudhari, A. D., & Ramnani, S. P. (2004). Radiation processing of natural polymers: Achievements & trends. *Radiation Processing of Polysaccharides, 1422*, 29.

Shabbir, A., Khan, M., Akhtar, M., Sadiq, Y., Jaleel, H., Ahmad, B., & Uddin, M. (2017). Regulation of functional activities and essential oil production in *Vetiveria zizanioides* L. Nash after γ-irradiated sodium alginate elicitation. *Turkish Journal of Biology, 41*(4), 661–672.

CHAPTER 13

Diseases Associated with Fenugreek

ULFAT JAN[1,2] and AMIR HUSSAIN SHEIR[2]

[1]*Division of Agriculture Entomology SKUAST Kashmir, Jammu and Kashmir, India*

[2]*Department of Electrical Engineering, Chandigarh University, Jammu and Kashmir, India*

ABSTRACT

Fenugreek (*Trigonella foenum-graecum* L.) is a legume crop which is grown once in a year and is grown all around the world, from India to Mediterranean region, China, Africa, Europe, Australia, and North America. Not only is it used as a spice in cooking, but it also has medicinal and nutraceutical features and also cultivated as a fodder crop in many regions of the world. Not only this crop offers can be grown in new areas, but it also has the potential to expand traditionally developed areas. Unfortunately, this crop is susceptible to a number of biological and physiological diseases which can reduce productivity and causes the serious damage. Physiological diseases are caused due to lack of essential minerals, while biological diseases are caused by fungal, bacterial, viral, and insect infections. Common fungal, bacterial, and viral diseases that affect the crop include cercospora leaf spot, root-rot/collar/foot-rot, powdery mildew, downy mildew, rust, fusarium wilt, bacterial leaf spot, bacterial blight, bacterial wilt, citrus canker, scab, yellow mosaic and curly top. Each of these diseases has its own causal organism and symptoms. Among these diseases, two major fungal diseases that affect the crop are Cercospora leaf spot caused by *Cercospora traversiana* and Powdery mildew caused by *Erysiphe polygoni.* A number of chemical, biological, cultural, and plant quarantine methods for disease control are available, but disease management through an Integrated approach such as crop rotation, disease

Fenugreek: Improvement in Yield Potential Through Plant Growth Regulators (PGRs) and Depolymerized Polysaccharides. Tariq Ahmad Dar, Bilal Ahmad, Moinuddin, and Khalid Rehman Hakeem (Eds.)

forecasting, changing planting practices, monitoring environmental conditions (temperature, moisture, soil pH, nutrients, etc.), modifying the environment through drainage, irrigation, and so on, and developing new varieties that are resistant or tolerant to diseases is essential in order to improve crop productivity and importance of the crop in the near future.

13.1 INTRODUCTION

Fenugreek, scientifically known as *Trigonella foenum graecum*, is a type of aromatic plant which comes under the Leguminosae family. The fenugreek plant typically grows upright and has a strong, sweet scent. Its leaves are small and divided into three leaflets, ranging in color from green to violet. It gives single, soft milky or purple florets and a straight or sometimes round, golden capsule containing the seeds. This plant usually grows to 60 cm (23.6 inch) in height and is annual, meaning it only gives for a single season. The exact place of source of fenugreek is unclear; however, it is native to the western Mediterranean region. Fenugreek leaves can be consumed either fresh or dried from a culinary herb, as its seeds are used as a spice or flavoring agent. Each pod of fenugreek produces between 10 and 20 minute, flat and brown seeds, with each seed divided into two parts. The seeds of the plant which are usually yellow brown in color and are often used as a spice. The seeds and green leaves of Fenugreek have been used in food as well as in remedial purposes that is an ancient practice of human study. Fenugreek spice seeds have remedial properties such as Hypocholesterolemic, lactation support, antibacterial, gastric stimulant, for anoxic, antidiabetic agent, hypatoprotective effect and cancer. Little amount of volatile and fixed oils was found in fenugreek seeds (Sowmya and Rajyalakshmi, 1999). These beneficial physiological effects, fenugreeks antidiabetic and cholesterol-lowering effects, are mainly due to its dietary fiber content, which has promising functional food value (Srinivasan, 2006). Nowadays, it is used as food stabilizer, adhesive, and emulsifier due to its high fiber, protein, and gum content. Fenugreek proteins have been found more soluble at alkaline pH (Meghwal and Goswami, 2012). Fenugreek contains 23–26% protein, 6–7% fat and 58% carbohydrates of which about 25% is dietary fiber (US Department of Agriculture, 2001). According to a study by Shabbeer et al. (2009), treatment with fenugreek inhibited the growth of breast, pancreatic and prostate cancer cell lines, but primary prostrate or immortalized prostrate cells remained unaffected. Recently, fenugreeks antibacterial and antifungal has been discovered. It has been proposed that fenugreek is an important source of bio-active compounds useful for the development of better and new antifungal agents (Haouala et al., 2008).

13.2 MAJOR DISEASES OF FENUGREEK

The major diseases of fenugreek and their control are discussed in subsections.

13.2.1 FUNGAL DISEASES OF FENUGREEK AND THEIR MANAGEMENT

The fungal diseases of fenugreek and their management are discussed in succeeding sections.

13.2.1.1 CERCOSPORA LEAF SPOT: CERCOSPORA TRAVERSIANA

Signs of the disorder include circular sunken spots on the foliage with yellowish halos around them, dead patches on the foliage, and discoloration on pods.

1. **Survivorship and Expansion:** Disease causing organism exists particularly in plant matter such as desiccation-resistant pseudo-stromata but can also remain as conidia in trash or seeds. When there is an adequate amount of moisture, new conidia are created and can spread through droplets of water or air currents to other plants or plant parts.
2. **Favorable Conditions:** Ideal conditions include temperature between 25°C and 35°C during the day, and temperatures not lower than 16.1°C at night, accompanied by a relative humidity of 90–95%. If temperatures drop below 15°C or leaves remain wet for more than 11 hours, the risk of infection is greatly reduced.
3. **Management:** Blitox spray should be applied with 0.3% concentration.

13.2.1.2 POWDERY MILDEW: ERYSIPHE POLYGONI AND LEVEILLULA TAURICA

Symptoms of the disease include white powdery spots that spread across the leaves, as well as yellow spots on the underside of the leaf.

1. **Survivorship and Expansion:** The mildew is found on plant parts and plant waste, such as leaflets that have dropped to the ground, throughout the cold winter months.

2. **Favorable Conditions:** Humidity is involved in the growth and spread of powdery mildew. When humidity is high, spores form more easily, while as lower humidity favors spore dispersal that is why powdery mildew is more common when temperatures are lower, and the air is moist.
3. **Management:** The disease can be managed by applying inorganic sulfur at 25 kg/ha or spraying 0.25% Wettable sulfur at 20 days interval.

13.2.1.3 ROOT ROT/COLLAR/FOOT ROT: RHIZOCTONIA SOLANI AND SCLEROTIUM ROLFSII

Disease signs of *Rhizoctonia* root rot include:

1. **Contamination:** Infection begins at the tips of small lateral roots and the base of the main root, with the "spear-point" symptom appearing on the rootlets.
2. **Leaf Symptoms:** Small, pale, and watery lesions develop randomly on both sides of the leaflets. These spots enlarge, become rough in shape, and change to light brown with darker brown edges. Eventually, the center of each spot turns pale and fades away.
3. **Survivorship and Spread:** The pathogen can survive in soil for extended periods without a host, with inoculum levels gradually increasing over several growing seasons.
4. **Favorable Conditions:** High humidity and warm weather create ideal conditions for disease spread.
5. **Management:** Root rot can be managed by applying Neem cake at 150 g/hectare and treating seeds with *Trichoderma viride* at 4 g/kg. Additionally, soil application of Carbendazim at 0.5 g/L or Copper oxychloride at 2 g/L is recommended.
6. **Crop Rotation:** Implementing a crop rotation of 3–4 years helps reduce disease incidence.

13.2.1.4 DOWNY MILDEW: PERONOSPORA TRIGONELLAE

1. **Disease Signs:** Symptoms of downy mildew differ from those of other diseases
    - Upper leaf surfaces display tiny pale chlorotic patches, particularly around the edges.

- The underside of leaves shows a white, cottony fungal growth that often appears as ashy purple.
- Infection also results in reduced plant development.

2. **Survivorship and Spread**: The mildew persists in plant material and decaying vegetation, such as fallen leaves.
3. **Favorable Conditions**: The disease is likely to occur in environments with temperatures between 18–24°C and 4–10°C and moisture levels above 80%.
4. **Control**: Downy mildew can be managed by applying 20 g of Difoltan or another copper-based fungicide.

13.2.1.5 RUST: UROMYCES TRIGONELLAE

1. **Disease Symptoms:**
   - During the late spring or early summer of the disease cycle, rust is visible as a colored dust, made up of tiny aeciospores that come down on the fenugreek plant, creating Uredinales lying on underside of the leaves. This powder-like substance is known as rust and consists of hair-like (ligulate) structures called telia.
   - Dark brown pustules, which can be either round or oblong in shape, are often seen on the leaves and petioles.
2. **Survivorship and Expansion:**
   - The rust remains alive primarily by producing teliospores, which are thick bordered, dormant spores that can remain on the leaf debris or on the ground.
   - The disease spread by wind-borne uredospores from infected crop.
3. **Favorable Environment:** A temperature between 20 and 25°C and moisture of 86–92% are ideal conditions for a high amount of rust.

13.2.1.6 DAMPING OFF: PYTHIUM APHANIDERMATUM

1. Disease Signs:
   - The sprouted seedlings show discoloration and soft rot at the bottom of the shoot as a result of water damage, leading to the death of the tissue in the cortex.

- The rotted seedlings give off an unpleasant smell and cause the entire seedling to fall over.

2. **Survivorship and Expansion:** The fungus thrives in soil and can spread through both direct contact with the soil and air-borne spores transported by wind or rain.

3. **Favorable Conditions:**

   - Ideal conditions for infection and development of disease include high soil moisture, fogginess, and temperatures less than 24°C for a period of time.
   - Congested seedlings, wetness due to excessive rainfall, inadequate drainage, and extra soil salts inhibit plant development and multiply the risk of damping off disease.

13.2.1.7 FUSARIUM WILT: FUSARIUM OXYSPORUM

1. Disease Symptoms:
   - The disease signs are initially seen as small discolorations in veins on the outer side of blooming leaves, which are then followed by the mature leaves drooping downwards.
   - At the early stages of development, plants can wither and die quickly after the emergence of the signs.
   - In older plants, the leaves start to droop downward and vein clearing are observed. This is then followed by stunted growth, lower leaves become pale in color, drooping of the leaves and young stems.
   - The diseased leaves show marginal necrosis, which is followed by rapid defoliation, dehydrating of conducting tissue, and eventually death of plant.
   - The signs develop more in older plants between the blossoming and mature of fruits.

2. **Survivorship and Expansion:** The infection is caused by soil-dwelling organisms as well as by the pathogens present in soil.

3. **Favorable Conditions:** Soil moisture and temperature that are moderately high are favorable for the spread of infection.

13.2.2 VIRAL DISEASES OF FENUGREEK

13.2.2.1 YELLOW MOSAIC DISEASE: BEAN YELLOW MOSAIC VIRUS

1. **Disease Symptoms:**
    - The symptoms of a mosaic virus infection are characterized by vein clearing, extreme shedding of leaves, leaf margins curling inwards, and reduced leaf size.
    - The seeds inside pods are very small, curled, and shriveled.
2. **Favorable Environment:**
    - The infection spreads by aphids in a semi-persistent way.
    - Aphids become more aggressive in humid situations, resulting in a larger population and spread of viruses.

Other viral diseases associated with fenugreek are bean yellow mosaic virus, alfa-alfa mosaic virus, cowpea mosaic virus, potato virus A and Y, and clover vein mosaic virus. These viral diseases cause average loss of seed and fodder yield.

13.2.2.2 BACTERIAL DISEASES ASSOCIATED WITH FENUGREEK

1. **Bacterial Blight:** *Pseudomonas syringae.*
2. **Symptoms:** Symptoms of the infection initially becomes visible on under-side of the leaf, shaded parts of plant and consisted of small round, yellow brown or black coloration, and circular patterns around them. Infected areas drop out and formed holes in the leaves with drop off. The disease becomes more prevalent at maturity of crop.

13.2.2.3 INTEGRATED DISEASE MANAGEMENT TECHNIQUES

The integrated disease management techniques for fenugreek crop involve:

- Avoiding monoculture and rotating crops to reduce the amount of pathogens in soil;
- Solarizing nursery beds can help to lower the inoculum amount of pathogens in the soil;
- Using tolerant varieties for cultivation;

- Weeds can be alternate host for pathogens, so they can be destroyed;
- Plant density and row width can be adjusted to help crops to escape from diseases;
- Destruction of infected plant material should be done during the field scouting;
- For seed-borne fungal diseases use healthy seeds as planting material provide valuable control.

13.3 CONCLUSION

Fenugreek is an important economical crop, but disease management through an integrated approach such as crop rotation, disease forecasting, changing planting practices, monitoring environmental conditions (temperature, moisture, soil pH, nutrients, etc.), modifying the environment through drainage, irrigation, and so on, and developing new varieties that are resistant or tolerant to diseases is essential in order to improve crop productivity and importance of the crop in the near future.

KEYWORDS

- **crop rotation**
- **cultivation**
- **disease management**
- **fenugreek**
- **nutrients**
- **pH**
- **resistant varieties**
- **seed-borne fungal diseases**

REFERENCES

Gupta, S. K., & Sharma, R. C. (2017). *Diseases of Vegetables, Ornamentals and Spice Crops*. Scientific Publishers.

Haoala, R., Hawala, S., El-Ayeb, A., Khanfir, R., & Boughanmi, N. (2008). Aqueous and organic extracts of *Trigonella foenum-graecum* L. inhibit the mycelial growth of fungi. *Journal of Environmental Science, 20*, 1453–1457.

Meghwal, M., & Goswami, T. K. (2012). A review on the functional properties, nutritional content, medicinal utilization, and potential application of fenugreek. *Journal of Food Processing Technology, 3*(9).

National Institute of Plant Health Management. (n.d.). Integrated Pest Management Package. Retrieved from: https://niphm.gov.in/IPMPackage/.

ResearchGate. (n.d.). Retrieved from: https://www.researchgate.net (accessed on 25 September 2024).

Shabber, S., Sobolewski, M., Anchoori, R. K., Kachaap, S., Hidalgo, M., Jimeno, A. M., & Khan, M. A. (2009). Fenugreek: A naturally occurring edible spice as an anticancer agent. *Cancer Biology & Therapy, 8*, 272–278.

Sowmya, P., & Rajyalakshmi, P. (1999). Hypocholesterolemic effect of germinated fenugreek seeds in human subjects. *Plant Foods for Human Nutrition, 53*, 359–356.

Srinivasan, K. (2006). Fenugreek (*Trigonella foenum-graecum* L.): A review of health beneficial physiological effects. *Food Reviews International, 22*(2), 203–224.

Tamil Nadu Agricultural University. (n.d.). Horticulture. Retrieved from: https://agritech.tnau.ac.in/horticulture/ (accessed on 25 September 2024).

U.S. Department of Agriculture. (2001). *Nutrient Database for Standard Reference, Release 14*. Washington, DC: U.S. Department of Agriculture.

Index

B

C

D

E

F

G

H

L

M

O

P

T

U

V

W

X

Y

Z